D1800387

MONOGRAPHS ON THE PHYSICS AND CHEMISTRY OF MATERIALS

General Editors

MONOGRAPHS ON THE PHYSICS AND CHEMISTRY OF MATERIALS

General Editors

WILLIS JACKSON H. FRÖHLICH N. F. MOTT

MULTIPLE-BEAM INTERFEROMETRY OF SURFACES AND FILMS. By S. TOLANSKY

THEORY OF DIELECTRICS, DIELECTRIC CONSTANT AND DIELECTRIC LOSS. By H. FRÖHLICH

PHYSICS OF RUBBER ELASTICITY. By L. R. G. TRELOAR

PHYSICAL PROPERTIES OF GLASS. By J. E. STANWORTH

NEUTRON DIFFRACTION. By G. E. BACON, *Second edition*

THE INTERFERENCE SYSTEMS OF CROSSED DIFFRACTION GRATINGS. By J. GUILD

DIELECTRIC BREAKDOWN OF SOLIDS. By S. WHITEHEAD

THE PHYSICS OF ELECTRICAL CONTACTS. By F. LLEWELLYN JONES

THE PHOTOGRAPHIC STUDY OF RAPID EVENTS. By W. D. CHESTERMAN

DETONATION IN CONDENSED EXPLOSIVES. By J. TAYLOR

GRAIN BOUNDARIES IN METALS. By D. MCLEAN

THE DETECTION AND MEASUREMENT OF INFRA-RED RADIATION. By R. A. SMITH, F. E. JONES, and R. P. CHASMAR

EXPERIMENTAL TECHNIQUES IN LOW TEMPERATURE PHYSICS. By G. K. WHITE

MOLECULAR DISTILLATION. By G. BURROWS

DIRECT METHODS IN CRYSTALLOGRAPHY. By M. M. WOOLFSON

OXIDE MAGNETIC MATERIALS. By K. J. STANDLEY

PHYSICAL PRINCIPLES AND APPLICATIONS OF JUNCTION TRANSISTORS. By J. H. SIMPSON and R. S. RICHARDS

THE THEORY OF DIELECTRIC BREAKDOWN OF SOLIDS. By J. J. O'DWYER

ULTRASONIC ABSORPTION. By A. B. BHATIA

CONTACT AND FRICTIONAL ELECTRIFICATION. By W. R. HARPER

An Introduction to the Theory of Superconductivity

BY

CHARLES G. KUPER

DEPARTMENT OF PHYSICS,
TECHNION: ISRAEL INSTITUTE OF TECHNOLOGY,
HAIFA, ISRAEL

CLARENDON PRESS · OXFORD
1968

Oxford University Press, Ely House, London W.1
GLASGOW NEW YORK TORONTO MELBOURNE WELLINGTON
CAPE TOWN SALISBURY IBADAN NAIROBI LUSAKA ADDIS ABABA
BOMBAY CALCUTTA MADRAS KARACHI LAHORE DACCA
KUALA LUMPUR HONG KONG TOKYO

Made in Great Britain at the Pitman Press, Bath

PREFACE

THERE is a considerable and rapidly growing literature dealing with the theory of superconductivity. The appearance of yet another book calls for, if not an apology, at least an explanation. With the development of superconducting solenoids for producing high magnetic fields, superconductivity has become a subject of considerable technological importance. Moreover, other applications are likely to follow. But the current theoretical literature is mostly very 'highbrow', written by theoreticians for other theoreticians; it is often unintelligible to engineers and applied physicists.

The present book is attempt to bridge this gulf, and to explain the theory to the *users* of superconductivity. The reader is assumed to be familiar with classical thermodynamics and with Fourier transforms. No knowledge of second quantization is assumed, and only a moderate knowledge of wave mechanics. The concepts of quantum field theory are explained and developed at the stage where they are required.

This volume has grown out of courses of lectures given by the author at the University of the Witwatersrand, the University of St Andrews, and the Hebrew University of Jerusalem. These courses were designed to give a general introduction to superconductivity to graduate students of experimental physics. It is hoped that the book will be useful to students at this level. But I have also been very conscious of the somewhat different requirements of metallurgists, applied physicists, and electrical engineers, and have stressed those aspects of superconductivity which are technologically important. In particular, the Ginzburg–Landau theory and Type II superconductivity are discussed at some length.

The material is divided into two parts. Part I is an account of the phenomenological theory, as developed particularly by F. and H. London, Gorter and Casimir, Ginzburg and Landau, Pippard, and Abrikosov. This part should be sufficient for most engineering and design purposes. No quantized-field techniques are used in Part I. In Part II, the Bloch–Fröhlich model Hamiltonian is introduced, and the underlying concepts are briefly explained. An elementary account follows of the microscopic theory of Bardeen, Cooper and Schrieffer, Bogoliubov, and Gor'kov. The microscopic theory is shown to lead to the equations postulated by the phenomenological theory. The *ad hoc* parameters of the Ginzburg–Landau theory acquire new significance. In particular the effective charge ε of the superconducting charge carriers is twice the electronic charge—a manifestation of the electron-pairing aspects of superconductivity.

In a book of this kind, electromagnetic units inevitably present a problem. Almost all the published work in the field has used either Gaussian cgs units or a system of natural units in which $\hbar = 1$. Most physicists (including the author) are firmly prejudiced in favour of one or other of these systems, and against the MKS system. But most engineers have the opposite prejudice (sanctified by international agreement). As a compromise between the conflicting requirements of different readers, Gaussian units have been used in the text, and superscript numerical references direct the reader to the appropriate item in Appendix F, where the necessary changes are listed to convert all formulae to their MKS equivalents.

It is with great pleasure that I thank my many colleagues who have assisted me by discussion and criticism. I am especially indebted to Professor H. Fröhlich, F.R.S., Dr B. D. Josephson, and Dr J. R. Waldram for numerous and penetrating comments on the entire subject matter. I have also benefitted greatly from comments on parts of the manuscript from Dr R. C. Clark, Dr M. Cook, Professor R. B. Dingle, Dr D. M. Finlayson, Dr L. Gunther, Professor A. B. Pippard, F.R.S., Dr A. W. B. Taylor, and Professor M. Weger. I am grateful to Mr D. Sarid for his assistance in preparing Appendix F, listing the MKS equivalents of all Gaussian electromagnetic formulae in the book, and to Miss Rachel Ostrovsky for typing the manuscript. I thank Drs A. G. Every, H. E. Price, and

Barbara D. Rothberg for making their notes of my Witwatersrand lectures available to me, and the Hebrew University of Jerusalem for the facilities they provided during the year 1965–66. And above all I thank my wife for her help in repeated revisions of the text, in checking cross-references and equation numbers, in compiling the bibliography and index, and in correcting proofs.

Jerusalem C. G. K.
June 1966

CONTENTS

PART I

PHENOMENOLOGICAL THEORY

PART II
MICROSCOPIC THEORY

LIST OF SYMBOLS

Symbols introduced *ad hoc* in a specific context are not necessarily included in this index. Unfortunately it has not proved practicable to entirely avoid the use of symbols with several meanings, but no confusion is likely to arise. Where one of the meanings is restricted to a small part of the book, this fact is noted below. As usual, vectors in three-dimensional space are represented by bold-face symbols, and their magnitudes by the corresponding light-face symbols. Quantum-mechanical operators are denoted by symbols bearing a circumflex accent. (The Hamiltonian $\hat{H}$ bears this accent even in classical contexts, to help to distinguish it from the magnetic field **H**.)

$\mathbf{A}$, A	vector potential	
A	surface area	§ 6.2
$\boldsymbol{\mathscr{A}}$, $\mathscr{A}$	vector potential in Ginzburg–Landau dimensionless units	Ch. 6, 7
$\mathscr{A}_1$, $\mathscr{A}_2$	limiting values of Taylor's material functions	§ 4.3
$\hat{a}$, $\hat{a}^+$	annihilation and creation operators for Fermions	
a	arbitrary coefficient (with numerical subscripts)	Ch. 3, 4
	size of specimen	Ch. 5
	lattice spacing	§ 9.1, § 10.3
	isotope-effect exponent	Ch. 15
$\mathbf{a}_1$, $\mathbf{a}_2$, $\mathbf{a}_3$	lattice vectors	§ 10.3
Av	(subscript) average over grand canonical ensemble	Ch. 12, 14
$\mathbf{B}$, B	magnetic induction	
$\boldsymbol{\mathscr{B}}$, $\mathscr{B}$	magnetic induction in Ginzburg–Landau dimensionless units	Ch. 6, 7

$\hat{b}$, $\hat{b}^+$	annihilation and creation operators for Bosons	
$\mathbf{b}_1$, $\mathbf{b}_2$, $\mathbf{b}_3$	reciprocal lattice vectors	§ 10.3
$\mathbf{C}$	arbitrary vector	Ch. 4
C	arbitrary coefficient	§ 2.2, § 7.1, § 9.2, §14.5
	Bloch constant	Ch. 10
$\mathscr{C}$	matrix element for emission of phonon	§ 9.5
c	speed of light	
	(with subscripts) specific heat	Ch. 2, § 10.1, § 12.2
$\mathbf{D}$	electric displacement vector	Ch. 4
D ($\mathbf{r}$)	diffusion coefficient	§ 14.5
$D_{\mathbf{q}}$	matrix element for phonon emission	§ 10.5, § 10.6
D/Dt	convective derivative	
d	thickness of layer	§ 4.3, Ch. 5, § 14.5
$\mathbf{E}$, E	electric field	
E	energy level	
E ($\mathbf{k}$)	energy of quasiparticle excitation of wave vector $\mathbf{k}$	
$\mathscr{E}$	energy per unit area	§ 5.1
	energy per unit length	§ 7.2
e	electronic charge	
e	base of Naperian logarithms	
$\mathbf{e}$	polarization vector	§ 9.1, § 9.3
F	Helmholtz free energy	§ 2.3, § 12.2
	Gor'kov's anomalous propagator	Ch. 14
$\mathscr{F}$	Fröhlich's dimensionless electron–phonon interaction constant	
$\mathscr{F}$	Lorentz force per unit length of fluxon	§ 7.3
f	Fermi–Dirac occupation number	
	arbitrary function	§ 4.3, § 6.2
	amplitude of Ginzburg–Landau wave function	§ 7.2

f	density of body force	§ 6.3
$\mathscr{f}$	Helmholtz free energy per unit volume	
G	Gibbs free energy	
$\mathscr{G}$	Green function	Ch. 14
g	Gibbs free energy per unit volume	
H, H	magnetic field	
H_e	external magnetic field	
H_c	critical magnetic field	
$H_{\mathrm{c}1}$, $H_{\mathrm{c}2}$	lower and upper critical fields	
$H_{\mathrm{c}3}$	surface critical field	§ 5.3, § 6.5, § 7.3
$\boldsymbol{\mathscr{H}}$, $\mathscr{H}$	magnetic field in Ginzburg–Landau dimensionless units	Ch. 6, § 7.1
$\hat{H}$	Hamiltonian	
$\hat{\mathscr{H}}$	'grand' Hamiltonian, $\hat{H}-\mu\hat{N}_\mathrm{e}$	
$\hbar$	Planck's constant/2π	
h	$(B_x-\mathrm{i}B_y)/H_\mathrm{c}$	§ 5.2
h.c.	Hermitian conjugate of preceding expression	
I	electric current	
i	$\sqrt{-1}$	
i	(subscript) running suffix	
J, J	electric current density	
$\mathscr{J}$	current density including 'magnetization current'	§ 4.3
j	(subscript) running suffix	
K, $\mathscr{K}$	integral kernel	
K	spring constant	§ 9.1
K_0	modified Bessel function of the second kind	§ 7.2
k, k	wave vector	
k_0	Debye cut-off	
k_F	Fermi-surface wave vector	
k	{Abrikosov's scaling parameter	§ 7.1
	(subscript) running suffix	

Symbol	Meaning	Reference
k	Boltzmann's constant	
$\bar{k}$	combined wave-vector and spin symbol	
L	latent heat	§ 2.1
	size of body	§ 5.2
	size of Born–Karman box	Ch. 10
$L(\eta_1, \eta_2, \Delta)$	function defined in equation (13.20)	§ 13.2
L	left-hand side of barrier	Ch. 8
$\mathscr{L}$	Lagrangian	§ 9.1
	Taylor's transverse material function	§ 4.3
l	(superscript) longitudinal	§ 4.3
l	mean free path	
	distance between pinning centres	§ 7.3
	(subscript) running suffix	§ 3.5, § 4.1
$\boldsymbol{l}$, l	wave vector	
d$\boldsymbol{l}$	element of length in line integral	§ 3.4
$\bar{l}$	combined wave-vector and spin symbol	
M, M	magnetic moment	§ 1.3, § 5.2, § 6.2
	magnetic moment per unit length	§ 5.1, § 7.2
	magnetization (i.e. magnetic moment per unit volume)	§ 4.3, § 6.4, § 7.1
M	mass of system	§ 10.1
	atomic or ionic mass	Ch. 9, § 10.6, § 12.2
$\mathscr{M}$	transition matrix element	§ 8.1
m	electronic mass	
m^*	effective mass	Ch. 6, 10
m_s	effective mass of Ginzburg–Landau charge carriers	§ 6.3
N	number of atoms	
N_e	number density of conduction electrons	

$\hat{N}$, $\hat{N}_e$	total number operator for electrons	
$\hat{n}$	number operator (with subscripts)	
n	number density; arbitrary integer; demagnetizing coefficient/4π	§ 1.3, § 5.2.1
n	(subscript) normal	
$\hat{O}$	arbitrary operator	App. C
O	order of magnitude of	
P	transition probability	Ch. 8, § 12.3
$\hat{\mathbf{P}}$, $\mathbf{P}$	total momentum operator, eigenvalue	§ 13.1
$\mathscr{P}$	principal value	
$\mathbf{p}$, p	momentum	Ch. 3
	wave vector	§ 9.5
$\hat{\mathbf{p}}$	momentum operator	Ch. 3, § 14.3
$\bar{p}$	combined wave-vector and spin symbol	§ 9.5
p	pressure	§ 2.1, § 6.3
	Lifshitz–Sharvin branching number	§ 5.2
	efficacy of pinning centres	§ 7.3
p_F	Fermi momentum	
Q	$\lvert \mathscr{A} - \kappa^{-1}\nabla\phi \rvert$	§ 7.2
$\mathbf{Q}$	momentum transfer/$\hbar$	Ch. 9, 10
	wave vector of current-carrying Cooper pairs	§ 13.1
Q_k, q_k	normal coordinates	Ch. 9
$\mathbf{q}$	wave vector	
$\tilde{\mathbf{q}}$	unit vector in direction of $\mathbf{q}$	§ 10.4
$\bar{q}$	combined wave-vector and spin symbol	§ 9.5
R	right-hand side of barrier	Ch. 8
R	resistance	Ch. 1, § 5.2.2
	radius	§ 5.2
	surface resistance	§ 6.1
$\mathbf{R}$	relative coordinate, $\mathbf{r}_1 - \mathbf{r}_2$	§ 13.2

Symbol	Meaning	Reference
$\mathscr{R}$	rate of flux creep	§ 7.3
$\mathbf{r}$, r	radius vector, polar coordinate	
S	entropy	
	unitary transformation matrix	§ 3.2, § 11.1
$\mathrm{d}S$	element of surface area	
$\mathscr{S}$	Taylor's longitudinal material function	§ 4.3
s	acoustic velocity	
	entropy per unit volume	§ 2.1
s	(subscript) superconducting	
T	Wick's time-ordering operator	Ch. 14
	(superscript) transformed	
T	temperature	
T_c	superconducting transition temperature	
$\mathscr{T}$	tunnelling probability	Ch. 8
t	time	
t	(subscript) transverse	§ 4.3
U	internal energy	
$U(\mathbf{r})$	Bloch's deformation potential	§ 10.4
$\mathscr{U}$	factor of BCS factorizable interaction Hamiltonian	§ 11.2, § 12.2
u	internal energy per unit volume	§ 2.1
$u_{\mathbf{k}}(\mathbf{r})$	factor of Bloch function with periodicity of lattice	Ch. 10
$u_{\mathbf{k}}$	Bogoliubov coefficient	Ch. 11, 12, 13, 14
V	potential energy	Ch. 3, 9, 10
	matrix element of BCS interaction	Ch. 11, 12, 14, 15
$\tilde{V}$	average self-consistent potential	§ 14.1
$\mathscr{V}_{\mathbf{Q}}$	Fourier transform of $V(\mathbf{r})$	§ 9.5
$\mathbf{v}$, v	velocity	
v_F	Fermi velocity	
$v_{\mathbf{k}}$	Bogoliubov coefficient	Ch. 11, 12, 13, 14

Symbol	Meaning	Reference
W	condensation energy	Ch. 2, § 11.2
	complex function of $B_x - \mathrm{i}B_y$, defined in eq. (5.18)	§ 5.2
$\hat{W}^+$	Blatt's pair-creation operator	§ 13.3
$\mathscr{W}$	matrix element of electron–electron interaction, relative to its value on the energy shell	Ch. 11
w	$\phi - \mathrm{i}A_z$	§ 5.2
X	surface reactance	§ 6.1
$\mathbf{x}$	displacement from equilibrium position	Ch. 10
x_s	fraction of superconducting phase	
x	coordinate	
y	coordinate	
Z	surface impedance	§ 6.1
z	coordinate	
$\mathscr{z}$	$x+\mathrm{i}y$	§ 5.2
	z, in Ginzburg–Landau dimensionless units	Ch. 6
	$-\kappa\mathrm{i}\sqrt{(2\pi)}\,(x+\mathrm{i}y)$	§ 7.1
α	*ad hoc* coefficient	
	Ginzburg–Landau coefficient	§ 2.2, § 6.3
$\hat{\alpha}$, $\hat{\alpha}^+$	annihilation and creation operators for Bogoliubons	
β	$1/kT$	
	Ginsburg–Landau coefficient	§ 2.2, § 6.3
Γ	matrix element for external perturbation	§ 12.3
γ	Sommerfeld specific heat coefficient	Ch. 2
	$\hbar/m$	§ 6.3
	line width	§ 14.1
Δ	half energy gap	
	Gor'kov's gap function	
	increment	
δ	Kronecker, Dirac symbols	
	skin depth	§ 4.1, Ch. 6
	$\xi - \lambda$	Ch. 5, Ch. 7

Symbol	Meaning	Reference
$\delta/\delta\phi$	functional derivative	
ε	single-particle energy effective charge of Cooper pair	
ε_{ijk}	fundamental skew tensor	§ 3.5, § 4.1
$\boldsymbol{\epsilon}$	dielectric constant	§ 4.3, § 15.3, App. F
ζ	order parameter Riemann ζ-function	Ch. 2, 6 § 14.3
η	single-particle energy relative to Fermi surface arbitrary function	 § 4.1
η_{C}	Coulomb cut-off	§ 15.1
$\tilde{\eta}$	single quasiparticle energy, in Hartree–Fock approximation	§ 14.2
Θ	Debye temperature	
$\boldsymbol{\Theta}$	arbitrary vector field (gauge transformation)	Ch. 4
θ	arbitrary function angle positive infinitesimal	Ch. 3 Ch. 5, § 7.1 Ch. 14, App. E
ϑ_1, ϑ_3	ϑ-functions	§ 7.1
κ	inverse Thomas-Fermi screening length Ginzburg–Landau parameter integer multiple of flux quantum Φ_0 high-frequency penetration depth	§ 10.2, § 11.2, Ch. 15 § 3.4 § 4.1
Λ	London parameter	
Λ_{ij}	von Laue tensor	§ 3.5, § 4.1, § 6.1
λ	penetration depth	
λ_{D}	Debye length	§ 14.3.1
$\boldsymbol{\mu}$	magnetic permeability	§ 1.1, § 4.3, App. F
μ	chemical potential, Fermi energy	
μ_{B}	Bohr magneton	§ 13.4

Symbol	Meaning	Reference
ν	e/m	§ 6.3
	number of Cooper pairs	§ 13.3
	valency	§ 15.2
$\hat{\nu}$	number operator $\hat{\alpha}^+\hat{\alpha}$ for Bogoliubons	§ 12.2
Ξ	density of states	
Ξ_0	normal density of states	
ξ	Pippard's coherence length	
	arbitrary scalar function	§ 3.3
	normal coordinate	Ch. 9, § 10.4
$\prod$	product	
Π_k, π_k	canonical momenta	Ch. 9
ρ	number density	
	mass density	§ 2.1
	charge density	Ch. 3, 4
$\boldsymbol{\rho}$	resistivity	§ 1.1
$\tilde{\rho}$ ($\mathbf{k}$, ω)	spectral density function	Ch. 14
Σ	sum	
σ	electron spin	
$\boldsymbol{\sigma}$	electrical conductivity	Ch. 3, 4, § 5.3, § 6.2
τ	parameter of ϑ-functions	§ 7.1
	imaginary time	§ 14.1
$\mathrm{d}\tau$	volume element	
Φ	magnetic flux, fluxoid	
	arbitrary function	Ch. 4
	Lifshitz–Sharvin function	§ 5.2
Φ_0	fluxoid quantum	
$\|\Phi_0\rangle$	BCS ground state	
$\|\Phi_N\rangle$	Bloch ground state	
$\|\Phi_Q\rangle$	BCS current-carrying state	§ 13.1
ϕ	electrostatic potential	
	Cooper-pair wave function	§ 13.3
	harmonic-oscillator wave function	§ 9.2
φ	phase of Ginzburg–Landau function	

Symbol	Meaning	Reference
χ	arbitrary function (gauge transformation)	Ch. 3, 4
	spin wave function	§ 9.4, Ch. 10
	magnetic susceptibility	§ 13.3
$\hat{\Psi}$, $\hat{\Psi}^{+}$	annihilation and creation operators	
Ψ	Ginzburg–Landau function	
ψ	Ginzburg–Landau function in dimensionless units	Ch. 6, 7
	general wave function	Ch. 3, 9
	Bloch wave function	Ch. 10
Ω	normalization volume	
	volume of specimen	
ω	frequency	
$\omega_{\mathbf{q}}$, $\omega_{\mathbf{k}}$	phonon frequency	
ω_{D}	Debye cut-off frequency	§ 10.6
$\bar{\omega}$	average phonon frequency	
ω_{e}	absorption edge	§ 2.3
$\|0\rangle$	vacuum state	

Note: Superscript numerical references in the text refer to Appendix F.

PART I

PHENOMENOLOGICAL THEORY

CHAPTER 1

MACROSCOPIC ELECTROMAGNETIC PROPERTIES OF SUPERCONDUCTORS

SUPERCONDUCTIVITY was discovered by Kamerlingh Onnes as long ago as 1911, but for many years it remained a laboratory curiosity. The reasons were twofold: (a) superconductivity was an academic study, with no apparent prospect of application, and (b) for almost half a century the phenomenon defied explanation. Very recently—within little more than the last fifteen years—both these aspects have changed dramatically.

Although at first the prospect of practical application of superconductivity appeared remote, the phenomenon itself has fascinated both experimental and theoretical physicists ever since its discovery. At an early stage in the history of the subject it was recognized that the superconducting state represented a distinct thermodynamic phase. But while this clarification made some aspects of superconductivity less mysterious, it also raised new and difficult questions. One important puzzle was the smallness of the condensation energy. If, as seemed likely, the condensation was purely electronic in nature, why was the condensation energy not comparable in magnitude with the Fermi energy?

An even more fundamental problem was the nature of the interactions responsible for superconductivity. Before 1950 many theoreticians had speculated about these interactions. Almost the only common feature of these speculations was agreement about the *irrelevance* of the crystalline lattice! Indeed, the crucial experiments on the isotopic-mass dependence of the superconducting transition temperature could easily have been done several years earlier—and surely would have been but for the prevailing climate of opinion.

In 1950 Fröhlich observed that Bloch's (1928) theory of metallic conduction could be formulated as a field theory. In field-theoretical language the mechanism responsible for electrical resistance is scattering of electrons, with associated emission or absorption of quanta ('phonons') of the acoustic vibration field of the medium. This electron–phonon interaction implies an attractive electron–electron interaction. Fröhlich suggested that this was the interaction responsible for superconductivity. His suggestion immediately received striking experimental confirmation with the discovery of the isotope effect (see § 10.6).

Following the discovery of the isotope effect, superconductivity was transformed from a subject of speculation to a well-formulated problem. But early attempts to construct a perturbation-theoretical model based on Fröhlich's interaction encountered severe mathematical difficulties. It gradually became clear that, in the words of Casimir (1953),

> . . . it may be just as difficult to find the actual [superconducting] state by perturbation methods as it would be to find the crystal structure of solid argon starting from the ideal gas law. In this latter case we can obtain a good approximation to the thermodynamic functions of both solid and gas if we start from the right model. In a similar way I believe that one should try to make a good guess as to what the state of affairs is and then calculate the energy.

In 1951 Onsager conjectured that conduction electrons might form quasi-molecular pairs bound by the Fröhlich interaction. These pairs would have Boson properties and might undergo a Bose–Einstein condensation. Such conjectures were motivated by the widely-accepted belief (e.g. Ginzburg 1952) that a condensed Bose gas of charged but non-interacting particles would have superconducting properties. In 1955 Schafroth rigorously demonstrated this property of the charged Bose gas. Much effort was spent on the search for a Bose-gas model (e.g.

Schafroth *et al.* 1957). But the 'good guess' hoped for by Casimir actually came in a form somewhat different from Onsager's conjecture; although in the new theory the electrons are paired, the pairs now overlap considerably and the condensate is not a simple Bose condensate.

In 1956 Cooper showed that the Fermi sea is unstable against pair formation when there exists any attractive interaction—however small—between electrons. In the light of Cooper's result, Bardeen, Cooper, and Schrieffer (1957)† constructed a variational wave function for a ground state with complete electron pairing, and orthogonal functions for low-lying excited states having only a few such pairs broken. Since then many papers have been published, making the mathematics more elegant and powerful (but often more complicated!) until now it is possible to claim that the microscopic theory of superconductivity is as well-founded as accepted theories in other branches of solid-state physics.

In parallel with the growth of the systematic theory of superconductivity, there has been an equally spectacular growth in experimental work. Twenty years ago some twenty elements and fifty to sixty alloys and compounds were known to become superconducting at low temperatures. Superconductivity could then reasonably be regarded as an esoteric phenomenon, but there are now more than a thousand known superconducting alloys and compounds, and superconductivity promises to become a useful technique of the metallurgist.

An important motivation for studying these numerous alloys has been the hope of finding a high-temperature superconductor. However, this hope still remains unfulfilled—the highest known superconducting transition temperatures‡ remain obstinately below 20°K. Nevertheless, the development of cryogenics has been so great that extensive technological applications of

† In conformity with common usage 'the theory of Bardeen *et al.*' will be abbreviated to 'the BCS theory'.

‡ *Added in proof:* Matthias *et al.* (1967) have recently found that the solid solution $Nb_3(Al_{0\cdot 8}\,Ge_{0\cdot 2})_1$ has a superconducting transition at 20·05°K.

superconductivity are now feasible. Amongst the enormous number of new materials, Kunzler *et al.* (1961) found that some, for example Nb_3Sn, remain superconducting in high magnetic fields ($\sim 10^5$ gauss). Already superconducting solenoids are extensively used as sources of high magnetic fields; it is hardly an exaggeration to say that conventional electromagnets are obsolescent. Two other applications of superconductivity which promise to become important are superconducting cables for loss-free power transmission, and the cryotron (Buck 1956). The latter is a small and very simple device which exploits the metastability of superconducting persistent currents. The cryotron can serve as a logic or memory element in a computer circuit.

Long before the physics of superconductivity was understood, a number of phenomenological theories had been proposed. These describe most aspects of the behaviour of superconductors very well. In fact the phenomenological theories—slightly improved in the light of the microscopic theory—should usually suffice for those whose primary interest in superconductivity is in possible applications. By analogy, an electronic engineer devising transistor circuits uses a phenomenological picture of 'electrons' and 'holes', each with its own effective mass. He need not concern himself with all the intricacies of, for example, the theoretical justification of the band model.

Accordingly, the programme in this book will be to develop the phenomenological concepts in Part I and the microscopic theory in Part II. Part I needs only a very modest acquaintanceship with wave mechanics. With the microscopic theory it is not possible to proceed at such an elementary level—the formalism of second quantization is essential for even a 'simple' account. However, the author hopes that no reader will be deterred by this need. No previous knowledge of second quantization is assumed. The necessary aspects of the formalism of creation and annihilation operators will be developed *ab initio* in Chapter 9.†

† Chapter 9 is conceived partly as a demonstration that second quantization is not really as difficult as is sometimes believed.

1.1. Electrical resistance at low temperatures

At and above room temperature, the electrical resistance of most metals is approximately a linear function of temperature. This familiar fact provides the basis for resistance thermometry. At much lower temperatures, Matthiessen's rule is obeyed to a good approximation, i.e.

$$R = R_{\text{res}} + R_{\text{ideal}}. \tag{1.1}$$

The residual resistance R_{res} is independent of temperature, but depends on the purity and previous thermal and mechanical history of the specimen. R_{res} is high for impure or cold-worked specimens and low for very pure and well-annealed ones. The residual resistance arises from the scattering of conduction electrons by randomly-distributed imperfections in the crystalline lattice (see, for example, Wilson 1953 § 10.4).

The ideal resistance R_{ideal}, in contrast, is insensitive to the purity and history of the specimen. It does, however, depend on the temperature. At sufficiently low temperatures ($T \ll \Theta$, the Debye temperature) it has the Bloch form

$$R_{\text{ideal}} \propto T^5. \tag{1.2}$$

The constant of proportionality in (1.2) is characteristic of the material. The ideal resistance arises from the scattering of conduction electrons by the thermal vibrations of the lattice (Bloch 1928).

In 1908 Kamerlingh Onnes succeeded in liquefying helium. This achievement made the temperature region around 4°K accessible to experiment. Onnes, using liquid helium as a cryogenic fluid, undertook a systematic study of the electrical resistance of metals at low temperatures. In 1911 he discovered a startling anomaly in the resistance of mercury. At 4·2°K mercury has an abrupt drop in its electrical resistance. In fact, below that temperature Onnes was unable to detect any resistance at all. This new phenomenon was soon discovered in other metals and is nowadays known to be rather widespread; over a thousand superconducting metals and alloys are now known, thanks largely to the work of Matthias and his associates.† In Fig. 1.1 the electrical resistivity of a typical non-superconducting metal (platinum) and a superconducting one (mercury) are contrasted.

Below its transition temperature, the electrical resistance of a superconductor is immeasurably small. If a current in a superconducting ring is established, it is found to persist without

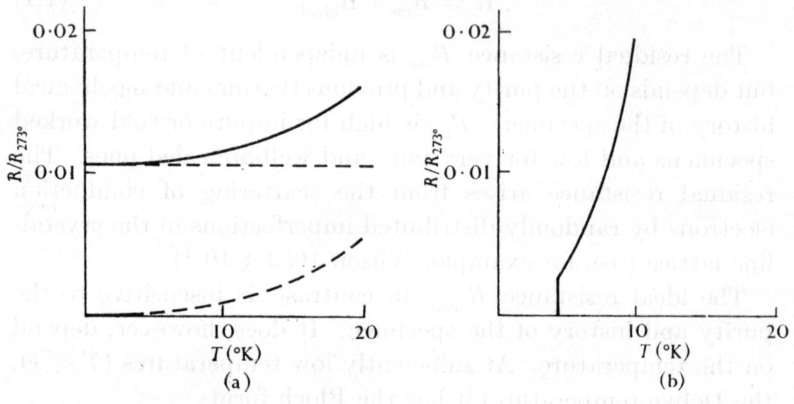

FIG. 1.1. The temperature-dependence of the electrical resistance of (a) a typical normal metal (platinum) and (b) a superconductor (mercury). From Shoenberg (1952 Fig. 2).

measurable attenuation for long periods of time. At M.I.T., Collins‡ has maintained a persistent current in a superconducting ring for $2\frac{1}{2}$ years. During that period he was unable to detect any diminution of current. From the failure of this and similar experiments to show any decay of the current, an upper bound for the resistance in the superconducting state can be established. If R_n is the resistance of the specimen when normal, then

$$R/R_n < 10^{-15}. \tag{1.3}$$

Thus, for all practical purposes, we may say that the resistance of a superconducting specimen is zero.

† See, for example, Matthias *et al.* (1963).

‡ Unpublished, but quoted by Crowe (1957), Lynton (1962), and Blatt (1964).

The early measurements referred to above were made with small currents. At first it was assumed that a superconducting specimen could carry arbitrarily large currents. If so, the phenomenon would have obvious technological importance, since in the absence of resistance there is no Joule energy dissipation. But all early attempts to construct high-field solenoids failed—usually destroying the apparatus!

The situation was gradually clarified in further experiments.† It was found that at any temperature T below the transition

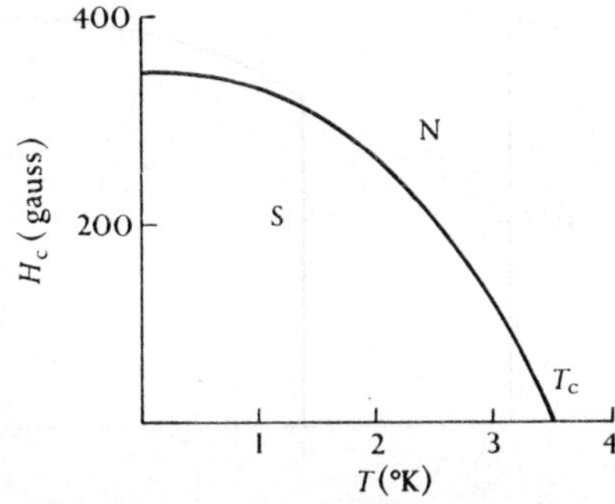

FIG. 1.2. The temperature-dependence of the critical field in tin.

temperature T_c, there is a 'critical' magnetic field H_c. When the applied field exceeds H_c, superconductivity is destroyed. To within a few per cent, the dependence of H_c on T is parabolic:

$$H_c = H_0(1-T^2/T_c^2). \tag{1.4}$$

For most superconducting elements, $H_0 \sim 300$ gauss.[(1)]

In Fig. 1.2 the critical field H_c is plotted as a function of temperature for a typical superconducting element (tin). Fig. 1.2 resembles a phase diagram for a thermodynamic system which has two phases. We shall see that this is indeed a useful interpretation of Fig. 1.2, i.e. *the superconducting state can be regarded as a distinct thermodynamic phase.*

† Onnes (1914); Tuyn and Onnes (1926).

(1) Numerical superscript references such as this refer to Appendix F.

Similarly the current I carried by a superconducting wire cannot be increased without limit. There is a critical value I_c, above which the electrical resistance becomes appreciable. Fig. 1.3 shows the dependence of the resistance on the current. As I increases through I_c, most of the resistance (often as much as 80 per cent) is abruptly restored. This discontinuity is followed by a gradual restoration of the remaining normal resistance as the current is increased further. Silsbee (1916) suggested that the phenomena of critical current and critical field are related.

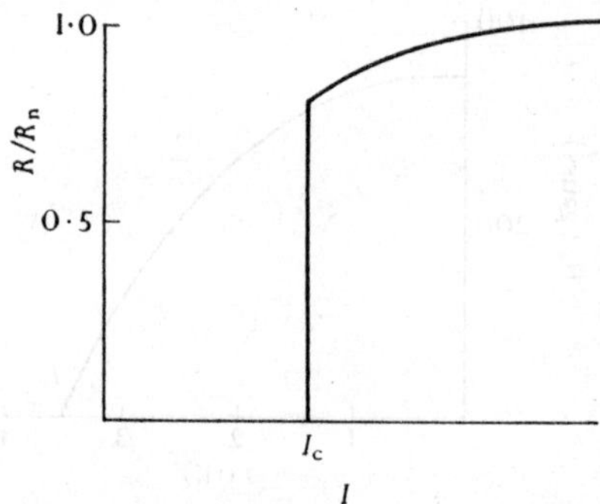

FIG. 1.3. The destruction of superconductivity by large currents.

More precisely, he postulated that destruction of superconductivity by a current should occur when the magnetic field produced by the current reaches H_c. In a straight cylindrical wire of radius a, the critical current I_c is given by the condition that the magnetic field at the surface of the wire is critical:

$$H_c = 2I_c/ac. \qquad (1.5)^{(2)}$$

Silsbee's hypothesis agrees well with experiment for most superconducting elements and for some other materials. However, it fails for many other materials—the so-called type II superconductors to be discussed in Chapter 7.

Let us now study the implications of zero electrical resistivity. For simplicity, the dielectric constant ϵ and magnetic permeability μ are both assumed to be unity. If then the resistivity

$\rho = 0$ (i.e. the conductivity $\sigma = \rho^{-1}$ is infinite), Ohm's law implies that

$$\mathbf{E} = 0. \tag{1.6}$$

It follows that curl $\mathbf{E} = 0$, and hence by one of Maxwell's equations that

$$\dot{\mathbf{B}} = 0, \tag{1.7}$$

i.e. the magnetic induction $\mathbf{B}$ is constant in time. To illustrate this result, there is a well-known and rather spectacular demonstration experiment (Arkadiev 1947) in which a permanent magnet 'floats' above a saucer of a superconducting metal (Fig. 1.4(a)). Equation (1.7) provides an explanation for

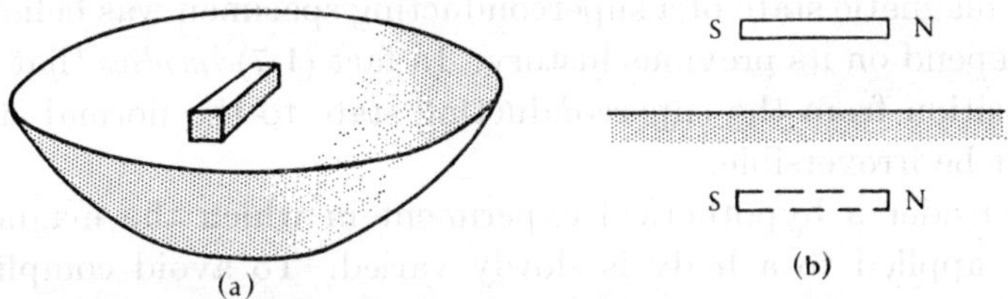

FIG. 1.4. The 'floating magnet' experiment: (a) over a concave saucer (stable equilibrium); (b) over a plane surface (neutral equilibrium). In (b) the image magnet is shown.

this effect. For simplicity consider first the case of an infinite plane surface instead of the concave saucer (Fig. 1.4(b)). Since $\mathbf{B} = 0$ in the absence of the magnet, it follows from (1.7) than $\mathbf{B}$ remains zero in the superconductor while the magnet is brought up to it. But the normal component $B_{\perp}$ has to be continuous, so that $B_{\perp}$ must vanish at the surface of the superconductor. We can satisfy this boundary condition by putting an image magnet below the surface, as shown. It is clear that there will be an equilibrium height where the repulsive force between the magnet and its image will just balance its weight. The equilibrium is stable against vertical displacements, but neutral to horizontal displacements; replacing the plane surface by the concave surface of Fig. 1.4(a) will stabilize the equilibrium.

Exercise. Find the equilibrium height of a small magnet (mass m and magnetic moment M) floating above a superconducting saucer of radius of curvature R.

1.2. The Meissner effect

In the next chapter we shall derive some thermodynamic relations for the superconducting transition. The derivations depend on the assumption that the superconducting state is a distinct thermodynamic phase, and that the transition between the normal and superconducting states is, at least in principle, reversible. When these thermodynamic formulae were first derived this was a bold assumption, since it was generally believed at that time that the transition was *not* reversible. The magnetic state of a superconducting specimen was believed to depend on its previous history. In fact (1.7) *implies* that the transition from the superconducting state to the normal state must be irreversible.

Consider a hypothetical experiment in which the magnetic field applied to a body is slowly varied. To avoid complications arising from the distortion of the field outside the body (cf. § 5.1), let the body have zero demagnetizing coefficient. For example, consider a long needle with its axis parallel to the field. Let the field be initially greater than the critical field H_c. While the body is in the normal state a static magnetic field will produce no electric current. (*Changes* in the field produce eddy currents, but they will decay Ohmically.) When the field falls to the value H_c, we would expect the body to become superconducting, with the magnetic induction taking the value H_c of the applied field. Equation (1.7) implies that the flux is now locked in the body and will persist even if the external field is removed (Fig. 1.5(a)). But on the other hand if the original cooling of the specimen were performed in the absence of a field, the body would have $\mathbf{B} = 0$ when it became superconducting. And now *this* value would be maintained by (1.7), even if an external field $H_e < H_c(T)$ were now imposed. Fig. 1.5(b) shows how the magnetic moment of the specimen would vary with the applied field if this argument were correct.

The figure (illustrating this hypothetical behaviour at constant temperature $T < T_c$) shows a typical hysteresis loop.

If the above argument were valid, it would clearly be wrong to regard superconductivity as a thermodynamic phase—the states with different amounts of trapped flux could not all be in thermodynamic equilibrium. Long-lived metastable states cannot be rejected *a priori* as impossible. Diamond at room temperature shows no tendency to graphitize, nor glass to

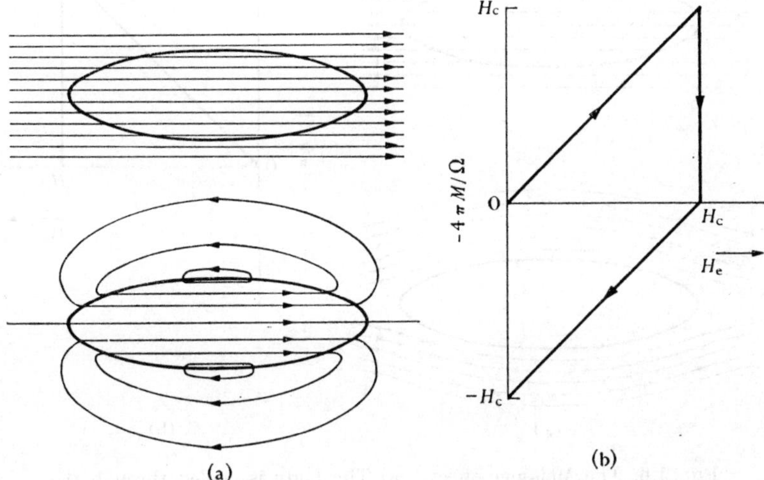

FIG. 1.5. The magnetic properties of a superconductor without a Meissner effect. (a) The body is cooled in a magnetic field which is subsequently removed; the flux in the body is trapped. (b) The magnetization curve, showing hysteresis.[8]

crystallize. But in all cases like these the excess entropy is frozen in by the *immobility* of the atoms at low temperatures. It seems hard to accept a similar metastability of supercurrents—characterized as they are by their extreme mobility.

The difficulty was resolved experimentally by Meissner and Ochsenfeld (1933). They showed that a body entering the superconducting state *expels* any magnetic flux in it (Fig. 1.6(a)). Equation (1.7) is replaced by the stronger condition

$$\mathbf{B} = 0. \tag{1.8}$$

The magnetic moment curve is now reversible (Fig. 1.6(b)), and the idea that the superconducting state is a thermodynamic phase is, *a posteriori*, justified.

Thus 'perfect conductivity' ($\rho = 0$) is not sufficient to describe superconductors; we also need 'perfect diamagnetism' ($\mathbf{B} = 0$). One is tempted to ask whether (1.8) alone might be

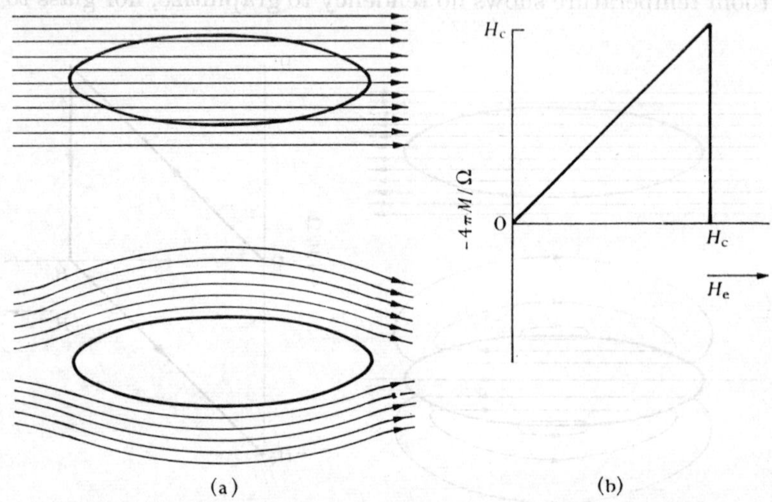

FIG. 1.6. The Meissner effect. (a) The body is cooled through the transition in the presence of a magnetic field; the field is expelled from the body when it becomes superconducting. (b) The magnetization curve[8] which is reversible (in contrast with Fig. 1.5(b)).

sufficient—does perfect diamagnetism imply perfect conductivity? This is a difficult question to answer because it is difficult to formulate precisely. Its resolution will be deferred until § 4.3: the formal result is that the two conditions (1.6) and (1.8) are independent and both are needed for a phenomenological description. However, Evans and Rickayzen (1964), in a careful study of how (1.6) and (1.8) arise in the microscopic theory, conclude that they are necessarily linked in *any* microscopic theory which includes scattering of electrons.

1.3. Shape-dependent phenomena is simply-connected bodies

In § 1.2 the body was assumed for simplicity to be a long needle, aligned parallel to the applied magnetic field. With this geometry, the magnetic field at the surface of the needle is equal to the applied field, provided that the demagnetizing coefficient of the needle is negligible. When the demagnetizing effect of the body is not negligible, the situation is more complicated. There is then a range of fields for which the equilibrium configuration is a finely-divided two-phase structure, known as the intermediate state.

Let us consider the problem of a sphere of radius a in an applied field $\mathbf{H}_e$. Outside the sphere, Laplace's equation is satisfied,

$$\nabla^2 \mathbf{B} = 0. \tag{1.9}$$

The boundary condition at large distances from the sphere is

$$\mathbf{B} = \mathbf{H}_e, \qquad (1.10)^{(3)}$$

while inside the sphere (1.8) holds. This field distribution is to be maintained by a surface current on the sphere.† However, such a surface current cannot compensate a discontinuity in the normal component $B_\perp$ of the field, so that $B_\perp$ must be continuous, i.e. from (1.8),

$$B_\perp = 0. \tag{1.11}$$

It is readily verified that the solution of (1.9) with boundary conditions (1.10), (1.11) is

$$\mathbf{B} = \mathbf{H}_e + \nabla(\mathbf{M}.\mathbf{r}/r^3), \qquad (1.12)^{(4)}$$

where the origin of the radius vector $\mathbf{r}$ is chosen to be at the centre of the sphere and the magnetic moment $\mathbf{M}$ of the sphere is given by

$$\mathbf{M} = -\tfrac{1}{2}\mathbf{H}_e a^3. \qquad (1.13)^{(5)}$$

† For many purposes it is convenient to fulfil the condition $\mathbf{B} = 0$ by putting the magnetic permeability $\mu = 0$, but allowing a magnetic field $\mathbf{H} \neq 0$ to persist. In this way of looking at the problem, the surface currents are 'free' currents rather than 'true' currents, and need not be explicitly considered. For a full discussion of this formalism, see Shoenberg (1952).

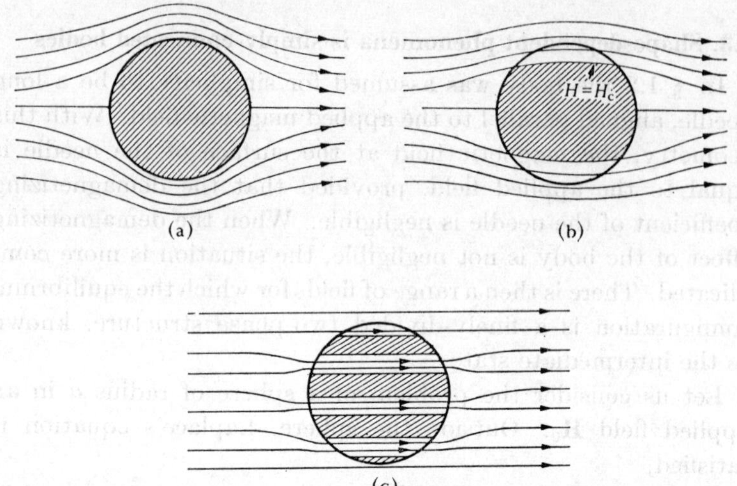

Fig. 1.7. Entry to the intermediate state: (a) shows that the field will first become critical on the equator; (b) shows that an equatorial band of normal material cannot be an equilibrium configuration; in (c) the laminar structure proposed by Peierls (1936) and London (1936) is illustrated. The superconducting regions are shaded.

Thus a superconducting sphere behaves like a magnetic dipole of moment **M** provided that the field is nowhere critical (Fig. 1.7(a)). The field will be greatest at $r = a$, on the equator of the sphere (the direction of the applied field $\mathbf{H}_e$ defining the polar axis); there it will have the magnitude

$$B_{\max} = \tfrac{3}{2} H_e. \tag{1.14}$$ [6]

Hence it follows that when

$$H_e = \tfrac{2}{3} H_c \tag{1.15}$$

superconductivity must begin to be destroyed. More generally, for an ellipsoid of demagnetizing coefficient $4\pi n$ $(0 < n < 1)$, equation (1.15) is replaced by

$$H_e = n H_c. \tag{1.16}$$

For fields greater than nH_c but less than H_c, the normal (n-) and superconducting (s-) phases must coexist. One's first guess would be that the superconductivity of an equatorial ring is gradually 'eroded' as the field is increased. But it is easy to see that this configuration is not possible, for as such hypothetical erosion proceeds, the sphere becomes effectively a prolate spheroid with n smaller than 2/3. It follows from (1.16) that for a given value of H_e, the field on the (shrunken) equator would be reduced. However, the n-s boundary would remain convex towards the n-phase. Hence the field in the newly-established ring of n-phase around the equator would be lower than the field at the boundary between the n- and s-regions. We are forced to conclude that no macroscopic distribution of n- and s-phases is stable when $nH_c < H_e < H_c$. The escape from the apparent paradox is that the *whole* body forms a finely-divided laminar structure of alternate n- and s-domains, the intermediate state proposed by London (1936) and Peierls (1936). The factors determining the grain size will be discussed in § 5.2. For the present we shall assume that these domains are so small that the intermediate state is effectively homogeneous. The field is then precisely H_c everywhere in the material and all n-s boundaries are in equilibrium. The fraction of superconducting material is

$$x_s = (H_c - H_e)/H_c(1-n), \tag{1.17}$$

and the magnetic moment M is

$$M = -\Omega n(H_c - H_e)/4\pi(1-n). \qquad (1.18)^{(7)}$$

Here Ω is the volume of the specimen. Fig. 1.8 illustrates the magnetization curve for a sphere ($n = \frac{2}{3}$).

Experiments agree tolerably well with the above macroscopic theory for large enough specimens; however there is often some hysteresis. The normal regions tend to 'supercool', and the superconducting regions to 'superheat'. For smaller specimens there are more systematic departures from the above ideal behaviour, associated with the detailed structure of the intermediate state (see § 5.2).

For non-ellipsoidal bodies one cannot define a demagnetizing coefficient, and the situation is still more complicated—in

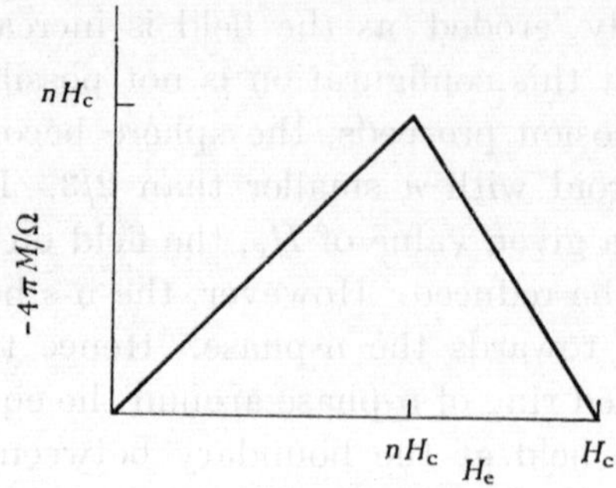

FIG. 1.8. The magnetization curve[8] for a superconducting ellipsoid; the falling part corresponds to the intermediate state. The curve shows the ideal behaviour of the London–Peierls theory. Departures from this ideal curve are discussed in § 5.2.1 and illustrated in Fig. 5.5.

general it is possible to have some regions of intermediate state and other regions of pure n- or s-character.

1.4. Persistent currents

In a multiply-connected body, equation (1.8) fails to uniquely define the state of the body. To see this, we replace the simply-connected body of Fig. 1.6(a) by a torus (Fig. 1.9). When the body becomes superconducting the flux in the *material* is expelled, but there is no reason why the flux in the *hole* should be expelled too. In fact it is not expelled; instead it is trapped inside the hole by the infinite conductivity of the ring. If we remove the external field the torus will carry just that current necessary to preserve the flux through the ring. This is a 'persistent' current of the type described on p. 6.

Of course, such persistent-current-carrying states are not true equilibrium states. In the absence of an external field the state of lowest energy is the one with no trapped flux. But the rate of decay of such a persistent current is determined by the electrical resistance of the superconductor. In principle it is possible for a body to show a virtually complete Meissner effect, while nevertheless a persistent current in it decays to zero with a finite lifetime. In fact, no measurable decay has been found.†

† Collins's experiment; see footnote, p. 6.

The possibility of trapped flux in a multiply-connected superconductor explains why many experiments have not shown the ideal behaviour described above. For example, Onnes and Tuyn (1923) failed to discover the Meissner effect

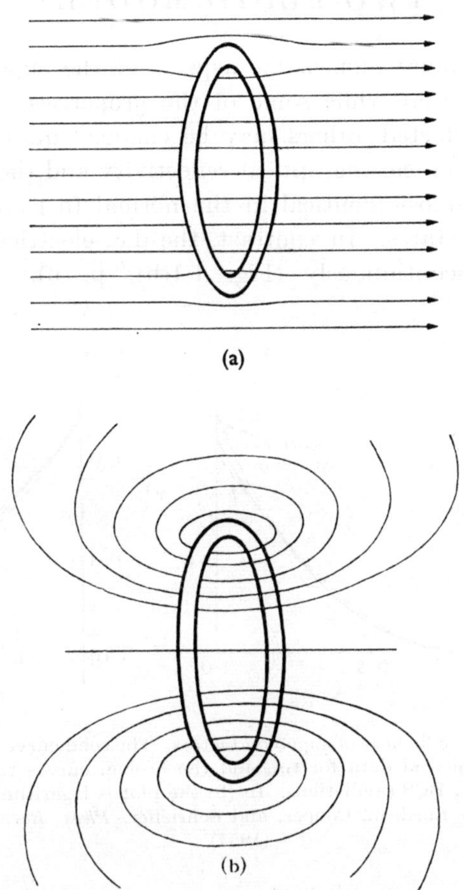

FIG. 1.9. Flux trapped by a superconducting torus.

because they used a *hollow* sphere in their work. When such a body enters the intermediate state, multiply-connected superconducting domains are inevitably formed and nearly all the flux can be trapped by them.

CHAPTER 2

TWO-FLUID MODELS

ONE of the more remarkable aspects of the superconducting transition is that while some of the properties of a material are barely affected, others may be changed quite drastically. Thus, for instance, the optical reflectivity and the elastic constants are almost identical in the normal (n-) and superconducting (s-) states. In contrast, the d.c. electrical resistivity vanishes discontinuously (Fig. 1.1(b), p. 6). Among the

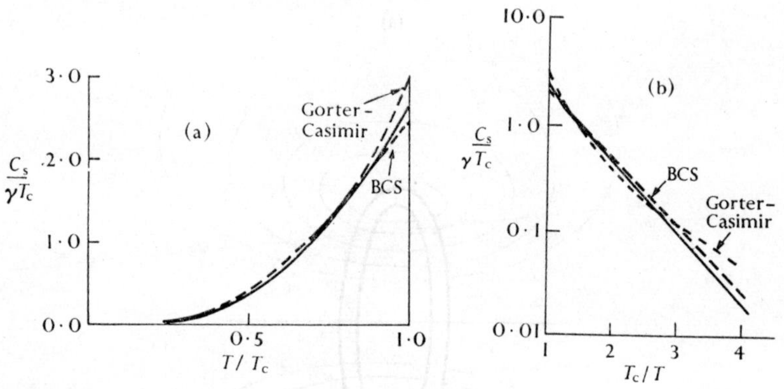

FIG. 2.1. Specific heat of superconductors. The solid curve represents the experimental data for tin, and the broken curves the Gorter–Casimir and BCS predictions. In (b) the plot is logarithmic against T^{-1}. From Bardeen, Cooper, and Schrieffer, *Phys. Rev.* **108,** 1175 (1957).

other properties which change drastically, we shall single out the specific heat for discussion at some length. Historically the specific heat anomaly was one of the earliest features of the superconducting transition to be discovered; it has remained important because of its crucial role in clarifying and testing theoretical concepts.

In normal metals at temperatures well below the Debye temperature, the specific heat is dominated by an 'electronic' contribution c_{el} of Sommerfeld form, with a relatively small Debye background contribution:

$$\left.\begin{aligned} c &= c_{\text{el}}+c_{\text{Debye}}, \\ c_{\text{el}} &= \gamma T, \\ c_{\text{Debye}} &\propto T^3. \end{aligned}\right\} \tag{2.1}$$

A superconducting metal obeys the law (2.1) above its transition temperature T_{c}. Moreover, superconductivity can be destroyed by a sufficiently large magnetic field ($> H_{\text{c}}$, cf. Fig. 1.2, p. 7) and the specific heat of the normal state can thus be measured below T_{c}. It continues to obey (2.1). The solid curves in Fig. 2.1(a), (b) illustrate the behaviour of superconducting tin. The ordinate is the ratio of the electronic specific heat in the n- and s-states. In calculating this electronic contribution for the s-state, it is assumed† that the lattice contribution does not change on passing through the transition. From the normal state data the Debye contribution c_{Debye} is extracted, leaving the electronic specific heat for the s-phase,

$$c_{\text{el}} = c-c_{\text{Debye}}. \tag{2.2}$$

At the transition temperature T_{c}, a finite discontinuity is observed in c_{el}, but there is no latent heat. I.e. when there is no magnetic field the transition is of second order in Ehrenfest's (1933) classification. Except at the lowest temperatures the specific heat of a superconductor can be approximated quite well by a T^3 curve (though with a coefficient very different from c_{Debye}!). But at very low temperatures this T^3 relation fails. Instead it is found (Corak *et al.* 1956) that

$$c_{\text{el}} \sim \exp(-b/T). \tag{2.3}$$

† In support of this assumption, note that the elastic constants are, to a good approximation, unaffected by the transition.

The solid curve in Fig. 2.1(b) shows the specific heat plotted logarithmically against the inverse temperature, and illustrates that (2.3) remains valid over the entire temperature range $0 < T < T_c$. We shall later (§ 2.3) that equation (2.3) provides strong evidence for the existence of an energy gap.

The presence of a subcritical magnetic field hardly modifies the specific heat of either the n- or the s-phase. But (cf. (1.4)) the transition to the superconducting state occurs at a lower temperature. This fact implies that the transition in the presence of a magnetic field must be associated with a latent heat.

A good deal of evidence suggests that the superconducting transition is primarily an electronic one—thus, for example, the 'lattice' contributions to the specific heat and thermal conductivity in the superconducting phase are only very slightly shifted from their normal values. The elastic constants, crystal structure, etc., are likewise unaffected.

In § 2.1 we shall discuss the thermodynamic implications of the above behaviour of the specific heat. As in other transitions with a specific heat discontinuity but with no latent heat, we are reminded of an Ehrenfest second-order process. Hence, following Landau (1937*a*), we look for a parameter ζ which is to be a measure of the long-range order in the system. A familiar example of such a long-range-order parameter is the molecular field in the Weiss theory of ferromagnetism.

In § 2.2 the 'two-fluid' concept is introduced. It provides a framework for introducing a Landau order parameter. It is convenient to think of the conduction electrons undergoing a progressive condensation as the temperature falls. At any temperature $T < T_c$ only a fraction ζ of the electrons are in the condensate ('superfluid' electrons) and the remainder are 'normal' electrons.

2.1. Simple thermodynamic considerations

In this section the superconducting transition is assumed to be reversible. We shall relate the thermodynamic properties of superconductors to the critical field curve (Fig. 1.2, p. 7).

The applied magnetic field $\mathbf{H}_e$ is treated as an intensive thermodynamic variable; the magnetic contribution to the internal energy of a system† may be written

$$dU_m = \mathbf{H}_e . d\mathbf{M}, \tag{2.4}$$

where $\mathbf{M}$ is the magnetic moment of the system. Equation (2.4) is analogous to the familiar hydrostatic term

$$dU_{\mathrm{Hyd}} = -p\, d\Omega, \tag{2.5}$$

where p is the hydrostatic pressure and Ω is the volume of the system. Combining (2.4) and (2.5), and adding a heat term $T dS$, the differential expression for the internal energy of a magnetic system is

$$dU = \mathbf{H} . d\mathbf{M} - p\, d\Omega + T\, dS. \tag{2.6}$$

In the application of thermodynamics to superconductivity, volume changes are however usually negligible and the hydrostatic term can be dropped.

To avoid complications, the superconducting specimen is taken to be a long needle aligned parallel to the applied field‡ so that its demagnetizing coefficient may be neglected. With the assumption of reversibility, the transition may be treated as a true thermodynamic phase change. By analogy with the Gibbs function of ordinary thermodynamic systems, a 'magnetic Gibbs function'

$$G = U - TS - \mathbf{H}_e . \mathbf{M} \tag{2.7}$$

† The magnetic energy is often written in the more fundamental form

$$du_m = \frac{1}{4\pi} \mathbf{H} . d\mathbf{B}, \tag{2.8}$$ [9]

where u_m is the magnetic energy *per unit volume* and where $\mathbf{H}$ and $\mathbf{B}$ are the *local* values of the magnetic field and induction respectively. The form (2.4) is equivalent to (2.8) apart from a term[10] $\frac{1}{4\pi}\mathbf{H}_e . d\mathbf{H}_e$, the magnetic energy associated with the applied field $\mathbf{H}_e$ *in vacuo*. For a proof of this equivalence, see Appendix A.

‡ This geometry is not essential, but for other configurations the destruction of superconductivity will be spread over a range of values of the applied field H_e—there will be an intermediate state (§ 1.3, § 5.2).

may be defined. A necessary condition for the two phases (conveniently denoted as s and n) to coexist is that the magnetic Gibbs function per unit volume $g = G/\Omega$ takes the same value in both phases. Thus (in an obvious notation) the critical field is determined by the condition

$$g_s(H_c) = g_n(H_c). \qquad (2.9)$$

Since the normal phase has been assumed to be nonmagnetic, g_n is independent of the field, i.e.

$$g_n(H_c) = g_n(0). \qquad (2.10)$$

However, the magnetic moment of the s-phase (regarded as a perfect diamagnetic region, cf. (1.8)) is

$$\mathbf{M} = -\Omega \mathbf{H}_e/4\pi. \qquad (2.11)^{(10)}$$

From (2.11) and (2.7) it follows that

$$\begin{aligned} G_s(H_e) &= G_s(0) - \int_0^{H_e} \mathbf{M}(\mathbf{H}_e).\mathrm{d}\mathbf{H}_e \\ &= G_s(0) + \Omega H_e^2/8\pi. \qquad (2.12)^{(11)} \end{aligned}$$

Hence, with (2.9),

$$g_n(0) - g_s(0) = H_c^2/8\pi. \qquad (2.13)^{(12)}$$

Differentiating (2.13) with respect to T, the difference in entropy between the two phases, per unit volume, is found to be

$$\begin{aligned} (S_n - S_s)/\Omega &= -\frac{\mathrm{d}}{\mathrm{d}T}(g_n - g_s) \\ &= -\frac{1}{4\pi} H_c \frac{\mathrm{d}H_c}{\mathrm{d}T}. \qquad (2.14)^{(10)} \end{aligned}$$

From this entropy difference, the latent heat of transition in a magnetic field follows (Keesom 1924):

$$\begin{aligned} L &= T\,(S_n - S_s)/\Omega \\ &= -\frac{TH_c}{4\pi}\frac{\mathrm{d}H_c}{\mathrm{d}T} \quad \text{per unit volume,}^{(10)} \end{aligned}$$

or, if ρ is the density,

$$L = -\frac{TH_c}{4\pi\rho}\frac{dH_c}{dT} \text{ per unit mass.}^{(10)} \qquad (2.15)$$

Equation (1.4) and Fig. 1.2 show that dH_c/dT is always $\leqslant 0$. Hence, from (2.14), the entropy of the s-phase never exceeds that of the n-phase. *In other words, the superconducting state is more ordered than the normal state.*

The difference in specific heat between the n- and s-phases is found by differentiating (2.14) with respect to temperature:

$$\begin{aligned}\Delta c &\equiv c_s - c_n\\ &= \frac{T}{\rho}\frac{\partial}{\partial T}\frac{S_s - S_n}{\Omega}\\ &= \frac{T}{4\pi\rho}\left\{H_c\frac{d^2H_c}{dT^2}+\left(\frac{dH_c}{dT}\right)^2\right\}.\end{aligned} \qquad (2.16)^{(10)}$$

In particular, at $T = T_c$, H_c vanishes and (2.16) simplifies to

$$(\Delta c)_{T_c} = \frac{T_c}{4\pi\rho}\left(\frac{dH_c}{dT}\right)^2_{T_c}, \qquad (2.17)^{(10)}$$

which is known as Rutger's formula. At this temperature, (2.15) gives zero latent heat, i.e. the transition in the absence of a magnetic field is of second order.

At $T = 0$, $(S_n - S_s)/\Omega$ must once more vanish, by the third law of thermodynamics. Hence, from (2.14), it follows that

$$\lim_{T\to 0}\frac{dH_c}{dT} = 0. \qquad (2.18)$$

All the above thermodynamic predictions agree well with experiment, and thus confirm the initial assumption that the superconducting transition is in principle reversible.

2.2. The order parameter and two-fluid models

As we have seen in § 2.1, the superconducting transition in the absence of a magnetic field is of second order (Ehrenfest

1933). In other words, the specific heat is discontinuous but there is no latent heat. The low-temperature phase can therefore be characterized by a parameter ζ of long-range order (Landau 1937*a*). This parameter is required to take the value unity at $T = 0$, i.e. the system at absolute zero is fully ordered. At the transition temperature T_c, ζ must vanish (cf. Landau and Lifshitz 1959 Chapter 14, pp. 430 ff.).

As well as having an explicit dependence on temperature, the Gibbs free energy G should also have a functional dependence on ζ. Minimization of G will then yield a functional dependence of ζ on T. Near the transition temperature, g may be expanded in a Taylor series:

$$g = g_0+\alpha\zeta+\tfrac{1}{2}\beta\zeta^2+\ldots. \tag{2.19}$$

At $T = T_c$, G must have its minimum at $\zeta = 0$. Hence

$$\alpha(T_c) = 0;\ \beta(T_c) > 0. \tag{2.20}$$

Below the transition the superconducting phase is the stable one. Hence the Gibbs free energy of the superconductor should be lower than that of the normal phase, i.e. for $T < T_c$, $\alpha(T) < 0$ to ensure that $G-G_0 < 0$.

When Landau developed his theory of Ehrenfest second-order transitions, it was thought that such transitions were widespread. However, it is now believed that the superconducting transition is unique in following the Ehrenfest scheme. Other transitions formerly classified as of second order are in fact Onsager-type λ-transitions. In these λ-transitions the specific heat has a logarithmic infinity at the transition, i.e. in the neighbourhood of the transition the specific heat is dominated by a term proportional to $\ln|T-T_c|$ (see Fig. 2.2). While there is no general theory of λ-transitions, the theory of the Ising ferromagnet (Onsager 1944) makes it clear that the logarithmic divergence arises from the behaviour of the short-range order. In particular as we approach a λ-transition from the high-temperature side, the specific heat rises sharply before we actually reach the transition. There is as yet no long-range order, yet the system has, as it were, some 'premonition' of the approaching transition. What happens is that although the long-range order has not yet become established, there is *some* order extending over several atomic distances.

In contrast to these λ-transitions, the superconducting transition appears quite sharp (Fig. 2.1); the normal phase shows no

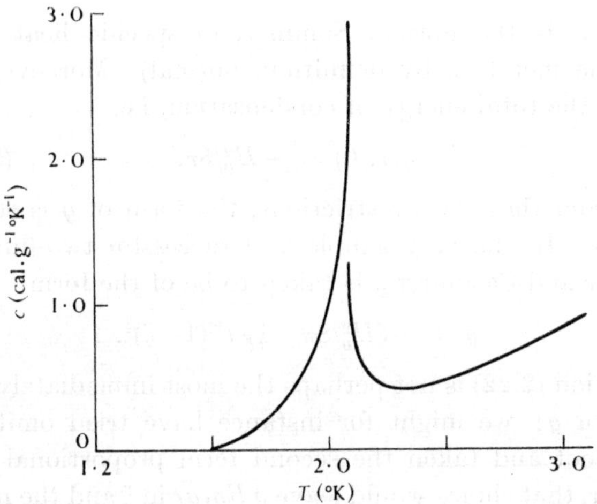

FIG. 2.2. A typical λ-transition: the specific heat of liquid He^4. From Atkins, *Liquid Helium*, p. 33. Cambridge University Press, 1959.

foreknowledge of the approaching transition. Thus superconductivity is characterized by the existence of long-range order, but *there is no short-range order* apart from that implied by the long-range order.†

† Thouless (1960) has shown that the above statement is not strictly true: there is some short-range order. In a microscopic-theoretical study of critical fluctuations, he finds that within a region $|T-T_c|<10^{-12}$ °K, the specific heat has a singularity of $|T-T_c|^{-\frac{1}{2}}$ character (see also Brout 1965 § 7.2). The smallness of the singular region is related to the Pippard coherence length ξ (§ 6.2). Since ξ is so large, a superconducting charge has $\sim 10^{12}$ 'neighbours', i.e. the 'short-range' order is really of fairly long range.

For many purposes it is convenient to picture the order parameter ζ as representing the fraction of electrons condensed into an ordered configuration, and correspondingly, $1-\zeta$ as the fraction of electrons not condensed. We will call these the fractions of 'super' and 'normal' electrons respectively.

This interpretation of ζ leads to the so-called two-fluid models—plural rather than singular because the functional form of $g(\zeta, T)$ is as yet unspecified. We require that

$$g(0, T_c) = -\tfrac{1}{2}\gamma T_c{}^2, \tag{2.21a}$$

where γT is the normal Sommerfeld specific heat (since if $\zeta = 0$ the metal is, by definition, normal). Moreover $G(1, 0)$ must be the total energy of condensation, i.e.

$$g(1, 0) = -H_0^2/8\pi. \qquad (2.21b)^{(11)}$$

Apart from these two restrictions, the form of g is somewhat arbitrary. In the very simple and successful two-fluid model of Gorter and Casimir,† g is taken to be of the form

$$g = -\zeta H_0^2/8\pi - \tfrac{1}{2}\gamma T^2(1-\zeta)^{\frac{1}{2}}. \qquad (2.22)^{(11)}$$

Equation (2.22) is not perhaps the most immediately obvious choice for g; we might for instance have tried omitting the square root and taken the second term proportional to $1-\zeta$. However, that choice would leave g *linear* in ζ and the minimum would occur at $\zeta = 0$ if the coefficient of ζ were positive, or at $\zeta = 1$ if the coefficient were negative. A transition would occur at the temperature T' at which this coefficient changed sign,[(11)]

$$\tfrac{1}{2}\gamma T'^2 = H_0^2/8\pi.$$

But, at the transition, ζ would change discontinuously from unity to zero, i.e. the model would have a *first*-order transition. To avoid this, Gorter and Casimir made the implicit assumption (in (2.22)) that the properties of the two fluids are not independent.

† Gorter and Casimir (1934) actually considered a more general class of models with

$$g = -\zeta H_0^2/8\pi - \tfrac{1}{2}\gamma T^2(1-\zeta)^C, \qquad (2.23)^{(11)}$$

and chose $C = \frac{1}{2}$ as the model in best agreement with experiment.

To find the properties of the Gorter–Casimir model, (2.22) must be minimized with respect to ζ:[11]

$$\tfrac{1}{4}\gamma T^2(1-\zeta)^{-\frac{1}{2}} = H_0^2/8\pi,$$

i.e.

$$\zeta = 1-(T/T_c)^4, \tag{2.24}$$

where we have made use of the fact that $\zeta = 0$ at $T = T_c$, i.e. that

$$\tfrac{1}{4}\gamma T_c^2 = H_0^2/8\pi. \qquad (2.25)^{(11)}$$

On substituting for ζ from (2.24) in (2.22), and differentiating twice with respect to temperature, the specific heat is found to be

$$c_s = 3\gamma T^3/T_c^2. \tag{2.26}$$

Thus the Gorter–Casimir model can reproduce the experimental specific heat quite well (Fig. 2.1), except at the lowest temperatures, $T \ll T_c$. At these very low temperatures the experimental values deviate significantly from the T^3-dependence of Gorter and Casimir (see equation (2.3)).

The critical magnetic field is easily calculated; from (2.16) it is related to the specific heat discontinuity Δc. The normal specific heat is of course $c_n = \gamma T$ and hence the critical field curve is parabolic:

$$H_c = H_0(1-T^2/T_c^2). \tag{2.27}$$

Equation (2.27) agrees with experiment to within about 3 per cent (cf. Fig. 2.3).

Exercise. Find the form which equations (2.24), (2.26), and(2.27) take in the more general Gorter–Casimir model (2.23).

2.3. The energy-gap model

As we have seen, the Gorter–Casimir T^3 specific-heat law fails at very low temperatures, and is replaced by the exponential dependence (2.3).† In Fig. 2.1(b), the specific heat (or

† The experimental data are insufficient to determine whether the dependence of the specific heat on temperature is a pure exponential rather than, say, $T^n \exp(-b/T)$, since for reasonably small n, whether positive or negative, the exponential will 'swamp' the other factor.

more precisely c_s/c_n) is plotted logarithmically against T^{-1}, to illustrate the exponential behaviour. Such an exponential specific heat is characteristic of a system with an energy gap. By that we mean that the ground state is separated by a finite energy 2Δ from all the excited states—or at least 'almost' all—in a sense to be made more precise later. The factor 2 is chosen for subsequent convenience, so that Δ will be the

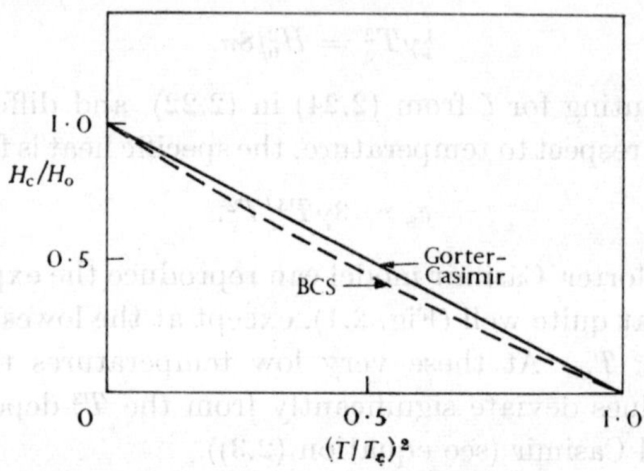

FIG. 2.3. The temperature dependence of the critical magnetic field. From Bardeen, Cooper, and Schrieffer, *Phys. Rev.* **108**, 1175 (1957).

threshold energy for a 'quasiparticle' excitation.† The idea that in superconductors there should be an energy gap $2\Delta \sim kT_c$ is not particularly new. It began to appear in the late 1940s in various theoretical speculations, of which the earliest were those of Ginzburg (1946) and Koppe (1947, 1950). Since then, an overwhelming mass of experimental evidence in support of the idea has accumulated.

The five main sources of experimental evidence for an energy gap in superconductors are:

(a) the low-temperature specific heat,
(b) the low-temperature thermal conductivity,
(c) the microwave absorption spectrum,
(d) ultrasonic attenuation, and
(e) tunnelling.

† In any process, only an *even* number of quasiparticles can be produced (cf. Appendix D).

(a) As remarked above, the specific heat measurements of Corak *et al.* (1956) show a low-temperature variation of the form (2.3) (and cf. Fig. 2.1(b)). It is very difficult to see how such a temperature dependence could arise, except as a consequence of an energy gap. But if there *is* an energy gap, the exponential variation is immediately explained.

(b) The thermal conductivity of tin, as measured by Goodman (1953), shows a temperature dependence similar to that of the specific heat. It is illustrated in Fig. 2.4. Historically

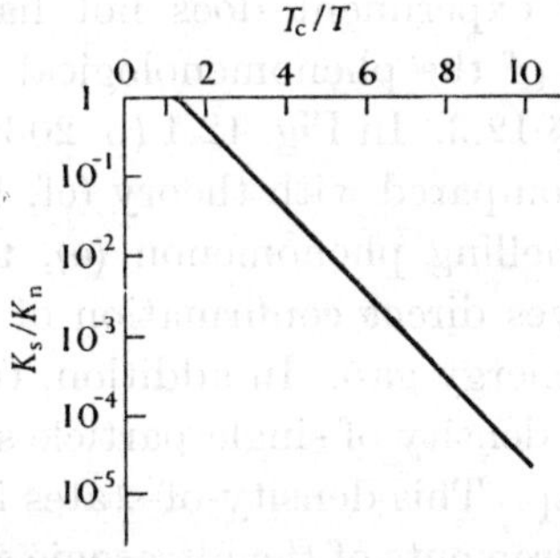

FIG. 2.4. The thermal conductivity of tin at low temperatures (after Goodman 1953).

this was an important experiment, in so far as it was probably the first to provide real evidence for an energy gap. Qualitatively, the interpretation of Goodman's results is straightforward: the electronic contribution to the conductivity will vary with the number of 'effective' electrons. The states above the energy gap form a quasi-continuum, and electrons excited into such states will behave in a more or less normal way. The number of such effective electrons will vary according to a Boltzmann factor $\exp(-\Delta/kT)$. This will be the dominant factor in the temperature dependence of the thermal conductivity. However, the quantitative theory of the thermal conductivity is somewhat more complicated than that of the specific heat, and will not be discussed in this book. For an account of the theory see Bardeen *et al.* (1959).

(c) One of the most clear-cut experiments showing the presence of an energy gap is the measurement by Glover and Tinkham (1957) of the microwave absorption spectrum.† In this experiment a sharp absorption edge is seen, whose angular frequency ω_e gives immediately the size of the energy gap $2\Delta = \hbar\omega_e$. The Glover–Tinkham experiment enables the energy gap to be measured in a very direct way, at any temperature $T < T_c$.

(d) Morse and Bohm (1957) have determined the energy gap from their measurements of the attenuation of ultrasound. Since this experiment does not have a simple intepretation in terms of the phenomenological theory, its discussion is deferred to § 12.3. In Fig. 12.1 (p. 206), their experimental energy gap is compared with theory (cf. Fig. 2.6).

Lastly, the tunnelling phenomenon (e), to be discussed in Chapter 8, also gives direct confirmation of the existence and magnitude of the energy gap. In addition, tunnelling provides information on the density of single-particle states in the neighbourhood of the gap. This density-of-states information is also contained in measurements of the ultrasonic attenuation (Morse and Bohm 1957, and see § 12.3) and the nuclear spin-lattice relaxation time (Hebel and Slichter 1959).

In view of the very strong evidence for an energy gap, the gap ought to find a place in the phenomenological theory. From time to time two-fluid models with a gap have been constructed (see, e.g. Ginzburg 1946, Koppe 1950, Bernardes 1957). It is assumed that there is a minimum energy necessary to excite an electron out of the superfluid into the normal fluid. In developing such models, various *ad hoc* assumptions are required. Instead of recapitulating the details of these models, we shall rather construct a *new* energy-gap model, in which the assumptions made are borrowed from the results of the microscopic theory of Bardeen *et al.* (1957), described in Chapters 11 and 12. This improved energy-gap model will be described as the 'BCS model'.

† The detailed calculation of the electromagnetic absorption is not so straightforward. It will be discussed later in terms of the microscopic theory; see § 12.3.

We postulate the existence of quasiparticle excitations, in one-to-one correspondence with the particle excitations of a normal Fermi gas. The thermally excited quasiparticles form the 'normal fluid'. These quasiparticles will be discussed in more detail in Chapter 12. By a 'quasiparticle' we merely mean an entity which has some particle-like attributes even though its detailed structure may be complicated. For example an electron in an ionic medium polarizes the medium and interacts with its own induced polarization. As the electron moves, it drags the polarization cloud around with it. At fairly low velocities this composite entity—the so-called polaron—behaves rather like a simple particle with a mass which can be very different from the mass of the bare electron. The phonons of a crystal lattice are another example of a system of quasiparticles. Indeed, from this point of view, the fundamental particles of physics should perhaps be regarded as quasiparticles. For they can never be observed in the absence of their interactions but are always clothed by them.

It is convenient to choose the zero of energy for the quasiparticles in a superconductor to be at the Fermi surface. The smallness of the superconducting energy[13] (typically $\sim 4 \times 10^3$ erg cm^{-3}) compared with the Fermi energy[14] (typically $\sim 10^{11}$ erg cm^{-3}) makes it plausible that only very few of the one-electron states are seriously affected by the transition. Hence, at energies far from the Fermi surface, the quasiparticle and normal Bloch states should coincide. But for every wave vector $\mathbf{k}$ there is both a quasiparticle state and a normal state, i.e. there is a one-to-one correspondence between quasiparticle and normal states. Since there are no quasiparticle states with energy $< \Delta$, the density of quasiparticle states near the Fermi surface (but further than Δ from it) is increased relative to the normal density of states; the states 'squeezed out' of the region $E < \Delta$ have to be accommodated in the region $E \gtrsim \Delta$.

The actual expression for the density of quasiparticle states will be taken from the BCS microscopic theory. The density of states $\Xi(E)$ per unit energy per unit volume is chosen to be (cf. Fig. 2.5)†

$$\Xi(E) = \frac{E\,\Xi_0(E)}{\sqrt{(E^2-\Delta^2)}}. \tag{2.28}$$

Here $\Xi_0(E)$ is the density of states in the normal metal. The energy gap 2Δ is temperature-dependent (Fig. 2.6): the BCS

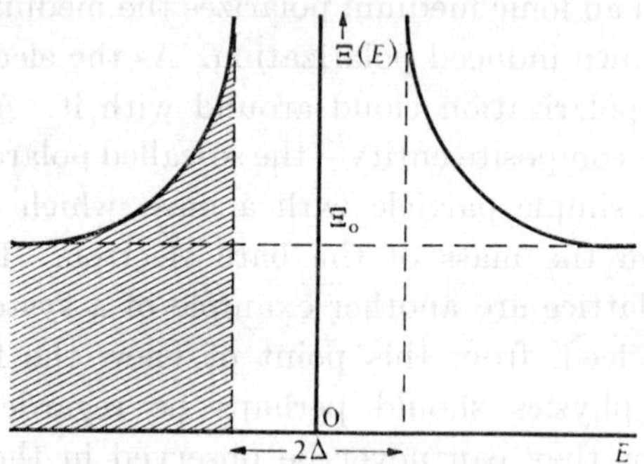

Fig. 2.5. The BCS density of states.

theory gives $\Delta(T)$ as the root of the transcendental equation (12.45). Near $T = T_c$ the equation reduces to

$$\Delta/kT_c = 3{\cdot}2(1-T/T_c)^{\frac{1}{2}}, \tag{2.29}$$

while for small T ($T \lesssim 0{\cdot}3T_c$), $\Delta(T)$ is very nearly constant, $= 1{\cdot}76\,kT_c$. The microscopic theory also predicts a condensation energy

$$W = -\tfrac{1}{2}\Xi_0\Delta^2, \tag{2.30}$$

† For the sake of familiarity, we use here a 'semiconductor' picture, in which the quasiparticle states are of two kinds—electron-like states with $p > p_F$, $E > 0$, and states with $p < p_F$, $E < 0$. The latter states are normally full, but when one is vacant it manifests itself as a second kind of particle, a 'hole'. In superconductors this point of view is sometimes misleading, but it will suffice in the present introductory text. The difficulty is that the Bogoliubov (1958) transformation (§ 12.1) superposes electrons and holes—the Bogoliubov quasiparticle does *not* have a well-defined charge. Adkins (1964) has discussed this point very fully, and has proposed an alternative representation, the 'electron excitation' picture. In problems where the charge of a quasiparticle is irrelevant, the semiconductor and Adkins pictures are equivalent. In tunnelling processes (Chapter 8) Adkins's picture is to be preferred, even though in the simplest case, of single-particle tunnelling, the semiconductor model is adequate.

which we may regard as the energy of the 'superelectrons'. From Fig. 2.6, we see that $\Delta(0)$ is finite and non-zero, and that as $T \to T_c$, $\Delta \to 0$ with infinite slope. Δ^2 vanishes with finite slope, so that the ratio $\Delta/\Delta(0)$ will serve as the order parameter ζ (cf. § 2.2).

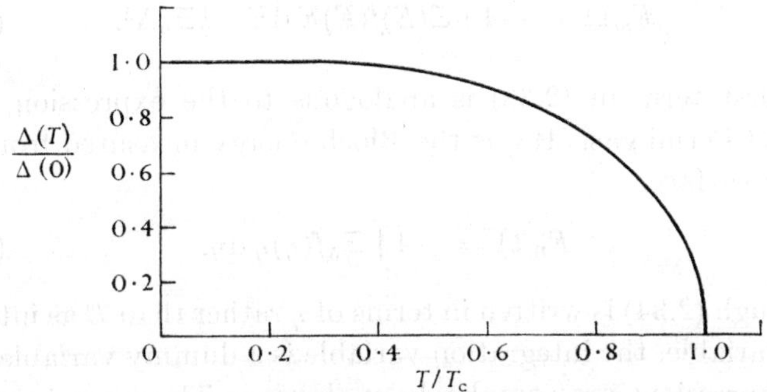

FIG. 2.6. The temperature-dependence of the BCS energy gap. From Bardeen, Cooper, and Schrieffer, *Phys. Rev.* **108**, 1175 (1957).

To construct the 'BCS model', we shall assume† that the quasiparticles (or 'normal' electrons) are statistically independent, and that the superelectrons, which contribute the energy (2.30), have no entropy. Then the distribution function for quasiparticles will be the normal Fermi-Dirac distribution

$$f(E) = \{1+\exp(\beta E)\}^{-1}, \tag{2.31}$$

where $\beta = 1/kT$. No chemical potential appears here since the quasiparticle energies were defined relative to the full Fermi sea. The entropy per unit volume is given by the usual expression,

$$S = 4k\int \Xi(E)\,dE\,\{f(E)\ln f(E) + (1-f(E))\ln(1-f(E))\},$$

or from (2.31),

$$S = 4k\int \Xi(E)\,dE\,\{\beta E f(E) + \ln(1+e^{-\beta E})\}. \tag{2.32}$$

† These assumptions are strictly valid only for systems of independent particles. They suffice because the BCS–Bogoliubov microscopic theory succeeds so well in decoupling the quasiparticles from one another.

The factor 4 in (2.32) arises from summing over spins and over quasielectrons and quasiholes (but see footnote, p. 32). The integrals are taken over positive E only.

Let us assume that the Helmholtz free energy per unit volume, in the absence of a magnetic field, is given by

$$F_{\mathrm{s}}/\Omega = -4\int \Xi(E)f(E)E\,\mathrm{d}E - \tfrac{1}{2}\Xi_0\Delta^2. \tag{2.33}$$

The first term in (2.33) is analogous to the expression for a normal Fermi gas. If η is the Bloch energy measured from the Fermi surface,

$$F_{\mathrm{n}}/\Omega = -4\int \Xi_0 f(\eta)\eta\,\mathrm{d}\eta. \tag{2.34}$$

Although (2.34) is written in terms of η rather than E as integration variable, the integration variable is a dummy variable, and the expressions are completely analogous. The second term of (2.33) is the contribution of the condensation energy (2.30) to the Helmholtz free energy.†

With the above postulates the thermodynamic properties of the system can be calculated. In particular the critical magnetic field can be found as follows. The magnetic Gibbs function G and the Helmholtz free energy F are equal when there is no magnetic field. Hence (2.13) may be rewritten as

$$H_{\mathrm{c}}^2/8\pi = f_{\mathrm{n}} - f_{\mathrm{s}}, \tag{2.35}$$ [11]

† The postulate (2.33) looks very natural, but it is really quite arbitrary. It would be just as natural to put

$$U_{\mathrm{s}}/\Omega = 4\int \Xi(E)f(E)E\,\mathrm{d}E - \tfrac{1}{2}\Xi_0\Delta^2$$

for the internal energy. However, by integrating (2.32) by parts, the entropy can be written

$$TS = 4\int \Xi(E)f(E)\,\mathrm{d}E\,(2E - \Delta^2/E).$$

It is evident that the above expression for the internal energy is consistent with (2.33) only if $\Delta = 0$. The real reason for preferring (2.33) is that it more or less agrees with the BCS microscopic theory (§ 12.2). If an additional small term $2\int \Xi(E)f(E)\,\mathrm{d}E\,\Delta^2/E$ is added to the right-hand side of (2.33), it agrees exactly with BCS.

where $\not f$ represents the Helmholtz free energy per unit volume. Combining (2.28), (2.33), (2.34), and (2.35), we have

$$H_c^2/8\pi = 4\Xi_0\left\{\int_0^\infty \eta f(\eta)\,\mathrm{d}\eta - \int_\Delta^\infty \frac{E^2 f(E)\,\mathrm{d}E}{\sqrt{(E^2-\Delta^2)}}\right\} - \tfrac{1}{2}\Xi_0\Delta^2. \tag{2.36}$$

(2.36)[11]

When the BCS expression for $\Delta(T)$ is introduced into (2.36) and the integrals are computed, the resulting critical-field curve (Fig 2.3) is fairly close to the Gorter–Casimir prediction (1.4). The slight deviations from the Gorter–Casimir parabola ($\sim$4 per cent) are qualitatively (though not quantitatively) in agreement with experiment.

The electronic heat capacity is readily calculated by differentiating (2.32):

$$\begin{aligned} c_s &= -\beta\,\mathrm{d}S/\mathrm{d}\beta \\ &= -4k\beta\int_\Delta^\infty \mathrm{d}E\,E\left\{\Xi(E)\frac{\mathrm{d}f}{\mathrm{d}\beta} + f\frac{\mathrm{d}\Xi(E)}{\mathrm{d}\beta}\right\} \\ &= 4k\beta^2\int_\Delta^\infty \mathrm{d}E\,\Xi(E)f(E)\{1-f(E)\}\left\{E^2+\beta\frac{\mathrm{d}}{\mathrm{d}\beta}(\tfrac{1}{2}\Delta^2)\right\}, \end{aligned} \tag{2.37}$$

after some manipulation. The electronic heat capacity is made up of two parts. There is a contribution

$$4k\beta^2\int \mathrm{d}E\,\Xi(E)f(E)\{1-f(E)\}\beta\frac{\mathrm{d}}{\mathrm{d}\beta}(\tfrac{1}{2}\Delta^2) \tag{2.37a}$$

arising from the temperature-dependence of the energy gap 2Δ. The other term

$$4k\beta^2\int \mathrm{d}E\,\Xi(E)f(E)\{1-f(E)\}E^2 \tag{2.37b}$$

is naturally interpreted as the specific heat of the quasiparticles; cf. the normal specific heat

$$c_n = 4k\beta^2\int \mathrm{d}\eta\,\Xi_0\eta^2 f(\eta)\{1-f(\eta)\}. \tag{2.38}$$

Only the former part (2.37a) contributes to the discontinuity at the transition temperature T_{c}, since as $\Delta \to 0$, $\Xi(E) \to \Xi_0$ and the quasiparticle term (2.37b) becomes identical with c_{n}. The magnitude of the specific heat discontinuity is

$$\left.\frac{c_{\mathrm{s}}-c_{\mathrm{n}}}{c_{\mathrm{n}}}\right|_{T_{\mathrm{c}}} = 1{\cdot}43. \tag{2.39}$$

For comparison, the Gorter–Casimir (1934) value is 2·00. Mostly, the experimental values (Table 2.1) are intermediate between the BCS and the Gorter–Gasimir predictions. At very low temperatures $T \ll T_{\mathrm{c}}$, the BCS specific heat is of course exponential:

$$c_{\mathrm{s}}(T)/c_{\mathrm{n}}(T_{\mathrm{c}}) \simeq 8{\cdot}5 \exp(-1{\cdot}44 T_{\mathrm{c}}/T), \tag{2.40}$$

in good agreement with experiment (Fig. 2.1(b)).

TABLE 2.1†

Superconductor	$\left.\frac{c_{\mathrm{s}}-c_{\mathrm{n}}}{c_{\mathrm{n}}}\right\|_{T_{\mathrm{c}}}$ *(experimental)*
Pb	2·65
Hg	2·18
Nb	2·07
Sn	1·60
Al	1·60
Ta	1·58
V	1·57
Zn	1·25
Tl	1·15

† This table is taken from Bardeen and Schrieffer (1961).

Exercise. Calculate $(\mathrm{d}H_{\mathrm{c}}/\mathrm{d}T)_{T_{\mathrm{c}}}$ using (2.36) and (2.29). Hence, by Rutger's formula (2.17), verify (2.39).

CHAPTER 3

THE LONDON ELECTRODYNAMIC EQUATIONS

THERE is at least one important respect in which the macroscopic theory outlined in Chapter 1 is inadequate. We have assumed the existence of mathematical surface currents. In other words, the skin depth of the superconducting currents has been neglected. Physically this skin depth must be non-zero. Nevertheless it is reasonable to hope that it will be an intrinsic property of the *material* and independent of the geometry of any particular specimen. If this conjecture is indeed correct then the description in Chapter 1 will be good for large enough bodies. However, deviations are to be expected in bodies whose size is comparable with the characteristic depth of penetration of magnetic fields. We shall see that this characteristic depth is $\lesssim 10^{-5}$ cm. Thus size-dependent effects will only be important for specimens with at least one exceedingly small dimension (e.g. thin films).

Becker *et al.* (1933) generalized the 'perfect conductivity' equation (1.6) to a form in which there is a finite skin depth for currents. It is assumed that the scattering processes which lead to the normal electrical resistivity are somehow suppressed in a superconductor. The charge carriers are, however, assumed to have the same *inertial* properties as in a normal metal. If the effective mass and charge of a carrier are m and ε respectively, then a uniform electric field $\mathbf{E}$ produces a uniform acceleration

$$\frac{d\mathbf{v}}{dt} = \frac{\varepsilon}{m}\mathbf{E}. \tag{3.1}$$

If n_s is the number density of superconducting charge carriers, the natural definition of the supercurrent density is

$$\mathbf{J}_s = n_s \varepsilon \mathbf{v}. \tag{3.2}$$

From (3.1) we then have†

$$\mathbf{E} = \Lambda\,\dot{\mathbf{J}}_\mathrm{s}, \tag{3.3}$$

where

$$\Lambda = m/n_\mathrm{s}\varepsilon^2. \tag{3.4}$$

There is no reason to assume that $\mathbf{J}_\mathrm{s}$ is the *total* current density. In general we might expect that some of the charge carriers are superconducting, while others remain normal. The total current density would then be

$$\mathbf{J} = \mathbf{J}_\mathrm{s}+\mathbf{J}_\mathrm{n}, \tag{3.5}$$

where $\mathbf{J}_\mathrm{n}$ is the normal or Ohmic component,

$$\mathbf{J}_\mathrm{n} = \sigma\mathbf{E}. \tag{3.6}$$

This assumption (3.5) with (3.6) is the natural extension of the two-fluid concept from a thermodynamic to an electrodynamic context. It greatly increases the field of applicability of the two-fluid models of § 2.2.

Under direct-current conditions the normal part $\mathbf{J}_\mathrm{n}$ will vanish. For, by (3.3), $\mathbf{J}_\mathrm{s}$ can be independent of time only if the applied electric field $\mathbf{E}$ vanishes. But if $\mathbf{E} = 0$, then from (3.6), $\mathbf{J}_\mathrm{n} = 0$ also; the perfect-conducting carriers form a path of zero resistance which can short circuit the normal carriers. The superconducting charge carriers present a non-zero impedance to an alternating applied field, but it is inductive

† The derivatives in (3.1) and (3.3) are strictly hydrodynamical or 'convective' derivatives $\mathrm{D}/\mathrm{D}t$ (e.g. Milne-Thomson 1955 p. 67), i.e. these time derivatives are taken with respect to a volume element moving with the electron fluid. It is often true, for geometrical reasons, that $\mathbf{v}.\mathrm{grad}\,\mathbf{v} = 0$; if this holds then the local derivative $\partial\mathbf{v}/\partial t = \mathrm{D}\mathbf{v}/\mathrm{D}t-(\mathbf{v}.\mathrm{grad})\mathbf{v}$ is identical with the convective derivative. I am indebted to Dr. J. R. Waldram for making me aware of the fallacy in many accounts of the London equations, where (3.3) is written with a local derivative in place of a hydrodynamic one. The difference is non-trivial; in terms of the local derivative, (3.3) reads

$$\mathbf{E} = \Lambda(\partial\mathbf{J}_\mathrm{s}/\partial t)+\Lambda\,\mathrm{grad}\,(\tfrac{1}{2}J_\mathrm{s}^2/\rho_\mathrm{s}),$$

where ρ_s is the superconducting charge density $n_\mathrm{s}\varepsilon$. The term in $\mathrm{grad}(\frac{1}{2}J_\mathrm{s}^2)$ leads to the Magnus force on Abrikosov's (1957) quantized flux lines, to be described in Chapter 7.

rather than resistive in character. Equation (3.3) exhibits the inductive nature of the response explicitly.

We assume (cf. p. 8) that the dielectric constant and magnetic permeability of the medium are both unity. Maxwell's equations then taken the form[15]

$$\operatorname{div} \mathbf{E} = 4\pi\rho; \quad \operatorname{curl} \mathbf{E} = -\frac{1}{c}\dot{\mathbf{B}}; \tag{3.7a}$$

$$\operatorname{div} \mathbf{B} = 0; \quad \operatorname{curl} \mathbf{B} = \frac{1}{c}\dot{\mathbf{E}} + \frac{4\pi}{c}\mathbf{J}. \tag{3.7b}$$

Combining (3.3) with (3.7a),

$$c \operatorname{curl}(\Lambda\dot{\mathbf{J}}_s) = -\dot{\mathbf{B}}. \tag{3.8}$$ [16]

Under steady conditions (i.e. $\dot{\mathbf{E}} = 0$; $\mathbf{J}_n = 0$) we may eliminate the current† between (3.7b) and (3.8):

$$\begin{aligned}\dot{\mathbf{B}} &= -c \operatorname{curl}(\Lambda\dot{\mathbf{J}}_s)\\ &= -c^2 \operatorname{curl}\{(\Lambda/4\pi) \operatorname{curl} \dot{\mathbf{B}}\}\\ &= -\lambda^2 \operatorname{curl} \operatorname{curl} \dot{\mathbf{B}}\\ &= \lambda^2\nabla^2\dot{\mathbf{B}},\end{aligned} \tag{3.9}$$ [17]

where

$$\lambda = (\Lambda c^2/4\pi)^{\frac{1}{2}} \tag{3.10}$$ [18]

is a parameter having the dimension of length. It has the significance of a depth of penetration of an applied field into the superconducting medium.

By a derivation similar to (3.9), eliminating $\dot{\mathbf{B}}$ instead of $\dot{\mathbf{J}}$, it is easily seen that

$$\dot{\mathbf{J}} = \lambda^2\nabla^2\dot{\mathbf{J}}, \tag{3.11}$$

and, from (3.11) and (3.3),

$$\mathbf{E} = \lambda^2\nabla^2\mathbf{E}. \tag{3.12}$$

† In deriving (3.9) it has been assumed that Λ is constant. In the improved phenomenological theory (§ 6.3) of Ginzburg and Landau (1950) a parameter equivalent to Λ can be defined, but is slightly field-dependent.

Equations (3.9), (3.11), and (3.12) are all of the form

$$\nabla^2\theta = \lambda^{-2}\theta. \tag{3.13}$$

von Laue (1949) has shown that the regular solutions of this equation decrease exponentially from the surface of the superconductor into its interior. Let us verify von Laue's result† for a simple geometry (Fig. 3.1). Consider an infinite plane

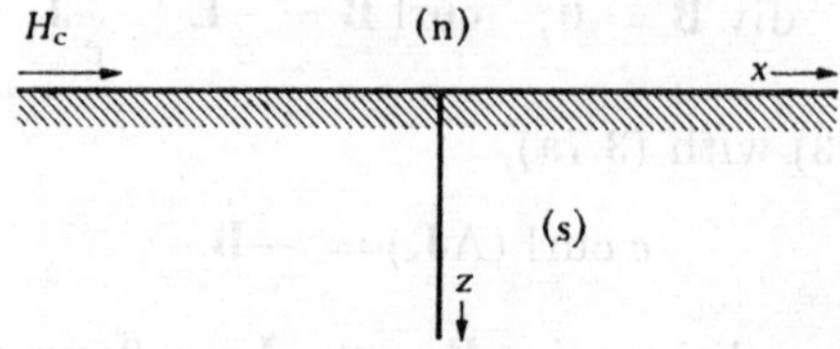

FIG. 3.1. Coordinate system for the discussion of field penetration.

boundary $z = 0$ with a superconducting medium filling the half-space $z > 0$. Under these condition θ will depend on z only. The equation (3.13) takes the form

$$\frac{d^2\theta}{dz^2} - \lambda^{-2}\theta = 0,$$

and has solutions

$$\theta = a_1 \exp(z/\lambda) + a_2 \exp(-z/\lambda).$$

The increasing exponential is not regular at $z = \infty$, so we require $a_1 = 0$, i.e.

$$\theta = \theta(0)\exp(-z/\lambda). \tag{3.14}$$

Thus we see that (3.12) and (3.9) are the desired generalizations of (1.6) and (1.7) to allow for the finite penetration depth λ. Deep in the material $\mathbf{E} = \dot{\mathbf{B}} = 0$, in agreement with (1.6) and (1.7).

† For a general proof, see London (1950 p. 51).

3.1. The Meissner effect

The 'acceleration' theory (p. 37) of Becker *et al.* (1933) does not contain the Meissner effect—(3.3) merely represents the generalization of (1.6) to allow for the finite skin depth of the supercurrents. But even in the macroscopic theory we have seen that perfect conductivity (1.6) is not sufficient. The additional postulate (1.8) of perfect diamagnetism is needed. The phenomenological theory of F. and H. London (1935) incorporates the Meissner effect into its description in essentially the same way. Equation (3.3) is assumed to hold, but is augmented by an additional equation:

$$c \operatorname{curl} \Lambda \mathbf{J}_S = -\mathbf{B}. \tag{3.15}$$[16]

We note that this second London equation (3.15) is of the same form as (3.8), after 'cancelling' the time derivatives on the two sides.

When the calculations leading to (3.9) and (3.11) are repeated, replacing (3.8) by (3.15), a new feature emerges: the time derivatives in (3.9) and (3.11) are replaced by the field and current density themselves. Thus it is $\mathbf{B}$ and $\mathbf{J}$, rather than $\dot{\mathbf{B}}$ and $\dot{\mathbf{J}}$, which are exponentially attenuated away from the boundary of the superconductor. In other words (3.15) does indeed contain a description of the Meissner effect.

The second London equation (3.15) is not a logical *consequence* of the first (3.3), nor vice versa. Nevertheless (3.15) is consistent with (3.3) in the sense that it prescribes a particular value (zero) for the integration constant in the time integral of (3.9). We therefore expect the actual penetration depth λ to be related to the microscopic parameters n_S, ε, m as in the theory of Becker *et al.* (p. 37). That is to say

$$\lambda^2 = mc^2/4\pi n_S \varepsilon^2. \tag{3.16}$$[19]

At this stage precise values of the parameters cannot be assigned, but we would expect to get a correct order of magnitude by assigning normal electronic values. This assignment gives

$\lambda \sim 0{\cdot}3\times10^{-5}$ cm, which is indeed of the right order of magnitude (see §§ 6.1, 6.2). Following the Londons, let us assume the validity of both (3.16) and the Gorter–Casimir expression for n_s (i.e. $n_s = \zeta N_e$, with ζ given by (2.24), and N_e the number density of conduction electrons). Then

$$\lambda = \lambda_0(1-T^4/T_c^4)^{-\frac{1}{2}}, \tag{3.17}$$

where λ_0 is the value of λ at zero temperature.

Thus far only steady conditions have been discussed, i.e. $\dot{\mathbf{E}} = 0$. Hence the normal component of the current is absent from the equations. If conditions are *not* steady, Ohm's law (3.6) is required to eliminate $\mathbf{J}_n$. Moreover, the displacement current in Maxwell's equations is no longer zero. Assuming that no space charges can occur (i.e. that the charge density $\rho = 0$), the elimination of $\mathbf{E}$ and $\mathbf{J}$ leads to

$$\nabla^2\mathbf{B} = \lambda^{-2}\mathbf{B}+(4\pi\sigma/c^2)\dot{\mathbf{B}}+c^{-2}\ddot{\mathbf{B}}, \qquad (3.18)^{(20)}$$

with similar equations for $\mathbf{E}$ and $\mathbf{J}$.

Exercises. (a) Find the current distribution in a cylindrical superconducting wire supplied with current through normal leads.

(b) Show that ρ satisfies[21]

$$\lambda^{-2}\rho+(4\pi\sigma/c^2)\dot{\rho}+c^{-2}\ddot{\rho} = 0,$$

and hence, solving, show that ρ decays to zero in a time $\sim 10^{-12}$ sec. (Take $\lambda^2 \sim 10^{-11}$ cm^2, $\sigma \sim 10^{19}$ e.s.u. as typical values.)[22]

To summarize, the London electrodynamic theory postulates the acceleration equation

$$\mathbf{E} = \Lambda\dot{\mathbf{J}}_s \tag{3.3}$$

and the additional London equation

$$c \operatorname{curl} \Lambda\mathbf{J}_s = -\mathbf{B}. \qquad (3.15)^{(16)}$$

The constant Λ is characteristic of the material, and of order $m/N_e e^2 \sim 10^{-31}$ sec^2. These equations imply that a static magnetic field penetrates into a superconducting medium to a depth[18] $\lambda = \sqrt{(c^2\Lambda/4\pi)} \sim 10^{-5}$ cm.

3.2. The vector potential and gauge invariance

To further study the implications of the London equations, it is useful to rewrite them in terms of a vector potential $\mathbf{A}$, instead of directly in terms of the electric and magnetic fields. The scalar potential ϕ and the vector potential $\mathbf{A}$ are auxiliary functions of position, defined to satisfy the relations

$$\left.\begin{aligned} \operatorname{curl} \mathbf{A} &= \mathbf{B}, \\ -\operatorname{grad} \phi - \frac{1}{c}\frac{\partial \mathbf{A}}{\partial t} &= \mathbf{E}. \end{aligned}\right\} \qquad (3.19)^{(23)}$$

However equations (3.19) are not sufficient to define $\mathbf{A}$ and ϕ uniquely—there is a great deal of 'gauge' freedom left to us in our choice of potentials. Maxwellian electrodynamics is greatly simplified by imposing the Lorentz condition

$$-\dot{\phi}/c - \operatorname{div} \mathbf{A} = 0. \qquad (3.20)^{(24)}$$

For then Maxwell's equations (3.7) reduce to

$$\left.\begin{aligned} \Box\phi &= -4\pi\rho, \\ \Box\mathbf{A} &= \frac{4\pi}{c}\mathbf{J}, \end{aligned}\right\} \qquad (3.21)^{(25)}$$

where the D'Alembertian operator $\Box \equiv \nabla^2 - \frac{1}{c^2}\frac{\partial^2}{\partial t^2}$.

Henceforth we shall assume that the potentials $\mathbf{A}$, ϕ have been chosen to satisfy the Lorentz condition (3.20). But even with the condition (3.20), $\mathbf{A}$ and ϕ are not unique. If $\chi(\mathbf{r}, t)$ is an arbitrary scalar function satisfying $\Box\chi = 0$, and if $\mathbf{A}$, ϕ are a set of potentials satisfying (3.19) and (3.20), then

$$\left.\begin{aligned} \mathbf{A}' &= \mathbf{A} + \operatorname{grad} \chi, \\ \phi' &= \phi - \frac{1}{c}\dot{\chi}, \end{aligned}\right\} \qquad (3.22)^{(26)}$$

are an equally good set of potentials.

Since the potentials are artificial constructions, and only the actual *fields* $\mathbf{E}$, $\mathbf{B}$ are observable physical quantities, a gauge transformation must leave invariant any equation which has physical content. By a 'gauge' transformation we mean a transformation (3.22) from one set of potentials to another.

The advantages—as well as the disadvantages!—of expressing physical relations in terms of the potentials will become clear when we study the behaviour of a mechanical system containing electric charges. If the Hamiltonian of the field-free system is known, the formalism requires only slight modification to describe the system when an electromagnetic field is present (see, e.g. Leech 1958). Let the Hamiltonian, in the absence of fields, be

$$\hat{\mathscr{H}}_0 = \sum_{i=1}^{n} p_i^2/2m_i + V(\mathbf{r}_1, \mathbf{r}_2, \ldots, \mathbf{r}_n) \tag{3.23}$$

and let the electric and magnetic fields at a point $\mathbf{r}$ be $\mathbf{E}(\mathbf{r})$ and $\mathbf{H}(\mathbf{r})$ respectively. Then the behaviour of the system (whether it is a classical or a quantum-mechanical one) is specified by the Hamiltonian

$$\hat{\mathscr{H}} = \sum_i \left\{\mathbf{p}_i - \frac{e}{c}\mathbf{A}(\mathbf{r}_i)\right\}^2 \Big/ 2m_i + V(\mathbf{r}_1, \mathbf{r}_2, \ldots, \mathbf{r}_n) - \sum_i e\phi(\mathbf{r}_i). \qquad (3.24)^{(26)}$$

However, the momenta $\mathbf{p}_i$ are not the Newtonian momenta $m_i\mathbf{v}_i$; instead we have

$$m_i\mathbf{v}_i = \mathbf{p}_i - \frac{e}{c}\mathbf{A}(\mathbf{r}_i). \qquad (3.25)^{(26)}$$

The kinetic-energy contribution to $\hat{\mathscr{H}}$ is thus $\sum \frac{1}{2}m_i v_i^2$.

Thus the momenta $\mathbf{p}_i$ depend on the gauge, although the eigenvalues of the Hamiltonian can be shown to be invariant. In a quantum-mechanical system the time-independent Schrödinger equation is obtained by making the replacement $\mathbf{p}_i \to \hat{\mathbf{p}}_i \equiv -i\hbar\nabla_i$, i.e.

$$\hat{\mathscr{H}}\psi \equiv \left[\sum_i \left\{\hat{\mathbf{p}}_i - \frac{e}{c}\mathbf{A}(\mathbf{r}_i)\right\}^2 \Big/ 2m_i + V(\mathbf{r}_1, \ldots, \mathbf{r}_n) - \sum_i e\phi(\mathbf{r}_i)\right]\psi(\mathbf{r}_1, \ldots, \mathbf{r}_n)$$
$$= E\psi(\mathbf{r}_1, \ldots, \mathbf{r}_n); \qquad (3.26)^{(26)}$$

the solutions ψ transform with the gauge. Under the transformation (3.22), the single-particle wave functions transform thus:

$$\psi'(\mathbf{r}) = \psi(\mathbf{r})\exp\{ie\chi(\mathbf{r})/\hbar c\}. \qquad (3.27)^{(26)}$$

In terms of the vector potential, the London equation (3.15) can be expressed in a particularly simple form. Let us adopt the natural convention that static fields have time-independent potentials. The time-independent Lorentz condition (from (3.20)) is just

$$\operatorname{div} \mathbf{A} = 0.$$

By (3.19), equation (3.15) takes the form

$$\operatorname{curl}\,(\Lambda \mathbf{J}_s + \frac{1}{c}\mathbf{A}) = 0 \qquad (3.28)^{(26)}$$

in terms of the vector potential $\mathbf{A}$. Hence it follows that

$$\Lambda \mathbf{J}_s + \mathbf{A}/c = \operatorname{grad} \theta, \qquad (3.29)^{(26)}$$

where θ is a divergence-free scalar field. In a simply-connected body, θ is single valued; it then follows that we can transform θ away by a gauge transformation. Our freedom of gauge allows us to add to $\mathbf{A}$ the gradient of an arbitrary divergenceless scalar field χ. We need merely choose[16] $\chi = -c\theta$ to define a new gauge—the so-called 'London gauge'—in which

$$c\Lambda \mathbf{J}_s = -\mathbf{A}. \qquad (3.30)^{(16)}$$

In this form the London equation is somewhat reminiscent of Ohm's law (3.6). But whereas Ohm's law is a local relation between a current density and the electric *field*, (3.30) relates the density of the supercurrent to the vector *potential*; its validity is restricted to the London gauge. In a general gauge all we can assert is (3.28).

Exercises. (a) Derive (3.21) from (3.7), using (3.19) and (3.20).

(b) Verify that the potentials $\mathbf{A}'$, ϕ' of (3.22) do indeed correspond to the same fields as $\mathbf{A}$, ϕ.

(c) Consider the Hamiltonian (3.24) for a single charged particle ($n = 1$). Show that the gauge transformation (3.22) is equivalent to the canonical transformation[26]

$$\hat{\mathscr{H}}^T = S^{-1}\hat{\mathscr{H}}S, \qquad S = \exp\{\mathrm{i}(e/\hbar c)\chi(r)\},$$

and hence verify (3.27).

3.3. 'Stiffness' of wave functions

In the light of equation (3.25), it is natural to introduce a supercurrent momentum density $\mathbf{p}_\mathrm{S}$ defined by

$$\mathbf{p}_\mathrm{S} = \varepsilon\Lambda\mathbf{J}_\mathrm{S} + \varepsilon\mathbf{A}/c. \tag{3.31}$$ [26]

In terms of this momentum density, the second London equation (3.15) takes the form

$$\mathrm{curl}\ \mathbf{p}_\mathrm{S} = 0. \tag{3.32}$$

Hence

$$\mathbf{p}_\mathrm{S} = \mathrm{grad}\ \xi, \tag{3.33}$$

where ξ is a scalar function of position.

Of course the definition (3.31) of $\mathbf{p}_\mathrm{S}$ is not gauge invariant; under the gauge transformation (3.22), we have

$$\mathbf{p}_\mathrm{S}' = \mathbf{p}_\mathrm{S} + \frac{\varepsilon}{c}\,\mathrm{grad}\ \chi, \tag{3.34}$$ [26]

since $\mathbf{J}_\mathrm{S}$ is a quantity with direct physical significance and is therefore invariant. Hence, from (3.33), it follows that

$$\xi' = \xi + \frac{\varepsilon}{c}\chi. \tag{3.35}$$ [26]

In order that the time-independent Lorentz condition $\mathrm{div}\ \mathbf{A} = 0$ shall be satisfied, we require

$$\nabla^2\chi = 0. \tag{3.36}$$

London (1950 p. 72) has shown that the gauge can be so chosen that $\mathbf{A}$ satisfies the boundary condition

$$A_{\perp} = 0 \tag{3.37}$$

on the surface of the superconductor. With this choice, the solution of (3.32) for a simply-connected superconductor is

$$\mathbf{p}_{\mathrm{s}} = 0, \tag{3.38}$$

i.e.

$$\Lambda c \mathbf{J}_{\mathrm{s}} = -\mathbf{A}. \qquad (3.38')^{(16)}$$

Equation (3.38′) is identical with (3.30), i.e. the present gauge is just the London gauge of § 3.2. This shows the significance of the London gauge: it is the gauge for which the supercurrent momentum density vanishes.

From this point of view, superconductivity represents a *long-range ordering* of the momenta of the superconducting charge carriers—all these carriers (Cooper pairs, § 11.1) have to have the same momentum (zero). This is in close analogy with the superfluidity of liquid helium; in Bogoliubov's (1947) weakly-interacting Bose-gas model, for example, there is a macroscopic occupancy of the zero-momentum single-particle quantum state.† Of course the particularly simple form (3.38) holds *only* in the London gauge. The phenomenon of long-range ordering of the momenta will in fact be present in any other gauge, but it will show not up in so simple a fashion (§ 13.3).

If $\mathbf{p}_{\mathrm{s}}$ is assumed to be the sum over individual charge carriers of the expectation values of their momenta, then the non-invariant character of $\mathbf{p}_{\mathrm{s}}$ merely reflects the non-invariant character of the wave function (cf. exercise (c), p. 46). To discuss the field-dependence of the wave function we must restrict ourselves to a particular gauge—for definiteness say the London gauge. In this gauge, $\mathbf{p}_{\mathrm{s}} = 0$ for a simply-connected

† It was this analogy that prompted Schafroth (1955) to develop the charged-Bose-gas model for superconductivity. However, although (cf. Blatt 1964) it is possible to cast the microscopic theory into a form close to that of Schafroth's model, it is not usually convenient to do so.

superconductor, whatever the magnetic field. This can be achieved if the wave function in the London gauge is independent of the field (so long as the field is weak enough not to destroy the superconductivity). This postulated 'stiffness' or rigidity of the wave function is in marked contrast with the behaviour of a free-electron gas. For free electrons, only the component of the momentum parallel to the applied-field direction z is unaffected; the motion perpendicular to the field is quantized in such a way that the angular momentum component L_z has integer eigenvalues.

3.4. Multiply-connected bodies: fluxoid conservation

It is rather unusual in physics for purely topological properties of a body to be important, but superconductivity provides an example. In a multiply-connected† superconducting body, although the Meissner effect ensures that the magnetic flux will be expelled from the material of the body, flux may be trapped in the holes. It is this possibility that allows a persistent current in a torus. From the macroscopic description of Chapter 1, it follows that the flux through a hole is *constant*—any attempt to change it by varying the applied magnetic field will induce surface eddy currents which, being undamped, will continue to maintain the flux.

London (1950 § 6) shows how a related variable, the 'fluxoid' through a closed curve may be introduced. An important property of the fluxoid is that it is constant in time, *even if the curve comes within the penetration layer at the surface of a hole.* In other words the fluxoid is the flux through the hole, corrected for the layer of superconducting material in which the flux is not zero (penetration layer).

† A multiply-connected region is defined as one in which there exist closed curves which cannot be deformed continuously to a point without passing outside the region. If there are no such curves (i.e. if all closed curves can be deformed to a point) the region is said to be simply connected. If the region can be made simply connected by introducing just one 'cut' surface, through which the curves are not allowed to pass, we say that initially it was doubly connected. An n-ply connected region requires $n-1$ cuts to make it simply connected. As examples, a sphere is a simply connected region, and so is a spherical shell. A torus is doubly connected.

Consider a superconducting body S (Fig. 3.2) with a hole N which may be vacuum or may be an inclusion of normal phase. Let C be a closed contour entirely in the medium S, and let Σ be a surface bounded by C. If C surrounds the hole and S

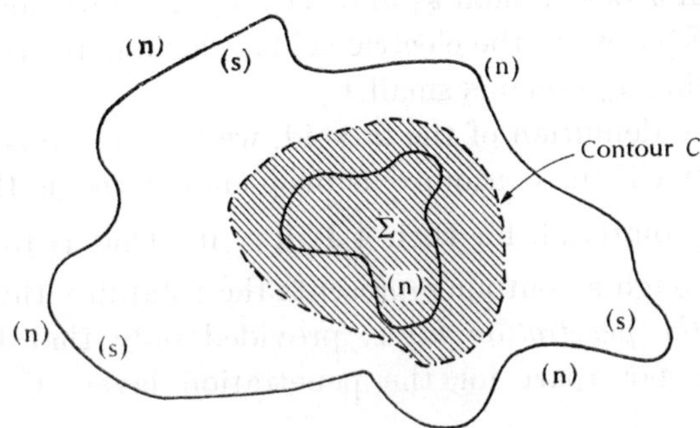

FIG. 3.2. Contour of integration for fluxoid conservation.

is multiply connected, then Σ cannot lie entirely in S. Integrating one of Maxwell's equations over the surface Σ,

$$-\frac{1}{c}\frac{\partial}{\partial t}\int_{\Sigma}\mathbf{B}.\mathbf{dS} = \int_{\Sigma}\text{curl }\mathbf{E}.\mathbf{dS}$$

$$= \oint_{C}\mathbf{E}.\mathbf{dl}, \qquad (3.39)^{(26)}$$

by Stokes's theorem. But from (3.3), the right-hand side is equal to

$$\frac{\partial}{\partial t}\Lambda\oint_{C}\mathbf{J}_{\mathrm{S}}.\mathbf{dl},$$

on interchanging the order of $\partial/\partial t$ and $\oint_{C}\mathbf{dl}$. Hence[16]

$$\frac{\partial}{\partial t}\left\{\int_{\Sigma}\mathbf{B}.\mathbf{dS}+c\Lambda\oint_{C}\mathbf{J}_{\mathrm{S}}.\mathbf{dl}\right\} = 0,$$

which may be written

$$\dot{\mathbf{\Phi}} = 0, \qquad (3.40)$$

where the *fluxoid* Φ is defined as

$$\Phi \equiv \int_{\Sigma} \mathbf{B}.d\mathbf{S}+c\Lambda\oint_{C}\mathbf{J}.d\mathbf{l}. \qquad (3.41)^{(16)}$$

In (3.40) there is actually a slight error, arising from the appearance of $\mathbf{J}$ rather than $\mathbf{J}_s$ in (3.41). In the static limit, (3.40) is exact. Even when the electric field is varying, the error made by neglecting $\mathbf{J}_n$ remains small.†

From the definition of the fluxoid, we see that Φ is identical with the flux if the contour C is deep enough inside the superconducting material, for then $\oint_C \mathbf{J}.d\mathbf{l} = 0$. That is to say, the fluxoid through a contour represents the total flux through the hole *and the penetration layer*, provided only that holes are sufficiently far apart for the penetration layers to have no overlap.

3.4.1. Fluxoid quantization

The conservation equation (3.40) is the only condition on Φ implied by the London equations (3.3), (3.15). However, F. London suggested that for entirely different reasons Φ should satisfy a further condition: it should be quantized, in units of[16] $2\pi\hbar c/\varepsilon$.

Consider a torus of superconducting material. If we neglect the penetration layer‡ (cf. Chapter 1) then $\mathbf{B} = 0$ in the superconductor. Hence, in the London gauge, curl $\mathbf{A}$ = div $\mathbf{A} = 0$ in the medium. Thus

$$\mathbf{A} = \text{grad}\,\chi, \qquad (3.42)$$

where χ is a scalar function of position; χ need not be single valued, since it is not defined except in the superconducting medium.

† From (4.18), the ratio J_n/J_s will be of order[27] $\lambda^2\sigma\omega/c^2$ where ω is a characteristic frequency of the varying field. Inserting numerical values, we find this ratio $\sim 10^{-9}\omega$, where ω is to be expressed in sec^{-1}. The neglect of $\mathbf{J}_n$ is therefore justified provided that the field does not vary appreciably in 10^{-9} sec.

‡ A more careful analysis, including correct treatment of the penetration layer (London 1950 § 26), reproduces the present results. The only difference in the final result is that Φ represents the fluxoid rather than the flux.

In general we may write

$$\int \mathbf{B}.\mathrm{d}\mathbf{S} = \oint \mathrm{grad}\ \chi.\mathrm{d}\mathbf{l} = \kappa, \tag{3.43}$$

where the integral is taken along a path embracing the hole once and only once. κ is the increase in χ on going once round the hole.

Confining our attention *only* to the material of the body makes (3.42) look like a gauge transformation of zero field (cf. (3.22)). But, in contrast to a real gauge transformation, (3.42) is *not* merely a formal redefinition of the potentials. It cannot be applied throughout space but holds only in the superconducting medium.

However, it would seem reasonable to assume that the superconducting charge carriers cannot 'see' what the vector potential is at points outside the medium. Granting this assumption we can find the influence of the field on the wave function† by regarding (3.42) as a true gauge transformation. According to (3.34), we must replace the momentum operator[26] $\hat{\mathbf{p}}$ by $\mathbf{p}-(\varepsilon/c)\,\mathrm{grad}\,\chi$. This is achieved by introducing a phase factor[26] $\exp(\mathrm{i}\chi\varepsilon/\hbar c)$ into the wave function, since

$$\begin{aligned}\hat{\mathbf{p}}\,\psi'(\mathbf{r}) &\equiv \hat{\mathbf{p}}\,\mathrm{e}^{\mathrm{i}\varepsilon\chi(\mathbf{r})/\hbar c}\psi(\mathbf{r}) \\ &= \mathrm{i}\hbar\left[\mathrm{e}^{\mathrm{i}\varepsilon\chi(\mathbf{r})/\hbar c}\nabla\psi(\mathbf{r})+\frac{\mathrm{i}\varepsilon}{\hbar c}\{\nabla\chi(\mathbf{r})\}\psi'(\mathbf{r})\right] \\ &= \mathrm{e}^{\mathrm{i}\varepsilon\chi(\mathbf{r})/\hbar c}\left(\hat{\mathbf{p}}-\frac{\varepsilon}{c}\nabla\chi\right)\psi(\mathbf{r})\end{aligned} \qquad (3.44)^{(26)}$$

(cf. (3.27)). But we must remember that the transformation (3.44) is not a true gauge transformation. We are not really redefining the potentials at all; as already remarked χ does not exist outside the superconductor although the electromagnetic field does. Thus the wave function ψ is uniquely defined (to within a *genuine* gauge transformation), and in

† The 'wave function of the superconducting charge carriers' is a concept which we are not strictly in a position to define yet. It will be introduced in § 6.3, in describing the Ginzburg–Landau theory. In the microscopic theory it represents the wave function of the Cooper pairs (see (13.30) and (14.45)).

particular it must be single valued. To guarantee single-valuedness, we must impose the flux quantization condition

$$\kappa = n\Phi_0,$$

where n is an integer and

$$\Phi_0 = 2\pi\hbar c/\varepsilon. \qquad (3.45)^{(16)}$$

For then[26]

$$\exp(i\varepsilon\kappa/\hbar c) = 1,$$

and (3.44) is satisfied with $\psi = \psi'$.

In London's account of flux quantization, the argument was substantially the one presented above, but the charge ε of the carriers was assumed to be the electronic charge e. Experimentally the quantization has been confirmed (Doll and Näbauer 1961, Deaver and Fairbank 1961), but the flux quantum Φ_0 was found to be[16] $\pi\hbar c/e$. Thus the charge ε must be taken to be $\varepsilon = 2e$, i.e. the charge carriers of superconductivity have *twice* the normal electronic charge. This is in accord with the ideas of the microscopic theory; in it the superconducting charge carriers are quasi-molecular electron pairs† (see Chapter 11).

As remarked above (footnote, p. 50), a more careful argumeut would show that it is really the fluxoid, rather than the flux, that is quantized. But for most purposes, especially when considering bodies of macroscopic dimensions, the difference between the fluxoid and the flux is negligible.

Fluxoid quantization helps to account for the phenomenon of persistent currents. Since the fluxoid can only change discontinuously and since the quantum of fluxoid is macroscopic[28] ($\Phi_0 \simeq 2\times 10^{-7}$ gauss cm²), it follows that transitions involving fluxoid changes will be exceedingly improbable. States with flux trapped in a non-superconducting inclusion will therefore be exceedingly long-lived. These arguments do not give any indication just *how* long-lived they are, but we

† It has been shown (Byers and Yang 1961, Bardeen 1961, Brenig 1961, Onsager 1961) that flux quantization follows rigorously from the BCS theory, with the charge value $\varepsilon = 2e$.

should expect something like the observed metastability of persistent currents. There have been some experiments (Galkin *et al.* 1957), where decay of a persistent current has been reported for temperatures very close to T_c. The likeliest explanation of this effect is that thermal fluctuations could destroy the doubly-connected topology for a time long enough to enable a fluxoid quantum to leak out.

3.5. Anisotropic superconductors

In an attempt to extend the London electrodynamics to real crystalline materials, Ginzburg (1944) and von Laue (1949) generalized London's equations (3.3), (3.15). They replaced London's scalar parameter Λ by a second-order symmetric tensor Λ_{ij}:

$$\left.\begin{aligned} E_i &= \sum_j \Lambda_{ij}(\dot{J}_s)_j, \\ B_i &= -c\sum_{j,k,l}\varepsilon_{ijk}\left[\frac{\partial}{\partial x_k}\{\Lambda_{jl}(J_s)_l\}-\frac{\partial}{\partial x_j}\{\Lambda_{kl}(J_s)_l\}\right]. \end{aligned}\right\} \quad (3.46)^{(16)}$$

Here $\varepsilon_{ijk} = 1$ if (i, j, k) is an even permutation of $(1, 2, 3)$;
$\varepsilon_{ijk} = -1$ if (i, j, k) is an odd permutation of $(1, 2, 3)$; and
$\varepsilon_{ijk} = 0$ otherwise. The usual tensor convention has been used, in which the coordinate axes are renamed x_1, x_2, x_3.

With a suitable choice of axes, the tensor Λ_{ij} becomes diagonal. For a tetragonal crystal (e.g. tin) the principal axes are obvious, and the tensor takes the form

$$\Lambda = \begin{pmatrix} \Lambda_1 & 0 & 0 \\ 0 & \Lambda_2 & 0 \\ 0 & 0 & \Lambda_2 \end{pmatrix}. \quad (3.47)$$

For the special case of a cubic crystal, the three diagonal components are all equal. The von Laue tensor—like all second-order tensors in a cubic crystal—is then isotropic.

Equation (3.47) makes two definite predictions about the properties of tin. Firstly, the penetration depth is independent of the azimuthal angle and, secondly, it is a monotonic function of the polar angle (where we use a system of polar coordinates based on the tetragonal axis). The second of these predictions has been examined by Pippard (see § 6.1), and appears to be invalid. Aluminium, which is a cubic metal, has also been examined (Williams 1962), and although the experimental evidence is far from conclusive, it seems that once more the penetration depth has some slight dependence on the crystallographic orientation. These experiments afford some evidence that the von Laue equations (3.46) may not be adequate to describe the superconductivity of crystals. The necessary modifications to the London theory will be discussed in § 6.2.

CHAPTER 4

SUPERCONDUCTING ELECTRODYNAMICS IN FOURIER SPACE

IN this chapter, the descriptive or phenomenological account of superconducting electrodynamics will be developed further. We shall be mainly concerned with the Fourier transforms of the field equations—both in space and time variables—as this facilitates the description of time-dependent and spatially-varying phenomena. We shall exploit the *linearity* of the field equations; since the solutions of linear equations are additive, general solutions can be built out of linear combinations of pure sinusoidal ones.

This programme is useful for two applications, one practical and the other a more abstract conceptual clarification. The behaviour of superconductors at high frequencies will be studied under the assumptions of the previous chapters—in particular the London equations (3.3) and (3.15). The high-frequency predictions are compared with experiment and do not fit well. The reasons for the poor agreement are twofold: (a) the breakdown of Ohm's law (3.6) under the conditions of the anomalous skin effect (even in a non-superconducting material), and (b) the inadequacy of the London theory. The modifications required will not be discussed here; they are deferred to Chapter 6.

The second application of the Fourier-transformed electrodynamics will be a careful study of some conceptual problems which arise in the limit of completely static and uniform fields. We should like to answer such questions as these: Are the two London equations (3.3) and (3.15) independent, or does (3.15) imply (3.3)? Can the Meissner effect be regarded as a consequence of the energy gap? These questions must be carefully reformulated before they can be answered, as there are some difficulties associated with the limiting processes $\mathbf{q} \to 0$, $\omega \to 0$.

If $\Phi(\mathbf{r}, t)$ is any function of position and time, we define the following Fourier transforms:

$$\Phi(\mathbf{r}, \omega) = (2\pi)^{-1}\int \mathrm{d}t\, \Phi(\mathbf{r}, t) \exp \mathrm{i}\omega t, \tag{4.1}$$

$$\Phi(\mathbf{q}, t) = (2\pi)^{-3}\int \mathrm{d}^3r\, \Phi(\mathbf{r}, t) \exp \mathrm{i}\mathbf{q}.\mathbf{r}, \tag{4.2}$$

$$\Phi(\mathbf{q}, \omega) = (2\pi)^{-4}\int \mathrm{d}t \int \mathrm{d}^3r\, \Phi(\mathbf{r}, t) \exp \mathrm{i}(\mathbf{q}.\mathbf{r}+\omega t), \tag{4.3}$$

where the integrals are over all space and time. Of course Φ need not be scalar; it could be, for example, a component of a vector. The inverse transformations are

$$\begin{aligned}\Phi(\mathbf{r}, t) &= \int_{-\infty}^{\infty} \mathrm{d}\omega\, \Phi(\mathbf{r}, \omega) \exp -\mathrm{i}\omega t \\ &= \int \mathrm{d}^3q\, \Phi(\mathbf{q}, t) \exp -\mathrm{i}\mathbf{q}.\mathbf{r} \\ &= \int \mathrm{d}\omega \int \mathrm{d}^3q\, \Phi(\mathbf{q}, \omega) \exp -\mathrm{i}(\mathbf{q}.\mathbf{r}+\omega t).\end{aligned} \tag{4.4}$$

It will also be convenient, for the subsequent development, to resolve the Fourier component $\mathbf{C}(\mathbf{q}, \omega)$ into its longitudinal and transverse components

$$\mathbf{C}(\mathbf{q}, \omega) = \mathbf{C}_{\parallel}(\mathbf{q}, \omega)+\mathbf{C}_{\perp}(\mathbf{q}, \omega). \tag{4.5}$$

Here $\mathbf{C}_{\parallel}$ and $\mathbf{C}_{\perp}$ are respectively in the direction parallel to $\mathbf{q}$ and in the plane normal to $\mathbf{q}$, i.e.

$$\left.\begin{aligned}\mathbf{q} \wedge \mathbf{C}_{\parallel}(\mathbf{q}, \omega) &= 0,\\ \mathbf{q}\,.\,\mathbf{C}_{\perp}(\mathbf{q}, \omega) &= 0.\end{aligned}\right\} \tag{4.6}$$

Equations (4.6) will serve as a formal definition of longitudinal and transverse vectors. The resolution of any vector $\mathbf{C}(\mathbf{q})$ into its longitudinal and transverse components $\mathbf{C}_{\parallel}$ and $\mathbf{C}_{\perp}$ is effected by taking

$$\left.\begin{aligned}\mathbf{C}_{\parallel} &= \mathbf{q}(\mathbf{q}\,.\,\mathbf{C})/q^2,\\ \mathbf{C}_{\perp} &= \mathbf{q} \wedge (\mathbf{q} \wedge \mathbf{C})/q^2.\end{aligned}\right\} \tag{4.7}$$

Note that *the distinction between longitudinal and transverse vectors breaks down in the limit* $q \to 0$. Hence the separation of a vector into longitudinal and transverse components becomes arbitrary in this limit.

The advantage of studying these Fourier-transformed quantities is that the differential operators ∇ and $\partial/\partial t$ are transformed into the multiplicative vector $\mathbf{iq}$ and the multiplicative scalar $i\omega$ respectively. To ensure the convergence of the Fourier integrals, all fields are assumed to vanish as $r \to \infty$, $|t| \to \infty$. Under these conditions Maxwell's equations take the form[29]

$$\left.\begin{aligned} &\mathbf{iq}.\mathbf{D}(\mathbf{q},\omega) = 4\pi\rho(\mathbf{q},\omega); \quad \mathbf{iq}\wedge\mathbf{E}(\mathbf{q},\omega) = -\frac{i\omega}{c}\mathbf{B}(\mathbf{q},\omega);\\ &\mathbf{iq}.\mathbf{B}(\mathbf{q},\omega) = 0; \qquad \mathbf{iq}\wedge\mathbf{H}(\mathbf{q},\omega) = \frac{4\pi}{c}\mathbf{J}(\mathbf{q},\omega)+\frac{i\omega}{c}\mathbf{D}(\mathbf{q},\omega),\end{aligned}\right\} \tag{4.8}$$

with the equation of continuity

$$i\omega\mathbf{q}.\mathbf{D}(\mathbf{q},\omega)-4\pi\mathbf{q}.\mathbf{J}(\mathbf{q},\omega) = 0. \tag{4.9}$$ [30]

In terms of the usual scalar and vector potentials ϕ, $\mathbf{A}$,

$$\left.\begin{aligned} \mathbf{E}(\mathbf{q},\omega) &= -\mathbf{iq}\,\phi(\mathbf{q},\omega)-\frac{i\omega}{c}\mathbf{A}(\mathbf{q},\omega),\\ \mathbf{B}(\mathbf{q},\omega) &= \mathbf{iq}\wedge\mathbf{A}(\mathbf{q},\omega).\end{aligned}\right\} \tag{4.10}$$ [26]

However the potentials are not uniquely defined; they admit of the gauge transformations (3.22). For the Fourier components of ϕ, $\mathbf{A}$, we may write the gauge transformation (3.22) as

$$\left.\begin{aligned} \mathbf{A}(\mathbf{q},\omega) \to \mathbf{A}'(\mathbf{q},\omega) &= \mathbf{A}(\mathbf{q},\omega)+\mathbf{\Theta}(\mathbf{q},\omega),\\ \phi(\mathbf{q},\omega) \to \phi'(\mathbf{q},\omega) &= \phi(\mathbf{q},\omega)+\chi(\mathbf{q},\omega),\end{aligned}\right\} \tag{4.11}$$

with the subsidiary conditions

$$\left.\begin{aligned} \mathbf{q}\wedge\mathbf{\Theta}(\mathbf{q},\omega) &= 0,\\ \omega\mathbf{\Theta}(\mathbf{q},\omega) &= c\mathbf{q}\,\chi(\mathbf{q},\omega).\end{aligned}\right\} \tag{4.12}$$ [16]

Exercise. Verify that the gauge transformation (4.11) leaves Maxwell's equations (4.8) invariant.

Since the potentials are not themselves measurable quantities, the gauge freedom does no harm provided the equations leave all physically-observable quantities gauge-invariant. Note that (4.11) allows **A** to be changed by a longitudinal vector only, i.e. $\mathbf{A}_\perp$ is an invariant quantity.

4.1. The London equations

For an isotropic medium† the London equations (3.3), (3.15) have the Fourier transforms

$$\left.\begin{aligned} \mathbf{E}(\mathbf{q}, \omega) &= \mathrm{i}\omega\Lambda\mathbf{J}_\mathrm{S}(\mathbf{q}, \omega), \\ ic\mathbf{q} \wedge \{\Lambda\mathbf{J}_\mathrm{S}(\mathbf{q}, \omega)\} &= -\mathrm{i}\mathbf{q} \wedge \mathbf{A}(\mathbf{q}, \omega). \end{aligned}\right\} \qquad (4.13)^{(16)}$$

On eliminating the electric field (using (4.10)), and separating the longitudinal and transverse components, (4.13) can be reduced to

$$\left.\begin{aligned} (\mathbf{J}_\mathrm{S})_\perp &= -\frac{1}{c\Lambda}\mathbf{A}_\perp, \\ (\mathbf{J}_\mathrm{S})_\parallel &= \frac{\mathbf{q}}{\omega\Lambda}\phi(\mathbf{q}, \omega) - \frac{1}{c\Lambda}\mathbf{A}_\parallel. \end{aligned}\right\} \qquad (4.14)^{(26)}$$

Exercise. Derive (4.14), using (4.10) and (4.13). Find explicitly the gauge transformation which reduces (4.14) to the London gauge (4.15).

The gauge may be chosen such that $\mathbf{A}_\parallel = 0$; the second of equations (4.14) then becomes

$$(\mathbf{J}_\mathrm{S})_\parallel = \frac{\mathbf{q}}{\omega\Lambda}\phi(\mathbf{q}, \omega). \qquad (4.15)$$

† Von Laue's generalization (3.46) for an anisotropic medium can similarly be transformed, to give[16]

$$E_i(\mathbf{q}, \omega) = i\omega \sum_{j=1}^{3} \Lambda_{ij}(J_\mathrm{S}(\mathbf{q},\omega))_j,$$

$$\mathrm{i}(\mathbf{q} \wedge \mathbf{A})_i \equiv B_i(\mathbf{q}, \omega) = -\mathrm{i}c \sum_{j,k,l} \varepsilon_{ijk}[\Lambda_{jl}\{J_\mathrm{S}(\mathbf{q}, \omega)\}_l q_k - \Lambda_{kl}\{J_\mathrm{S}(\mathbf{q}, \omega)\}_j q_l],$$

where $i = 1, 2, 3$. Since, however, Pippard's experimental data (cf. § 6.1) do not support the idea of a tensorial anisotropy, these equations have not proved useful.

This is just the London gauge of (3.30).† Evidently, in the London gauge, the transverse part of the supercurrent 'sees' the vector potential, while the longitudinal part 'sees' only the electric field.

The two-fluid concept has not yet been mentioned in the present context. But it is precisely in a.c. experiments that the normal fluid manifests itself. In any d.c. experiment the superfluid completely short circuits the normal part. Moreover, in order to measure the London penetration depth λ by d.c. methods it is necessary to perform experiments on specimens of dimensions comparable to λ ($\sim 3\times 10^{-6}$ cm). The first measurements (Shoenberg 1940) were made on a colloidal suspension of mercury particles. In order to measure the penetration depth in a macroscopic body, it is necessary to use high frequencies (H. London 1940).

To discuss the principles underlying London's method, consider the simple geometrical configuration of Fig. 3.1—a semi-infinite slab of superconducting material filling the half space $z > 0$. Let $\mathbf{B}$ be in the x-direction and of magnitude

$$B_x = H_{\mathrm{e}}\eta(z)\mathrm{e}^{\mathrm{i}\omega t}, \tag{4.16}^{(31)}$$

where $\eta = 1$ at the surface $z = 0$ and H_{e} is the amplitude of the field outside the medium. With this geometry, it is convenient to Fourier-transform only the time variable; in effect, this has already been done by choosing a single frequency in (4.16). Taking the single Fourier component (4.16) of (3.18), we find

$$\eta''(z) = (\lambda^{-2}+4\pi\mathrm{i}\omega\boldsymbol{\sigma}/c^2-\omega^2/c^2)\eta(z). \tag{4.17}^{(32)}$$

The three terms on the right-hand side of (4.17) are respectively the contributions from the London current (3.15), the Ohmic current (3.6), and the displacement current. Their magnitudes are in the ratios

$$1 : 4\pi\lambda^2\omega\boldsymbol{\sigma}/c^2 : \lambda^2\omega^2/c^2. \tag{4.18}^{(33)}$$

† Once again it should be stressed that all physical results must be independent of the gauge. Despite the simplification introduced by the London gauge (4.15), it is often desirable to preserve the explicit gauge-invariance of (4.14).

In particular the displacement current is negligible provided $\omega \ll 10^{16}\ \sec^{-1}$, i.e. it becomes important only at ultraviolet frequencies. Neglecting this term, we find that

$$\eta = \mathrm{e}^{-z/\kappa}$$

with

$$\kappa = \lambda(1+\mathrm{i}\lambda^2/\delta^2)^{-\frac{1}{2}}. \tag{4.19}$$

Here δ is the ordinary skin depth

$$\delta^2 = c^2/2\pi\omega\boldsymbol{\sigma}, \tag{4.20}$$ [34]

which characterizes the penetration of an alternating field into an Ohmic conductor of conductivity $\boldsymbol{\sigma}$.

From Maxwell's equations, the current density and electric field are easily found to be

$$\left.\begin{aligned} J_y &= (c/4\pi\kappa)H_\mathrm{e}\,\mathrm{e}^{\mathrm{i}\omega t}\eta(z),\\ E_y &= -(\mathrm{i}\omega\kappa/c)H_\mathrm{e}\,\mathrm{e}^{\mathrm{i}\omega t}\eta(z).\end{aligned}\right\} \tag{4.21}$$ [35]

Exercise. Derive (4.21) from (4.16), (4.17), and Maxwell's equations (4.8). Hence verify (4.23).

If we define the surface impedance† in the usual way:

$$Z = E_0\delta \Big/ \int\limits_0^\infty J_y(z)\,\mathrm{d}z, \tag{4.22}$$

then (4.21) leads to a value

$$Z = 4\pi\mathrm{i}\omega\kappa/c^2. \tag{4.23}$$ [36]

The frequency-dependent penetration depth κ reduces to the London depth λ at low frequencies, and to the normal skin depth δ at high frequencies. The transitional regime occurs when $\delta \sim \lambda$, i.e. at frequencies which are of the order of 10^3 to 10^4 Mc/s when we insert typical values for λ and $\boldsymbol{\sigma}$.

Equation (4.23) is readily tested experimentally, since the surface impedance is directly measurable. The most systematic

† This quantity is often called the 'impedance per unit square'. It is in fact the impedance of a square on the surface, of *arbitrary size*. To see this, note that the impedance of a rectangle on the surface is proportional to the dimension parallel to the field and inversely proportional to the dimension perpendicular to the field.

investigations are those of Pippard (1947, 1950*a*, *b*). In fact Pippard's experiments, which were carried out at the interesting range of frequencies $\sim 10^3$ to 10^4 Mc/s, do not confirm the theoretical predictions. However, the major source of disagreement is irrelevant to the phenomenon of superconductivity; even for a non-superconducting metal, (4.23) fails at frequencies $\sim 10^3$ Mc/s. This failure—the so-called anomalous skin effect—is a consequence of the fact that the formal skin depth δ (4.20) becomes smaller than the electronic mean free path. When this occurs the field is changing rapidly over one free path, and Ohm's law fails. H. London (1940) made an *ad hoc* correction to the conductivity to allow for this effect. A full treatment, by solution of the Boltzmann transport equation, has been given by Reuter and Sondheimer (1948). A simpler approximate treatment by Pippard (1947) gives essentially the same results, namely that at sufficiently high frequencies (4.20) breaks down and the skin depth δ eventually becomes proportional to $\omega^{-\frac{1}{3}}$ and independent of $\boldsymbol{\sigma}$.

When the skin depth (4.20) is replaced by the anomalous skin depth, (4.23) agrees qualitatively with Pippard's data for fairly impure materials near to the transition temperature. But for very pure Type I materials† well away from the transition point, (4.23) remains inadequate (Waldram 1964, Williams 1962). The significance of this disagreement will be discussed in Chapter 6; it will be necessary to modify the London equations, replacing the *local* relation (3.15) by Pippard's non-local expression (6.11).

4.2. Schafroth's criterion for a Meissner effect

In Chapter 6 we shall see that the detailed comparison of theory and experiment compels us to revise the London theory. London's second equation (3.15) has to be replaced by Pippard's more complicated non-local expression (6.11). With this development in mind, we shall derive Schafroth's condition that the medium should exhibit a Meissner effect. The London

† See Chapter 6 for a discussion of the classification of superconductors as Type I or II.

equation (3.15) need *not* be assumed; instead consider a system with the much more general linear relation between $\mathbf{J}$ and $\mathbf{A}$:

$$J_i(\mathbf{r}, t) = \sum_{j=1}^{3} \int \mathrm{d}^3 r' K_{ij}(\mathbf{r}-\mathbf{r}') A_j(\mathbf{r}', t). \tag{4.24}$$

This is still not the most general linear relation possible, in so far as it is still instantaneous although not necessarily local. Moreover the medium has been assumed to be uniform, although not necessarily isotropic; the kernel K is a function only of the vector displacement $\mathbf{r}-\mathbf{r}'$. It may seem strange at first sight to demand that K_{ij} should be a second-order tensor even in an isotropic medium. But with a scalar K it would not be possible to maintain the gauge invariance of (4.24) (cf. (4.26)).

Equation (4.24) possesses the spatial Fourier transform

$$J_i(\mathbf{q}, t) = \sum_j K_{ij}(\mathbf{q}) A_j(\mathbf{q}, t). \tag{4.25}$$

From (4.11) and (4.12) we see that, under gauge transformations, $\mathbf{A}(\mathbf{q}, t)$ may be changed by an arbitrary longitudinal vector. However $\mathbf{J}(\mathbf{q}, t)$ is a physically observable quantity, and must not change. Hence the gauge-invariance condition

$$\sum_j K_{ij}(\mathbf{q}, t)\, q_j = 0 \tag{4.26}$$

must be imposed. For a time-independent system, the equation of continuity is

$$\operatorname{div} \mathbf{J} = 0.$$

Applying this condition to (4.25) gives

$$\sum_i q_i K_{ij}(\mathbf{q}, t) = 0. \tag{4.27}$$

It may be shown that K_{ij} is a symmetric tensor.† It therefore follows that (4.26) and (4.27) are not independent; gauge

† The proof that K_{ij} is symmetric is analogous to the usual demonstration (e.g. Jeans 1925) that the dielectric tensor is symmetric.

invariance implies and is implied by continuity. For an isotropic medium K_{ij} is easily shown to be of the form

$$K_{ij} = \mathscr{K}_1(q^2)\, q_i q_j + q^2 \mathscr{K}_2(q^2)\, \delta_{ij}, \tag{4.28}$$

where, from (4.26) and (4.27),

$$\mathscr{K}_1(q^2) + \mathscr{K}_2(q^2) = 0. \tag{4.29}$$

If $\mathscr{K}_1(q^2)$ is taken to be a regular function near $q^2 = 0$, i.e.

$$\mathscr{K}_1(q^2) = a_0 + O(q^2),$$

then it follows from (4.25), (4.28), and (4.29) that

$$i\mathbf{q} \wedge \mathbf{J}(\mathbf{q}, t) = -a_0 q^2 \mathbf{B}(\mathbf{q}, t) + O(q^4)\mathbf{B}, \tag{4.30}$$

or in configuration space

$$\text{curl } \mathbf{J}(\mathbf{r}, t) = -a_0 \nabla^2 \mathbf{B}(\mathbf{r}, t) + O(\nabla^4 \mathbf{B}). \tag{4.31}$$

In other words, if $\mathscr{K}_1(q^2)$ is regular at the origin then there is no Meissner effect, since (4.31) is just the usual expression for a diamagnetic current.

Let us now consider what happens if $\mathscr{K}_1(q^2)$ has a singularity at the origin. A *simple pole* (simple in q^2, not in $|q|$!) is adequate for our purposes:

$$\mathscr{K}_1(q^2) = a_{-1} q^{-2} + O(q^0). \tag{4.32}$$

In this case, (4.25), (4.28), and (4.29) imply

$$i\mathbf{q} \wedge \mathbf{J}(\mathbf{q}, t) = -a_{-1} \mathbf{B}(\mathbf{q}, t) + O(q^2)\mathbf{B},$$

i.e.

$$\text{curl } \mathbf{J} = -a_{-1} \mathbf{B} + O(\nabla^2 \mathbf{B}). \tag{4.33}$$

Apart from corrections arising from the field inhomogeneities, this is just the second London equation (3.15). To test any model for a Meissner effect, all that is needed is to find out whether $\mathscr{K}_1(q^2)$ has a pole at the origin of q. The particular form of the kernel

$$\mathscr{K}_1^{\mathrm{L}} = 1/\Lambda c^2 q^2 \tag{4.34}$$ [37]

leads to precisely the London theory. Pippard's equation (6.11) follows from the more complicated kernel

$$\mathscr{K}_1^{\mathrm{P}} = \frac{1}{\Lambda c^2 q^2} \frac{\tan^{-1}\xi q}{\xi q}, \tag{4.35}^{(37)}$$

which also has the necessary pole at $q^2 = 0$ (Schafroth and Blatt 1956, Schafroth 1960).

In applying Schafroth's criterion (4.32), it is essential that the gauge invariance be preserved; Schafroth has shown that failure to preserve gauge invariance can lead to a spurious Meissner effect. In particular, if the starting point is a non-invariant approximate Hamiltonian, there is no unique way of defining the current **J**. Then by varying the definition of **J** the coefficient a_{-1} of (4.32) can be altered arbitrarily.

4.3. Dispersion relations for superconductors

The present section is a detailed study of the relation between 'perfect conductivity' and 'perfect diamagnetism' (cf. § 1.2). It will be shown that formally they are independent properties. However, the ways in which they manifest themselves are so closely related that one can hardly believe that they are really independent. Indeed, Evans and Rickayzen (1964) have shown that in any microscopic theory containing an electron-scattering mechanism the existence of the Meissner effect is both sufficient and necessary for perfect conductivity.

This section will be rather more mathematical than most of Part I, and will appeal less to physical intuition. A number of general sum rules will be derived, which admittedly can also be derived in a more intuitive manner (Ferrell and Glover 1958). However, the results of a non-rigorous calculation may be suspect. Since Taylor's (1961) approach leads to the same sum rules in a rigorous manner, we are able to accept these rules with confidence.

In the theory of dielectrics it is usual to define a complex dielectric constant

$$\boldsymbol{\epsilon}(\omega) = \boldsymbol{\epsilon}_1(\omega) + \mathrm{i}\boldsymbol{\epsilon}_2(\omega), \tag{4.36}$$

to relate the electric displacement vector to the electric field, which is assumed to vary sinusoidally with time. (Of course the electric field and displacement are real quantities, but it is convenient to write their time dependence as

$$\mathbf{E} = \mathbf{E}_0\,\mathrm{e}^{\mathrm{i}\omega t};\ \ \mathbf{D} = \mathbf{D}_0\,\mathrm{e}^{\mathrm{i}\omega t}, \tag{4.37}$$

leaving 'the real part of' to be understood.) The real part and the imaginary part of the dielectric constant respectively relate the in-phase and out-of-phase parts of the displacement to the electric field. The familiar dispersion relations of Kramers (1927) and Kronig (1926)

$$\left.\begin{aligned} \epsilon_1(\omega)-\epsilon_\infty &= \frac{2}{\pi}\mathscr{P}\int\frac{\epsilon_2(\omega')\omega'\,\mathrm{d}\omega'}{\omega'^2-\omega^2},\\ \epsilon_2(\omega) &= \frac{2}{\pi}\mathscr{P}\int\frac{\epsilon_1(\omega')\omega\,\mathrm{d}\omega'}{\omega^2-\omega'^2}, \end{aligned}\right\} \tag{4.38}$$

relate the real part ϵ_1 of the dielectric constant to the imaginary part (or the absorption). $\mathscr{P}$ denotes the principal value.

According to a theorem of Titchmarsh (1948), dispersion relations of the form (4.38) necessarily hold for any function $f(\omega)$ which satisfies the following conditions:

(a) $f(\omega)$ must be square-integrable, i.e.

$$\int_0^\infty |f(\omega)|^2\,\mathrm{d}\omega \text{ is finite,} \tag{4.39}$$

and (b) the Fourier transform

$$f(t) = \int_{-\infty}^{\infty} f(\omega)\exp(-\mathrm{i}\omega t)\,\mathrm{d}\omega \tag{4.40}$$

is real and vanishes for $t < 0$.

Following Taylor (1961), we shall define material functions for superconductors. These functions are constructed in such a way as to satisfy the Titchmarsh conditions and obey a pair of dispersion relations analogous to (4.38). It will be convenient to discuss the longitudinal and transverse functions separately. Exponential Fourier transforms (defined in (4.3)) are used, with the frequency variable ω ranging from $-\infty$ to ∞. When a function $f(t)$ is real, its transform satisfies

$$f(\omega) = f^*(-\omega). \tag{4.41}$$

We separate Maxwell's equations (4.8) into their transverse and longitudinal components, noting that **B** and **H** are entirely transverse:[38]

$$\left.\begin{aligned} \mathbf{q}\wedge\mathbf{E}^{\mathrm{t}}(\mathbf{q},\omega) &= -\frac{\omega}{c}\mathbf{B}(\mathbf{q},\omega),\\ \mathrm{i}\mathbf{q}\wedge\mathbf{H}(\mathbf{q},\omega) &= -\frac{4\pi}{c}\mathbf{J}^{\mathrm{t}}(\mathbf{q},\omega)+\frac{\mathrm{i}\omega}{c}\mathbf{D}^{\mathrm{t}}(\mathbf{q},\omega); \end{aligned}\right\} \tag{4.42}$$

$$4\pi\mathbf{J}^{\mathrm{l}}(\mathbf{q},\omega) = \mathrm{i}\omega\mathbf{D}^{\mathrm{l}}(\mathbf{q},\omega) = 4\pi\omega\,\rho(\mathbf{q},\omega)\,\mathbf{q}/q^2. \qquad (4.43)^{(30)}$$

These are two longitudinal relations (4.43) connecting the four fields $\mathbf{E}^{\mathrm{l}}$, $\mathbf{D}^{\mathrm{l}}$, $\mathbf{J}^{\mathrm{l}}$ and ρ, and two transverse relations (4.42) connecting the four fields **B**, **H**, $\mathbf{E}^{\mathrm{t}}$ and[30] $4\pi\mathbf{J}^{\mathrm{t}}-\mathrm{i}\omega\mathbf{D}^{\mathrm{t}}$. To determine the longitudinal fields, two constitutive equations are required:

$$\mathbf{J}^{\mathrm{l}}(\mathbf{q},\omega) = \boldsymbol{\sigma}^{\mathrm{l}}(\mathbf{q},\omega)\mathbf{E}^{\mathrm{l}}(\mathbf{q},\omega), \tag{4.44}$$

$$\mathbf{D}^{\mathrm{l}}(\mathbf{q},\omega) = \boldsymbol{\epsilon}^{\mathrm{l}}(\mathbf{q},\omega)\mathbf{E}^{\mathrm{l}}(\mathbf{q},\omega). \qquad (4.45)^{(39)}$$

The material functions $\boldsymbol{\epsilon}^{\mathrm{l}}(\mathbf{q},\omega)$ and $\boldsymbol{\sigma}^{\mathrm{l}}(\mathbf{q},\omega)$ have the significance of a complex longitudinal dielectric constant and of a complex longitudinal conductivity respectively. We may use the dielectric constant $\boldsymbol{\epsilon}^{\mathrm{l}}(\mathbf{q},\omega)$ to eliminate $\mathbf{D}^{\mathrm{l}}(\mathbf{q},\omega)$ from the equations; $\mathbf{E}^{\mathrm{l}}(\mathbf{q},\omega)$ must then satisfy

$$4\pi\boldsymbol{\sigma}^{\mathrm{l}}(\mathbf{q},\omega)\,\mathbf{E}^{\mathrm{l}}(\mathbf{q},\omega) = \mathrm{i}\omega\,\boldsymbol{\epsilon}^{\mathrm{l}}(\mathbf{q},\omega)\,\mathbf{E}^{\mathrm{l}}(\mathbf{q},\omega). \qquad (4.46)^{(30)}$$

When $\mathbf{D} = 0$, (4.46) reduces to the condition for plasma waves. However for superconductivity $\boldsymbol{\sigma}^{\mathrm{l}}(\mathbf{q},\omega)$ is the more interesting material function, since the displacement $\mathbf{D}^{\mathrm{l}}(\mathbf{q},\omega)$ will not play any further role in this development.

For the transverse equations, it is most expedient to eliminate the magnetic field $\mathbf{H}$ in favour of the magnetization $\mathbf{M}$:

$$\mathbf{H}(\mathbf{q}, \omega) = \mathbf{B}(\mathbf{q}, \omega) - 4\pi\mathbf{M}(\mathbf{q}, \omega). \tag{4.47}$$ [40]

The transverse equations now take the form[41]

$$\left.\begin{aligned} \mathbf{q} \wedge \mathbf{E}^{t}(\mathbf{q}, \omega) &= -\frac{\omega}{c}\mathbf{B}(\mathbf{q}, \omega), \\ ic\mathbf{q} \wedge \mathbf{B}(\mathbf{q}, \omega) &= -4\pi\left\{\mathbf{J}^{t}(\mathbf{q}, \omega) - ic\mathbf{q} \wedge \mathbf{M}(\mathbf{q}, \omega) + \frac{i\omega}{4\pi}\mathbf{D}^{t}(\mathbf{q}, \omega)\right\}. \end{aligned}\right\} \tag{4.48}$$

A single constitutive equation

$$\mathbf{J}^{t}(\mathbf{q}, \omega) - ic\mathbf{q} \wedge \mathbf{M}(\mathbf{q}, \omega) = \boldsymbol{\sigma}_{\mathrm{m}}(\mathbf{q}, \omega)\, \mathbf{E}^{t}(\mathbf{q}, \omega) \tag{4.49}$$ [42]

suffices to determine the three transverse fields $\mathbf{E}^{t}$, $\mathbf{B}$, and $\mathscr{J}^{t}$:

$$\mathscr{J}^{t}(\mathbf{q}, \omega) \equiv \mathbf{J}^{t}(\mathbf{q}, \omega) - ic\mathbf{q} \wedge \mathbf{M}(\mathbf{q}, \omega) + \frac{i\omega}{4\pi}\mathbf{D}^{t}(\mathbf{q}, \omega). \tag{4.50}$$ [43]

The complex conductivity $\boldsymbol{\sigma}_{\mathrm{m}}$ of (4.49) is not identical with the usual transverse conductivity, but reduces to it when the magnetic permeability $\mu(\mathbf{q}, \omega) = 1$, i.e. $\mathbf{M} = 0$. The total transverse current defined by (4.50) is the sum of the true current and the magnetization and displacement currents; the true current $\mathbf{J}^{t}$ may of course be Ohmic or not, according to the nature of the material. Note from (4.48) that $\mathbf{q} \cdot \mathscr{J}^{t}(\mathbf{q}, \omega) = 0$, or in configuration space that $\operatorname{div} \mathscr{J}^{t} = 0$. Hence, for any closed surface S,

$$\int_{S} \mathscr{J}^{t} \cdot d\mathbf{S} = \int_{\Omega} d\tau \operatorname{div} \mathscr{J}^{t} = 0,$$

i.e. the transverse current *does not transport any charge.*

The material functions $\boldsymbol{\sigma}^{l}$ and $\boldsymbol{\sigma}_{\mathrm{m}}$ need not satisfy the Titchmarsh conditions (4.39), (4.40), but Taylor (1961) shows that the new functions

$$\left.\begin{aligned} \mathscr{S}(\mathbf{q}, \omega) &= -i\omega\boldsymbol{\sigma}^{l}(\mathbf{q}, \omega), \\ \mathscr{L}(\mathbf{q}, \omega) &= c^{-2}\{-4\pi i\omega\, \boldsymbol{\sigma}_{\mathrm{m}}(\mathbf{q}, \omega) - \omega^{2}\}, \end{aligned}\right\} \tag{4.51}$$ [44]

do satisfy the conditions. The reason that $\boldsymbol{\sigma}^{l}$ and $\boldsymbol{\sigma}_{m}$ may fail to satisfy the conditions is that they may diverge in the static limit $\omega \to 0$; indeed we expect *a priori* that they *will* diverge in a superconductor. The functions defined in (4.51) are designed to remain finite even if the $\boldsymbol{\sigma}^{l}$, $\boldsymbol{\sigma}_{m}$ do diverge.

Since $\mathscr{S}(\mathbf{q}, \omega)$ and $\mathscr{L}(\mathbf{q}, \omega)$ satisfy Titchmarsh's conditions, the following dispersion relations are valid for both normal metals and superconductors:

$$\left.\begin{aligned} \operatorname{Re}\{\mathscr{S}(\mathbf{q}, \omega)-\mathscr{S}(\mathbf{q}, \infty)\} &= \frac{2}{\pi}\mathscr{P}\int_{0}^{\infty}\frac{\operatorname{Im}\mathscr{S}(\mathbf{q}, \omega')\omega'\mathrm{d}\omega'}{\omega'^{2}-\omega^{2}}, \\ \operatorname{Im}\mathscr{S}(\mathbf{q}, \omega) &= \frac{2}{\pi}\mathscr{P}\int_{0}^{\infty}\frac{\operatorname{Re}\mathscr{S}(\mathbf{q}, \omega')\omega\,\mathrm{d}\omega'}{\omega^{2}-\omega'^{2}}, \end{aligned}\right\} \tag{4.52}$$

$$\left.\begin{aligned} \operatorname{Re}\{\mathscr{L}(\mathbf{q}, \omega)-\mathscr{L}(\mathbf{q}, \infty)\} &= \frac{2}{\pi}\mathscr{P}\int_{0}^{\infty}\frac{\operatorname{Im}\mathscr{L}(\mathbf{q}, \omega')\omega'\mathrm{d}\omega'}{\omega'^{2}-\omega^{2}}, \\ \operatorname{Im}\mathscr{L}(\mathbf{q}, \omega) &= \frac{2}{\pi}\mathscr{P}\int_{0}^{\infty}\frac{\operatorname{Re}\mathscr{L}(\mathbf{q}, \omega')\omega\,\mathrm{d}\omega'}{\omega^{2}-\omega'^{2}}. \end{aligned}\right\} \tag{4.53}$$

In the limit $\omega \to 0$, (4.52) and (4.53) reduce to the sum rules

$$\int_{0}^{\infty}\operatorname{Re}\boldsymbol{\sigma}^{l}(\mathbf{q}, \omega')\mathrm{d}\omega' = \tfrac{1}{2}\pi\{\mathscr{S}(\mathbf{q}, \infty)-\mathscr{S}(\mathbf{q}, 0)\}, \tag{4.54}$$

$$\int_{0}^{\infty}\operatorname{Re}\boldsymbol{\sigma}_{m}(\mathbf{q}, \omega')\mathrm{d}\omega' = \tfrac{1}{8}c^{2}\{\mathscr{L}(\mathbf{q}, \infty)-\mathscr{L}(\mathbf{q}, 0)\}. \tag{4.55}$$ [45]

The difference between a normal conductor and a perfect conductor is that $\mathbf{E}^{l}(\mathbf{q}, 0) = 0$ for the perfect conductor, but $\mathbf{E}^{l}(\mathbf{q}, 0)$ is finite and nonzero for the normal conductor, i.e.

$$\left.\begin{aligned} \mathscr{S}(\mathbf{q}, 0) &= 0 && \text{for normal conductivity,} \\ \lim_{\mathbf{q}\to 0}\mathscr{S}(\mathbf{q}, 0) &= \mathscr{A}_{1} \neq 0 && \text{for perfect conductivity.} \end{aligned}\right\} \tag{4.56}$$

Also it can be shown that the condition for normal magnetism and the condition for a Meissner effect are respectively†

$$\left.\begin{aligned} \mathscr{L}(\mathbf{q}, 0) &= 0 \qquad \text{for normal magnetism,} \\ \lim_{\mathbf{q}\to 0} \mathscr{L}(\mathbf{q}, 0) &= \mathscr{A}_2 \neq 0 \text{ for Meissner effect.} \end{aligned}\right\} \tag{4.57}$$

Thus the conditions for perfect conductivity and the Meissner effect are formally independent (Taylor 1961). In real materials, however, there is always a relation between $\mathscr{S}(\mathbf{q}, 0)$ and $\mathscr{L}(\mathbf{q}, 0)$:[46]

$$\lim_{\mathbf{q}\to 0} \left\{ \mathscr{L}(\mathbf{q}, 0) - \frac{4\pi}{c^2} \mathscr{S}(\mathbf{q}, 0) \right\} = 0,$$

or

$$\lim_{\omega\to 0} \lim_{\mathbf{q}\to 0} \omega\{\boldsymbol{\sigma}_{\mathrm{m}}(\mathbf{q}, \omega) - \boldsymbol{\sigma}^{\mathrm{l}}(\mathbf{q}, \omega)\} = 0. \tag{4.58}$$

Evans and Rickayzen (1964) have shown that (4.58) is a consequence of the existence of a scattering mechanism for electrons, irrespective of its details. Thus, although (4.58) cannot be derived from the phenomenological theory, it does follow from *any* microscopic theory. It is consistent with the phenomenological theory, and may therefore be adopted as a phenomenological postulate. The physical meaning of (4.58) is very simple. We noted on p. 56 that the distinction between longitudinal and transverse vectors breaks down in the limit as $q \to 0$. Consider an external field of small wave number $\mathbf{q}$ applied to a superconductor. If $\boldsymbol{\sigma}^{\mathrm{l}}(\mathbf{q}, \omega)$ and $\boldsymbol{\sigma}_{\mathrm{m}}(\mathbf{q}, \omega)$ were different, it would mean that either the charge carriers were so delocalized or that their interactions were of such long range that they could 'sense' whether the applied field was longitudinal or transverse. Equation (4.58) asserts that in the limit $q \to 0$ this does not happen. In other words the mutual forces between the charge carriers have a finite range and the carriers themselves have a finite spatial extension. In Chapter 6 we shall see that typically this localization distance or coherence length ξ is $\sim 10^{-4}$ cm, large but indeed finite.

† In fact $\mathscr{L}(\mathbf{q}, 0) \propto q^2\mathscr{K}_1(q^2)$ (cf. equation (4.28)) in the limit as $q \to 0$.

It is this equality of the longitudinal and transverse conductivities that ensures the identity of the inertial coefficient Λ of London's acceleration equation (3.3) and the penetration coefficient Λ of the magnetic field equation (3.15). When London's equations hold, the coefficient Λ is related to the coefficients $\mathscr{A}_1$, $\mathscr{A}_2$ as follows:

$$\left.\begin{aligned} \Lambda^{-1} = \operatorname{Im}(\omega\boldsymbol{\sigma}) &= \mathscr{A}_1 \\ &= c^2\mathscr{A}_2/4\pi. \end{aligned}\right\} \qquad (4.59)^{(47)}$$

For a normal conductor the sum rules (4.54), (4.55) may be simplified, using (4.56), (4.57), and (4.51), to give

$$\left.\begin{aligned} \int_0^\infty \operatorname{Re}\boldsymbol{\sigma}^{\mathrm{l}}(\mathbf{q}, \omega)\,\mathrm{d}\omega &= \tfrac{1}{2}\pi\mathscr{S}(\mathbf{q}, \infty) = \lim_{\omega\to 0} \omega \operatorname{Im}\boldsymbol{\sigma}^{\mathrm{l}}(\mathbf{q}, \omega), \\ \int_0^\infty \operatorname{Re}\boldsymbol{\sigma}_{\mathrm{m}}(\mathbf{q}, \omega)\,\mathrm{d}\omega &= \tfrac{1}{8}c^2\mathscr{L}(\mathbf{q}, \infty) = \lim_{\omega\to 0} \omega \operatorname{Im}\boldsymbol{\sigma}_{\mathrm{m}}(\mathbf{q}, \omega). \end{aligned}\right\} \qquad (4.60)^{(45)}$$

This simplification is tantamount to observing that when (4.56) and (4.57) hold, the conductivities $\boldsymbol{\sigma}^{\mathrm{l}}$, $\boldsymbol{\sigma}_{\mathrm{m}}$ themselves obey dispersion relations analogous to (4.52) and (4.53). But it is clear from (4.56) and (4.57) that the simplified sum rules (4.60) are not applicable to superconductors.

Ferrell and Glover (1958) showed that a study of the transmission of electromagnetic radiation through thin films can yield information about the energy gap. Their argument is related to another argument, by Bardeen (1956), which makes it plausible that the electromagnetic properties of superconductors are related to the energy gap.† These arguments are outlined below.

If the spectrum of low-lying states has an energy gap 2Δ, then phonons of energy $\hbar\omega < 2\Delta$ cannot be absorbed. This implies that $\operatorname{Re}\boldsymbol{\sigma}(\mathbf{q}, \omega) = 0$ for $\omega < 2\Delta/\hbar$. On the other hand, at high frequencies $\omega \gg 2\Delta/\hbar$ the conductivity takes the same

† However, Bardeen's argument has been criticized by Buckingham (1957).

value in the s-phase as in the n-phase. Theoretically this follows because at frequencies much above $2\Delta/\hbar$ electrons are unable to 'see' the very small energy gap 2Δ. Experimental confirmation is afforded by the fact that the *optical* reflectivity of the n- and s-phases are the same.

Thus the right-hand side of (4.60) is unaltered by passing from the normal to the superconducting state. But the left-hand side *is* altered; $\text{Im}\,\boldsymbol{\sigma}(\mathbf{q}, \omega)$ is suppressed for $\omega < 2\Delta/\hbar$.

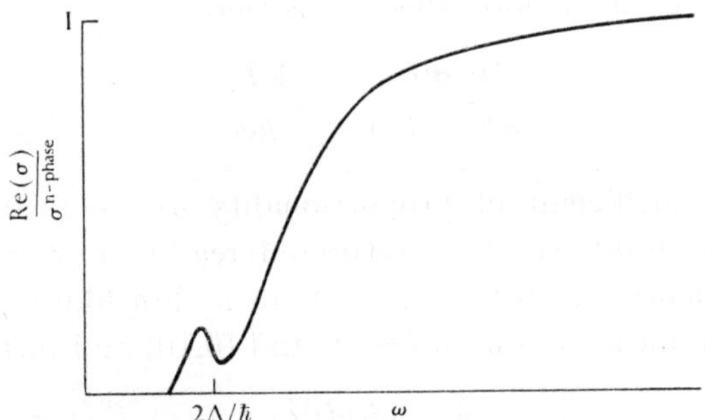

FIG. 4.1. $\text{Re}(\boldsymbol{\sigma})/\boldsymbol{\sigma}^{\text{n-phase}}$ for lead, as found by Ginsberg and Tinkham (1960) from infrared transmission through films.

The excess of the right-hand side over the left-hand side† is easily seen to be $\frac{1}{2}\pi\mathscr{A}_1$. The tacit assumption underlying this argument is that the conductivity as a function of frequency is not singular in the vicinity of $\omega \sim 2\Delta/\hbar$. Experimentally this assumption is confirmed— $\text{Re}(\boldsymbol{\sigma})$ is indeed zero for low frequencies. Figure 4.1 shows how it rises smoothly to its

† The actual calculation of Ferrell and Glover is non-rigorous, in so far as they assume that $\boldsymbol{\sigma}$ itself, rather than $\mathscr{S}$ or $\mathscr{L}$, obeys dispersion relations. While this assumption is correct in the n-phase, it breaks down in the s-phase. These authors patch up the dispersion relations by adding a real δ-function contribution $\frac{1}{2}\pi\mathscr{A}_1\delta(\omega)$ to $\boldsymbol{\sigma}(\omega)$; they remark that this is a natural description of the infinite d.c. conductivity. The δ-function then contributes, via the dispersion relations, to $\text{Im}\,\boldsymbol{\sigma}(\mathbf{q}, \omega)$ —in fact for $\omega \ll 2\Delta/\hbar$,

$$\text{Im}\,\boldsymbol{\sigma}(\mathbf{q}, \omega) \simeq \mathscr{A}_1/\omega. \tag{4.61}$$

In Taylor's rigorous formulation of the problem, (4.61) follows immediately from (4.51) and (4.56).

normal-phase value for $\omega \gg 2\Delta/\hbar$, apart from a minor peak just below $\omega = 2\Delta/\hbar$ (Ginsberg and Tinkham 1960)†.

From the experimentally-observed value of $\mathscr{A}_1$, the penetration depth λ follows from (3.10):

$$\lambda = c(\Lambda/4\pi)^{\frac{1}{2}} = c(4\pi\mathscr{A}_1)^{-\frac{1}{2}}. \tag{4.62}$$[48]

However, the values of λ found thus by Glover and Tinkham (1957) are larger than the London value λ_L (3.16), by up to a factor of 10. A further result of Glover and Tinkham—not predicted by the London theory—is that

$$\frac{\mathrm{Im}\,\boldsymbol{\sigma}(\omega)}{\boldsymbol{\sigma}^{\text{n-phase}}(\omega)} = \alpha\frac{\mathrm{k}T_c}{\hbar\omega}, \tag{4.63}$$

where the coefficient of proportionality $\alpha = 0{\cdot}27$ in tin and lead. This result can be interpreted readily in terms of the Pippard theory (§§ 6.1, 6.2). For, in a thin film of thickness $d \ll \lambda_L$, the mean free path $l \sim d$, and (6.10) and (6.15) give

$$\lambda = (\xi_0/d)^{\frac{1}{2}}\lambda_L.$$

Moreover, if we interpret Pippard's coherence length ξ as an uncertainty in the position of an electron, Heisenberg's uncertainty principle gives $\xi_0 \sim \hbar/\Delta p$. The momentum uncertainty Δp is expected to be $\sim \hbar v_F/\mathrm{k}T_c$, where v_F is the velocity of an electron at the Fermi surface of the metal. Thus

$$(\lambda/\lambda_L)^2 = \xi_0/d = \alpha\hbar v_F/\mathrm{k}T_c\, d, \tag{4.64}$$

and (4.63) follows. The dimensionless constant α is of course not fixed by this argument, but the BCS microscopic theory predicts the value 0·18, in fair agreement with the above experimental values.

† *Added in proof:* Recent experiments by Palmer and Tinkham (D. M. Ginsberg, private communication) suggest that the 'precursor' peak of Fig. 4.1 may be much smaller or even non-existent.

CHAPTER 5

SURFACE ENERGY AT A NORMAL-SUPERCONDUCTING INTERFACE

At first sight, the combination of the London electrodynamic equations of Chapter 3 with any of the thermodynamic two-fluid models of Chapter 2 seems to give an excellent phenomenological description of superconductivity. But on closer study it will be shown that there is a serious inconsistency. The macroscopic theory of Chapter 1 is fully consistent, but with the introduction of the concept of a penetration layer a paradox emerges. To resolve the paradox a new assumption is required. A surface energy, associated with normal-superconducting phase boundaries, is postulated. This assumption will be shown to have important implications in the theory of the intermediate state.

5.1. Parasitic solutions

Solutions of the London equations will be constructed which do not show a Meissner effect. As a preliminary, let us recalculate the magnetic Gibbs function for the superconducting phase, taking account of the penetration layer. In the first instance consider an isolated superconductor, large compared with the penetration layer. The geometry is chosen to avoid any effects arising from the distortion of the applied field: a long thin ellipsoid parallel to the applied magnetic field direction (cf. Fig. 1.5, p. 11). To simplify the discussion, the two axes perpendicular to the field will be taken to be equal, though this assumption is not essential.

Since, by hypothesis, the minor axis a is much greater than the penetration depth λ the induction and the current density inside the body are

$$\left.\begin{aligned}\mathbf{B} &= \mathbf{H}_e \exp\{-(a-r)/\lambda\},\\ \mathbf{J} &= \mathbf{J}_0 \exp\{-(a-r)/\lambda\}.\end{aligned}\right\} \qquad (5.1)^{(49)}$$

At a radial distance r from the axis, the induction $\mathbf{B}$ differs from[50] $\mathbf{H}_e$ by the induction due to the solenoidal supercurrent outside r. In other words,

$$\mathbf{B} = \mathbf{H}_e - \frac{4\pi}{c}\int_r^a \mathbf{J} \wedge d\mathbf{r}, \qquad (5.2)^{(51)}$$

or

$$\mathbf{B} = \mathbf{H}_e + \frac{4\pi}{c} J_0 \lambda\, (\mathbf{H}_e/H_e)\, [1 - \exp\{-(a-r)/\lambda\}], \qquad (5.3)^{(52)}$$

which agrees with (5.1) when we set[53] $4\pi J_0 \lambda/c = H_e$. (Note that $\mathbf{J}_0$ is perpendicular to $\mathbf{H}_e$.)

The magnetic moment per unit length is

$$\mathbf{M} = \pi a^2 \int_0^a \mathbf{J} \wedge d\mathbf{r}/c$$

$$= \frac{\mathbf{H}_e}{4\pi} \pi a^2 (1 - 2\lambda/a) \qquad (5.4)^{(54)}$$

neglecting $\exp(-\lambda/a)$ compared to unity. Hence the magnetic energy per unit length is

$$\tfrac{1}{2}\mathbf{M}.\mathbf{H}_e = \frac{1}{8\pi} H_e^2 (\pi a^2 - 2\pi\lambda a). \qquad (5.5)^{(11)}$$

The first term of (5.5) follows directly from (2.4), if the body is *uniformly* magnetized. The second term is the correction arising from the non-uniformity; it may be interpreted as a surface energy[11] $-\lambda H_e^2/8\pi$ per unit area.

In particular, if the superconducting cylinder is in equilibrium with normal metal outside, then $H_e = H_c$. In this case the penetration of the field leads to an effective surface energy

$$\alpha_\lambda = -\lambda H_c^2/8\pi \qquad (5.6)^{(11)}$$

per unit area. Although this result is exact only in the limit $\lambda/a \ll 1$, it will be adequate to exhibit an apparent failure of the Meissner effect.

Consider a lamina of n-phase introduced into a superconducting body. Let the n-lamina be oriented parallel to the external magnetic field. Let its thickness $2d_{\mathrm{n}}$ be small, $d_{\mathrm{n}} \ll \lambda$, and let the external magnetic field be $H_{\mathrm{e}} < H_{\mathrm{c}}$. The change in the free energy, on introducing the n-lamina, is made up of two parts. The condensation energy (2.13) of the n-s transition contributes[11] $\frac{2}{8\pi} H_{\mathrm{c}}^2 d_{\mathrm{n}}$ per unit area; it is positive since energy has to be *supplied* to destroy the superconductivity of the lamina. The other contribution is a lowering of the electromagnetic energy, because the n-lamina allows some magnetic flux to enter the body. By (5.6), the electromagnetic contribution is[11] $-\frac{1}{8\pi} H_{\mathrm{e}}^2(2d_{\mathrm{n}}+2\lambda)$. The net change of free energy is[11]

$$-\frac{1}{8\pi}\{H_{\mathrm{e}}^2(2d_{\mathrm{n}}+2\lambda)-2H_{\mathrm{c}}^2\, d_{\mathrm{n}}\},$$

and can clearly be negative for small enough d_{n}.

It is therefore energetically favourable to construct 'parasitic' configurations in which fine laminae (or perhaps threads) of n-phase penetrate through the superconductor. Indeed if (5.6) did not break down, it would be possible to make the free energy arbitrarily large and negative! All that would be needed is a sufficiently large number of these laminae.

This extreme conclusion is of course spurious. The free energy cannot be reduced *below* the value appropriate to complete penetration of the field. It will now be shown however that the energy can be lowered *to* that value. For this the approximation (5.6) is inadequate; the exponential term of (5.3) must be retained. An exact calculation is practicable for simple enough geometries. In particular for a superconducting lamina of thickness $2d_{\mathrm{s}}$, the magnetic energy per unit area can be shown to be

$$\mathscr{E}_1 = \frac{1}{8\pi} H_{\mathrm{e}}^2\{2d_{\mathrm{s}}-2\lambda \tanh(d_{\mathrm{s}}/\lambda)\}. \qquad (5.7)^{(11)}$$

(Note that (5.7) agrees with (5.5) if $d_{\mathrm{s}} \gg \lambda$.)

Exercise. Derive (5.7).

Consider now a fine-grained laminar structure of alternate layers of normal material (thickness $2d_{\mathrm{n}}$) and superconductor (thickness $2d_{\mathrm{s}}$), where $d_{\mathrm{n}} \ll d_{\mathrm{s}} \ll \lambda$. The former inequality implies that only a very small fraction of the material is in the normal state. The role of the second inequality is to ensure that the penetration of magnetic field into the superconductor is practically complete. In this limit $d_{\mathrm{s}} \ll \lambda$, (5.7) reduces to

$$\mathscr{E}_1 = \frac{1}{8\pi} H_{\mathrm{e}}^2 \frac{2d_{\mathrm{s}}^3}{3\lambda^2}. \tag{5.7$'$}$$ [11]

The total magnetization energy per unit volume of the material is just

$$\mathscr{E}_1/2(d_{\mathrm{n}}+d_{\mathrm{s}}),$$

since $\{2(d_{\mathrm{n}}+d_{\mathrm{s}})\}^{-1}$ is the number of pairs of n- and s-laminae per unit thickness. When d_{n} is neglected in comparison to d_{s}, the magnetization energy per unit volume is

$$\frac{1}{8\pi} H_{\mathrm{e}}^2 \frac{d_{\mathrm{s}}^2}{3\lambda^2}, \tag{5.8}$$ [11]

which can be made arbitrarily small. On the other hand the energy of condensation (2.13) is[11] $-H_{\mathrm{c}}^2/8\pi$ per unit volume of s-phase. But since d_{n} is negligible compared to d_{s}, this energy is also the condensation energy per unit volume of the laminar structure.

Thus we have constructed a configuration in which the magnetization energy is as close as we please to zero (its value in the n-phase), while the condensation energy has the full value[11] $-H_{\mathrm{c}}^2/8\pi$ per unit volume, characteristic of the s-phase. It is easy to satisfy oneself that this configuration minimizes the Gibbs free energy. Hence the state of lowest Gibbs free energy is one which does not show a Meissner effect!

The escape from the paradox was pointed out by H. London (1935). We must postulate the existence of an additional surface energy α'_{ns} *of non-magnetic origin*, at the n-s phase boundary. It is convenient to write

$$\alpha'_{\text{ns}} = \frac{1}{8\pi}\xi H_{\text{c}}^2, \tag{5.9}$$ (11)

where ξ has the dimension of a length. In § 6.4 it will be identified with Pippard's coherence length. A sufficient condition to suppress parasitic solutions is clearly

$$\xi > \lambda. \tag{5.10}$$

Superconductors for which (5.10) holds are known as 'Type I' superconductors. However, there are materials ('Type II' superconductors) for which (5.10) is not true.† For these materials, parasitic solutions can *and do* occur for fields well below H_{c}. There is a much lower field, above which there is no Meissner effect. By minimizing the Gibbs free energy for the laminar structure, the scale of the domain structure may be determined, although (cf. Chapter 7) the true configuration in a Type II superconductor is not a laminar one. The Gibbs function per unit volume of a laminar structure is

$$g = \frac{1}{8\pi}x_{\text{s}}\{H_{\text{c}}^2(1+2\xi/d_{\text{s}}) - 2H_{\text{e}}^2(\lambda/d_{\text{s}})\tanh(d_{\text{s}}/\lambda)\}, \tag{5.11}$$ (11)

where x_{s} is the superconducting fraction, i.e.

$$x_{\text{s}} = d_{\text{s}}/(d_{\text{s}}+d_{\text{n}}). \tag{5.12}$$

For fields near H_{c}, (5.11) has a minimum at a finite value of d_{s} provided $\xi < \lambda$, i.e. provided the material is Type II. For a Type I superconductor, in contrast, the minimum occurs for $d_{\text{s}} \to \infty$ (the pure superconducting state).

It is convenient to define the quantity

$$\delta = \xi - \lambda. \tag{5.13}$$

† For a more formal rigorous definition of Type I and Type II superconductivity, see p. 115.

The condition (5.10) for a superconductor to be of Type I may thus be written

$$\delta > 0.$$

In a Type I material, the field at an n-s interface must be very close to H_c, and the net surface energy is

$$\alpha_{ns} = \frac{1}{8\pi}\delta H_c^2. \qquad (5.14)^{(11)}$$

5.2. The intermediate state

Even in a Type I superconductor, it is possible to realize a finely-divided domain structure. In § 1.3 it was shown that a domain structure could be required in some circumstances, as a consequence of purely geometrical considerations. For example, consider a superconducting sphere in a magnetic field which at large distances from the sphere is uniform and takes the value $\mathbf{H}_e$. At the equator of the sphere, the field is $\frac{3}{2}\mathbf{H}_e$. It is clear that the pure superconducting state cannot be in thermodynamic equilibrium if $\frac{3}{2}H_e > H_c$. There must be a separation into two phases.

However, no *macroscopic* two-phase configuration can be in equilibrium (cf. Fig. 1.7, p. 14). If the field on an n-s boundary† is not precisely H_c, an infinitesimal displacement of the phase boundary in the appropriate direction will reduce the Gibbs free energy. But if the field at the interface *is* H_c, there must be regions where the interface is convex towards the n-phase. In that case the field inside the n-phase is less than H_c. The alternative—a configuration with sharp corners—is equally unacceptable; such corners are regions of infinite but still convex curvature.

If the interfacial surface energy were negligible, the only escape from the above dilemma would be for the domain structure to become *infinitely* fine-grained. In the limit, the field would be H_c everywhere in the n-phase. The macroscopic theory of the intermediate state (§ 1.3) corresponds to this limit. In a Type I superconductor, however, this infinitely

† It is assumed that the radius of curvature $R \gg \lambda$.

fine subdivision cannot be the equilibrium configuration. Its infinite surface area would endow it with an infinite Gibbs free energy. In fact the equilibrium structure will have a small but finite grain size; it will be determined by the balance between the field inhomogeneity on the one hand and the n-s surface energy on the other.

Landau (1937*b*, 1943) has described two very different models for the domain structure. In the first model the n- and s-domains are laminae, flat except very near the surface of the body. The laminar boundaries are allowed to curve near the surface, in such a way as to maintain the condition $H = H_c$ on the n-s interfaces (Fig. 5.1). However, just as in the hypothetical configuration with macroscopic n- and s-regions (Fig. 1.7, p. 14), the curvature of the surfaces would leave $H < H_c$ in the n-layers. To avoid this difficulty, Landau's second model has n-laminae which branch repeatedly as they approach the surface of the body (Fig. 5.2). Inside the body the field is uniform and of magnitude H_c. The branching becomes infinitely fine towards the surface of the body, so that the external field is homogeneous. Andrew (1948) has discussed a similar model with branching threads. All the models have their own difficulties and none is likely to represent the exact equilibrium state. In practice, however, the Gibbs function is rather insensitive to the details of the structure.

Lifshitz and Sharvin (1951) have discussed yet another configuration—probably the most realistic of all the theoretical models. In their model, the n-laminae are assumed to branch a finite number of times p. The number of branchings is regarded as one of the variational parameters and is optimized. For bodies of reasonable size, the result is that p is quite small, between 1 and 2. This means that Landau's unbranched model is quite a good one; the more complicated branching models will not be discussed here.

The principal assumptions of the model are:

(a) the domain structure is so coarse that the penetration layer of thickness $\sim\lambda$ may be neglected;

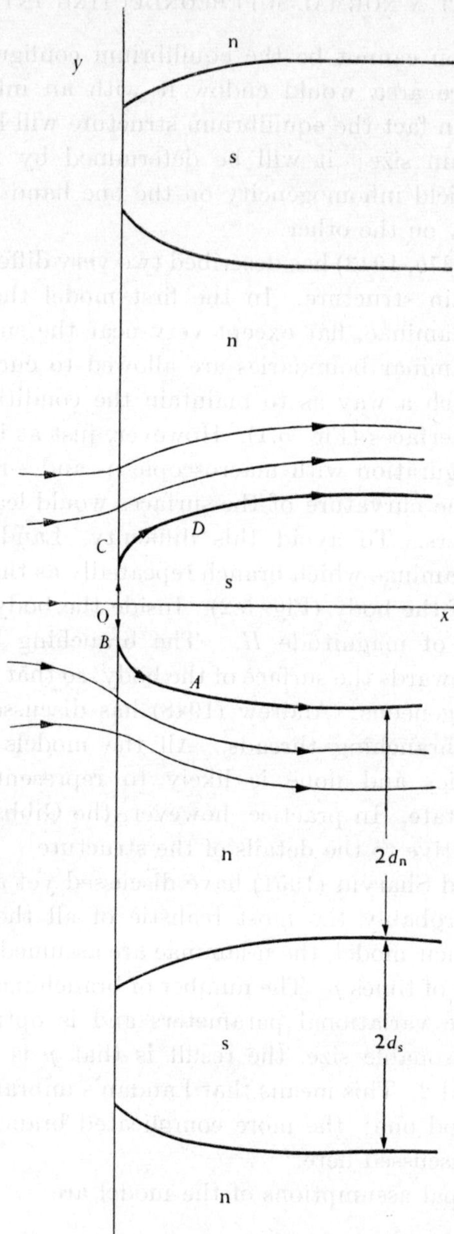

FIG. 5.1. Landau's (1937*b*) non-branched laminar model for the intermediate state.

(b) the domains are laminae which are curved only near the external surface of the body;

(c) the external surface is plane, and perpendicular to the applied magnetic field H_e; and

(d) the magnetic field is tangential to all n-s interfaces and has a constant magnitude H_c on them.

In § 5.2.1, the model will be used to calculate the magnetization of an ellipsoidal body. For this application, assumption (c)

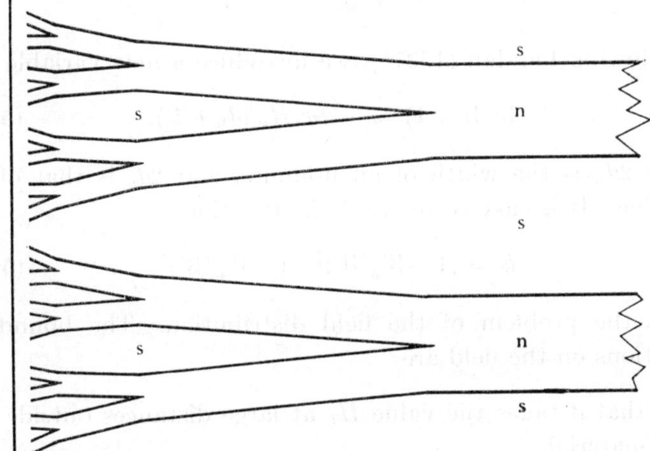

FIG. 5.2. Landau's (1943) branched model.

cannot be valid at all points on the external surface; it is therefore not surprising that the agreement with experiment is only *fairly* good.

The coordinates to be used are shown in Fig. 5.1. The problem is two-dimensional, in the sense that none of the variables depend on z, the direction perpendicular to the plane of the figure. The magnetic vector potential may be chosen to lie in the z-direction, so that for this calculation it behaves as a scalar. In terms of the magnetic scalar potential ϕ and of the vector potential A_z the magnetic field is

$$\left.\begin{aligned} B_x &= -\partial\phi/\partial x = \partial A_z/\partial y, \\ B_y &= -\partial\phi/\partial y = -\partial A_z/\partial x. \end{aligned}\right\} \tag{5.15}$$

Equations (5.15) have the form of Cauchy–Riemann equations for the function

$$h = \mathrm{d}w/\mathrm{d}z. \tag{5.16}$$

In (5.16), the variables are defined by

$$\left.\begin{aligned} h &= (B_x - \mathrm{i}B_y)/H_\mathrm{c}, \\ w &= \phi - \mathrm{i}A_z, \\ \text{and} \qquad z &= x + \mathrm{i}y. \end{aligned}\right\} \tag{5.17}$$

Following Landau (1937*b*), we introduce a new variable W:

$$\ln(W+1) = -2w/H_\mathrm{e}\,(d_\mathrm{n}+d_\mathrm{s}), \tag{5.18}$$

where $2d_\mathrm{n}$ is the width of an n-lamina and $2d_\mathrm{s}$ is that of an s-lamina. It is easy to see that the function

$$h = (1-W_0/W)^{\frac{1}{2}} - (-W_0/W)^{\frac{1}{2}} \tag{5.19}$$

solves the problem of the field distribution. The boundary conditions on the field are

(a) that it takes the value H_e at large distances outside the material,
(b) that it is constant in magnitude ($=H_\mathrm{c}$) at all points on the n-s boundary (including CD), and
(c) that it has no y-component at large distances inside the material (see Fig. 5.1).

In terms of the variable W, the boundary conditions may be written

$$\left.\begin{aligned} h &= H_\mathrm{e}/H_\mathrm{c} \text{ at } W = -W_0, \\ |h| &= 1 \qquad \text{for } W > W_0, \\ \mathrm{Im}(h) &= 0 \qquad \text{for } 0 < W < W_0. \end{aligned}\right\} \tag{5.20}$$

Provided the constant W_0 is chosen to be

$$W_0 = \tfrac{1}{4}(H_\mathrm{c}/H_\mathrm{e} - H_\mathrm{e}/H_\mathrm{c})^2, \tag{5.21}$$

equation (5.19) defines a field satisfying all these boundary conditions. Moreover, (5.18) and (5.19) together define h as a function of the complex variable z; the Cauchy–Riemann equations (5.15) are automatically satisfied.

To determine the form of the profiles CD (Fig. 5.1) it is necessary to integrate

$$\mathrm{d}z = -\mathrm{d}w/hH_{\mathrm{c}},$$

i.e.

$$z = \frac{d_{\mathrm{s}}+d_{\mathrm{n}}}{2\pi}\frac{H_{\mathrm{e}}}{H_{\mathrm{c}}}\int \frac{\mathrm{d}W}{h(W+1)}. \tag{5.22}$$

On performing the integration, parametric equations for the profile are obtained (Landau and Lifshitz 1960 p. 184).

In order to find the size of the domains, it is necessary to calculate the magnetic Gibbs function. Let us consider a parallel-sided slab of material, of thickness L. It will be assumed that L is so large that the second external surface of the slab has no influence on the profile of the region CD of Fig. 5.1. The Gibbs function per unit area is made up of the following three contributions:

(a) The surface energy of the n-s boundaries is

$$G_{\mathrm{surf}} = 2L\delta H_{\mathrm{c}}^2/8\pi(d_{\mathrm{n}}+d_{\mathrm{s}}), \qquad \text{(5.23a)}^{(11)}$$

where the correction arising from the curvature of the region CD has been neglected.

(b) The rounding of the 'corners' CD contributes a condensation energy per corner of magnitude

$$G_{\mathrm{cond}} = (H_{\mathrm{c}}^2/8\pi)\int_0^\infty (d_{\mathrm{n}}-y)\,\mathrm{d}x. \qquad \text{(5.23b)}^{(11)}$$

(c) Finally, the field-energy contribution is

$$G_{\mathrm{field}} = -\tfrac{1}{2}\mathbf{M}\,.\,\mathbf{H}_{\mathrm{e}}, \tag{5.23c}$$

where the magnetic moment **M** per corner is

$$\frac{1}{4\pi}\int (H_e\,d_n - H_c\,y)\,ds, \qquad (5.24)^{(10)}$$

and where

$$ds = \sqrt{\{(dx)^2+(dy)^2\}}. \qquad (5.25)$$

Exercise. Derive (5.24). Hint: Remember that the surface CD is a surface of discontinuity of the field; $\mathbf{B} = \mathbf{H}_c$ on the n-side of CD, and $\mathbf{B} = 0$ on the s-side.

Assembling the three contributions (5.23a, b, c), the total magnetic Gibbs function per unit area is

$$G = \frac{1}{4\pi(d_n+d_s)}\Bigg[H_c^2 L\delta + 2H_c^2\int\limits_{W=W_0}^{\infty}(y_\infty - y)\,dx + 2H_e\Bigg\{H_c\int\limits_{W=W_0}^{\infty}(y\,ds - d_n\,dx) + \int\limits_{W=0}^{W_0} H_y\,y\,dy\Bigg\}\Bigg]. \qquad (5.26)^{(10)}$$

The integrals in (5.26) cannot be expressed in closed form, but have to be evaluated numerically. Finally, in terms of the superconducting fraction x_s defined in (5.12), the Gibbs function per unit area is

$$G = \frac{1}{4\pi}H_c^2\left\{\frac{L\delta}{d_n+d_s} + (d_n+d_s)\Phi(x_s)\right\}. \qquad (5.27)^{(10)}$$

The function $\Phi(x_s)$ has been computed by Lifshitz and Sharvin (1951), and is tabulated in Table 5.1. G is minimized by choosing the domain size to be

$$(d_n+d_s)_{\min} = \sqrt{\{L\delta/\Phi(h)\}}. \qquad (5.28)$$

There is a considerable amount of experimental evidence showing that this domain structure really does occur in the intermediate state. Meshkovsky and Shalnikov (1947) first observed the domain structure by measuring the local magnetic field at the surface of a spherical tin specimen and on a cut plane inside it. They used a small bismuth probe as the field detector. Some of their results are illustrated in Fig. 5.3.

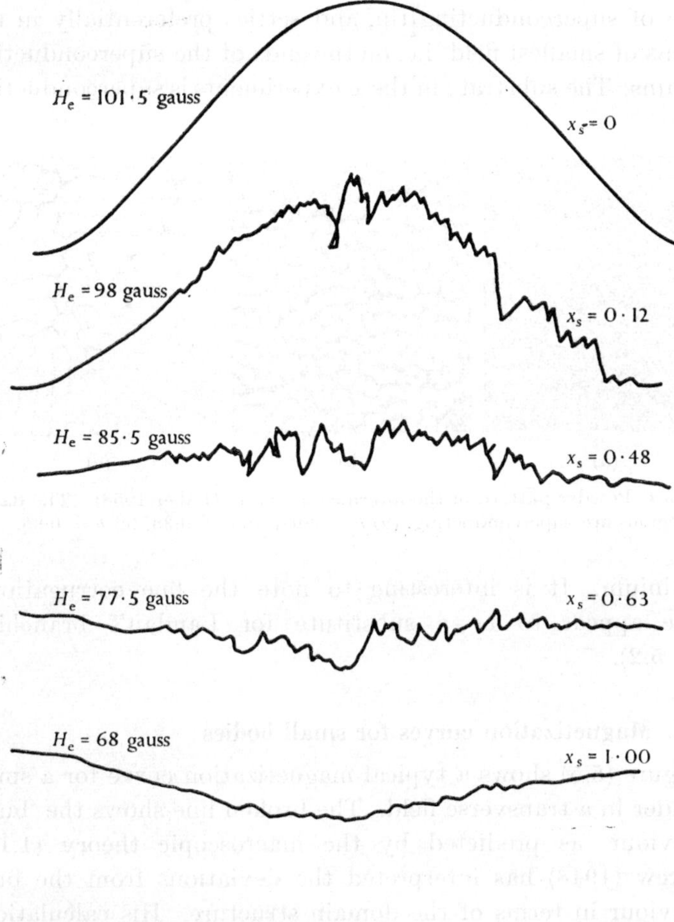

FIG. 5.3. Magnetic field on the surface of a tin sphere in the intermediate state, as found by Meshkovsky and Shalnikov (1947). The equator is at the ends of the curves, and a pole in the centre.[58]

More elegant, perhaps, are the powder photographs of Faber (1958), some of which are shown in Fig. 5.4. The powder is made of superconducting tin, and settles preferentially in the regions of smallest field, i.e. on the ends of the superconducting domains. The substrate in these experiments is superconducting

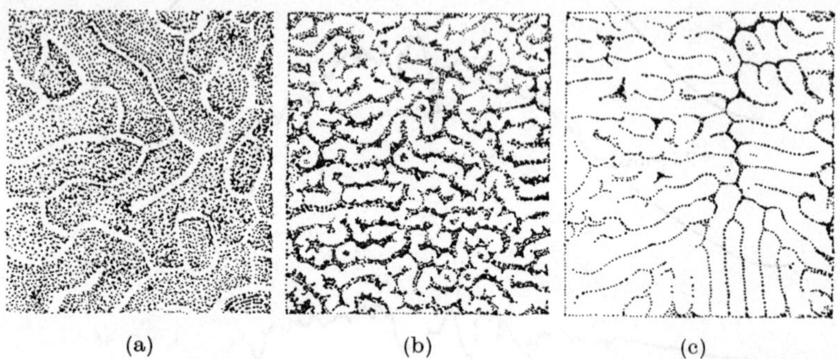

Fig. 5.4. Powder pattern of the intermediate state (Faber 1958). The dark regions are superconducting. (*a*) $h = 0{\cdot}08$, (*b*) $h = 0{\cdot}38$, (*c*) $h = 0{\cdot}65$.

aluminium. It is interesting to note the fine corrugations. These appear to be a substitute for Landau's branching (Fig. 5.2).

5.2.1. Magnetization curves for small bodies

Figure (5.5) shows a typical magnetization curve for a small cylinder in a transverse field. The broken line shows the 'bulk' behaviour, as predicted by the macroscopic theory (1.17). Andrew (1948) has interpreted the deviations from the bulk behaviour in terms of the domain structure. His calculations were based on the branching model (Fig. 5.2), but are presented here in terms of the unbranched model.

A difficulty arises in applying the domain theory to small specimens. The derivation of (5.27) presupposes an infinite parallel-sided slab of material. In applying the theory to a body of ellipsoidal or similar shape, it will therefore be necessary to find some interpretation of the length L. It must clearly be some average length of the s-laminae. Andrew has discussed the nature of the average, and has shown that for a cylinder of radius R in a transverse field, $L = \frac{3}{2}R$. A necessary assumption is that the domain structure is fine-grained compared with the size of the body. From (5.28), the grain size is proportional to $\sqrt{(L\delta)}$, and the assumption is therefore valid for large enough bodies. Assuming that $\delta \sim 5\times 10^{-5}$ cm, it is clear that 'large

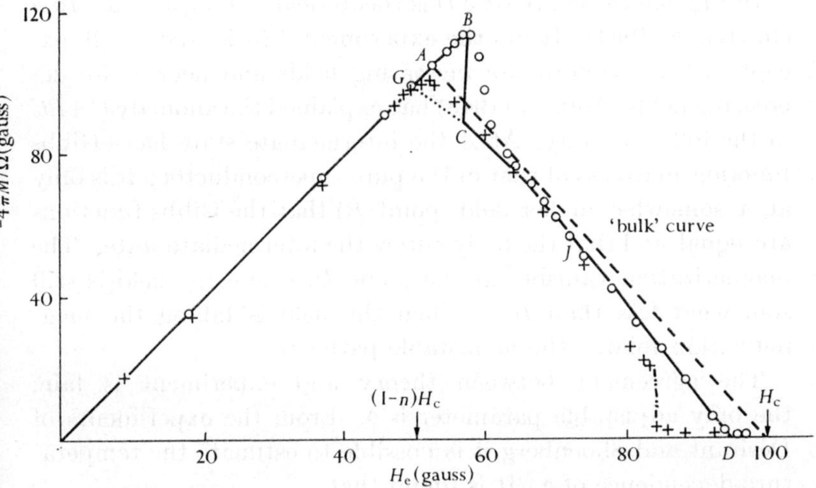

FIG. 5.5. Magnetization curve for cylinder in transverse field.(59). ○ increasing fields; + decreasing fields. The experimental points shown are those for Désirant and Shoenberg's (1948) tin specimen S9 at 3·0°K. The theoretical curve is drawn (full line) for $\delta = 5\times 10^{-5}$ cm.

enough' means $\gtrsim 10^{-2}$ cm. The theory should therefore be applicable to specimens such as those of Désirant and Shoenberg (1948).

For an ellipsoid, of demagnetizing coefficient $4\pi n$,

$$h = 1-nx_S, \tag{5.29}$$

and the magnetic Gibbs function is

$$G = -\frac{H_c^2\Omega}{8\pi n}\left\{(1-h)^2-2(1-h)\sqrt{\left(\frac{\Phi(x_S)\delta}{L}\right)}\right\}. \quad (5.30)^{(11)}$$

The magnetic moment of the specimen is

$$M = -\frac{\partial G}{\partial H_e}$$

$$= -\frac{H_c\Omega}{4\pi n}\left\{1-h-\left(\frac{\delta}{L}\right)^{\frac{1}{2}}\left(\Phi^{\frac{1}{2}}-\tfrac{1}{2}x_s\Phi^{-\frac{1}{2}}\frac{d\Phi}{dx_s}\right)_{x_s=(1-h)/n}\right\}. \quad (5.31)^{(10)}$$

In Fig. 5.5 the curve *GCJD* is theoretical, taking $\delta = 5\times10^{-5}$ cm (Kuper 1954). It fits the experimental (3°K) data well, except in the region *GC* for increasing fields and near *D* for decreasing fields. Andrew (1948) has explained the anomaly *GABC* in the following way. At *A* the intermediate state has a Gibbs function in excess of that of the pure superconductor; it is only at a somewhat higher field (point *B*) that the Gibbs functions are equal and that the body enters the intermediate state. The magnetization vanishes at the point *D*, where the field is still somewhat less than H_c. When the field is falling the magnetization follows the metastable path *CG*.

The agreement between theory and experiment is fair; the only adjustable parameter is δ. From the experiments of Désirant and Shoenberg it is possible to estimate the temperature-dependence of δ. It is found that

$$\delta = \delta_0(1-T^4/T_c^4)^{-\frac{1}{2}}, \quad (5.32)$$

with $\delta_0 \simeq 2{\cdot}2\times10^{-5}$ cm for tin.

This value is in fair agreement with that obtained by independent methods. From the domain size in large specimens, δ may be found by (5.28). Faber's (1952, 1954) experiments on nucleation and growth of s-phase are described in § 5.3. They give yet another independent estimate of δ. For tin, both these methods give $\delta_0 = 2{\cdot}5\times10^{-5}$ cm.

TABLE 5.1

x_s	0·0	0·1	0·2	0·3	0·4	0·5
Φ	0	0·0020	0·0065	0·0128	0·0182	0·0221
x_s	0·6	0·7	0·8	0·9	1·0	
Φ	0·0224	0·0195	0·0136	0·0055	0	

5.2.2. Destruction of superconductivity by a current

If a current I in excess of the Silsbee current (1.5) is passed through a cylindrical wire of radius a, some destruction of superconductivity must occur. Clearly the wire does not pass *entirely* into the normal state. If it did, the current density in it would be uniform. Inside the wire, the magnitude of the magnetic field H would then be proportional to the distance r from the axis of the cylinder,

$$H(r) = (r/a)H(a). \tag{5.33}$$

H would therefore be less than H_c close to the axis.

An alternative hypothetical configuration might be a core of s-phase along the axis, with a sheath of n-phase outside it. However this configuration must also be rejected. The reason is that the core would carry the *entire* current. Let the radius of the core be a'. Then $a' < a$, and

$$\begin{aligned} H(a') &= 2I/a'c \\ &> 2I/ac \\ &> H_c \end{aligned} \qquad (5.34)^{(55)}$$

a fortiori. As in the previous problem, of a body in an external magnetic field, the only possible equilibrium configuration is one with a fine-grained domain structure.

F. London (1937) showed that equilibrium could be established with a normal sheath covering a core in the intermediate state. Inside the intermediate-state core, he assumed that $H = H_c$ everywhere. A domain structure giving this field distribution is illustrated in Fig. 5.6. In a purely macroscopic description, the cone angle θ of the s-domains is not determined. But $\frac{1}{2}\pi - \theta$ should be as small as possible, to minimize the tendency of the current streamlines to spread out.

If $J(r)$ is the current density inside the core, the current $I(r)$ contained inside a cylinder of radius r is

$$I(r) = 2\pi \int_0^r J(r)\, r\, \mathrm{d}r. \tag{5.35}$$

This current has to produce a critical field everywhere inside the core, i.e.

$$2I(r)/cr = H_c. \tag{5.36}$$ [55]

Hence

$$J(r) = cH_c/4\pi r \tag{5.37}$$ [72]

and the total current in the core (of radius a') is

$$I_{\text{core}} \equiv I(a') = \tfrac{1}{2}H_c a'c. \tag{5.38}$$ [56]

As the domain structure of the core is such that n-domains reach its surface, they merge continuously into the sheath.

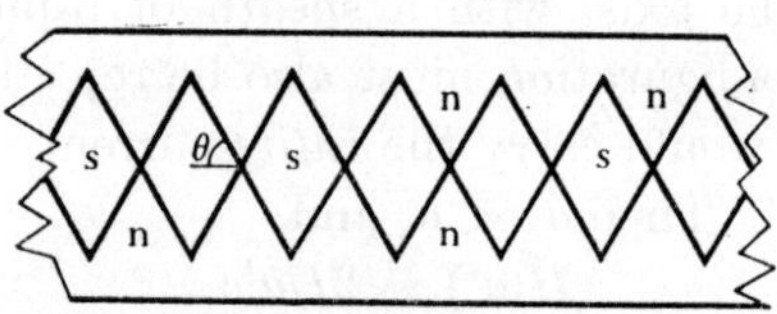

FIG. 5.6. Domain structure of the intermediate state of a wire carrying a current.

There cannot, therefore, be any discontinuity of J at $r = a'$. But in the sheath the current density is uniform. Hence

$$J_{\text{sheath}} = J(a'), \tag{5.39}$$

and the total sheath current is

$$I_{\text{sheath}} = \pi J(a')\,(a^2 - a'^2) = cH_c\,(a^2 - a'^2)/4a'. \tag{5.40}$$ [57]

Let R denote the resistance per unit length of the wire, R_n the resistance per unit length in the normal state, and R_{sheath} the resistance per unit length of the sheath. Then

$$R_{\text{sheath}} = R_n \frac{a^2}{a^2 - a'^2}. \tag{5.41}$$

The electric field E follows from Ohm's law. Since the sheath and core are in parallel, we may write E in the two alternative forms

$$\left.\begin{aligned} E &= R_{\text{sheath}} I_{\text{sheath}} = R_{\text{n}} a^2 H_{\text{c}} c/4a'; \\ E &= RI = R(I_{\text{sheath}} + I_{\text{core}}). \end{aligned}\right\} \qquad (5.42)^{(57)}$$

From (5.38), (5.40), and (5.42), it is readily shown that

$$R = \tfrac{1}{2}R_{\text{n}}\left[1+\sqrt{\left\{1-\left(\frac{acH_{\text{c}}}{2I}\right)^2\right\}}\right]. \qquad (5.43)^{(56)}$$

In Fig. 5.7, the above theoretical prediction (solid curve) is compared with the experimental data of Scott (1948). As

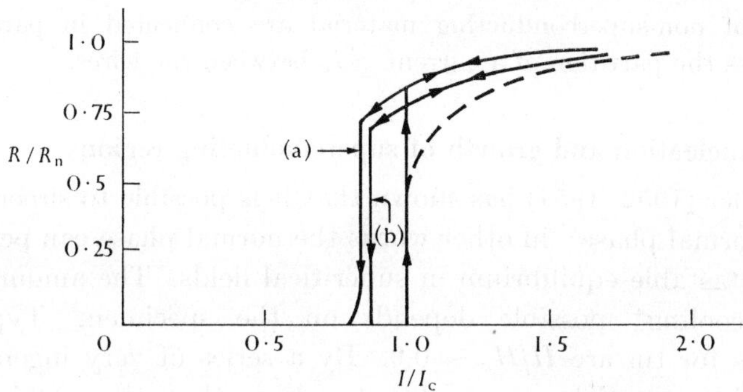

FIG. 5.7. Restoration of resistance by a current. The broken curve is theoretical (equation (5.43)); the full curves represent Scott's (1948) experimental results for indium wires of (a) 0·0106 cm and (b) 0·0357 cm diameter, at 3·34°K.

in the earlier experiments of Shubnikov and Alekseyevsky (1936), Scott confirms the existence of the theoretically-predicted discontinuity in the resistance at the Silsbee current[56] $I_{\text{c}} = \frac{1}{2}caH_{\text{c}}$. However, the observed magnitude of the discontinuity is not the predicted $\frac{1}{2}R_{\text{n}}$. Instead, it is a function of the radius. For wires of diameter $\sim 10^{-2}$ cm, $R(I_{\text{c}})/R_{\text{n}} \sim 0{\cdot}8$, and the discontinuity is smaller for thicker wires. It is a plausible extrapolation of the data that London's expression (5.43) will hold for very thick wires.

In the thin wires used in Scott's experiments, the electron mean free path is comparable with the radius. The discrepancy between the experiments and the London theory has been interpreted as a size effect, arising from the scattering of normal electrons by the n-s boundary (Kuper 1952). The role of the n-s surface energy is not important. Although the 'scattering' theory gives a resistance in good agreement with experiment, it leads to a cone angle $\theta \sim 1$ radian; the approximations made are therefore unreliable. Moreover no account has been taken of proximity effects (§ 14.5), which may modify the conduction properties of both phases. It is clear that the present state of the theory is unsatisfactory.†

Exercise. A thin wire of superconducting material and a thick wire of non-superconducting material are connected in parallel. Discuss the partition of a current $>I_c$ between the wires.

5.3. Nucleation and growth of superconducting regions

Faber (1952, 1954) has shown that it is possible to supercool the normal phase; in other words the normal phase can persist in metastable equilibrium in subcritical fields. The amount of supercooling‡ possible depends on the specimen. Typical values for tin are $H/H_c \sim 0{\cdot}9$. By a series of very ingenious experiments, Faber was able to show that the specimens contained 'flaws', which act as nucleation centres for the growth of the s-phase.

In Faber's experiments, a long rod was surrounded by a series of small coils, uniformly spaced along the rod. These coils could be used either as sources of magnetic flux or as flux detectors. The latter use of the coils makes it possible to observe the growth of s-phase along the rod. Faber first

† *Added in proof:* Baird and Mukherjee (1967) have constructed a new model in which the shape of the domains has been determined by computation, so that the current streamlines meet the n-s surfaces normally and H is nearly H_c everywhere in the n-phase. Their model does *not* give the London value $R(I_c)/R_n = 0{\cdot}5$, even for thick wires.

‡ Superheating of the superconducting phase has also been observed, but will not be discussed.

established that in a single specimen the s-phase always starts growing from the same place. He then applied a local magnetic field in the neighbourhood of this flaw, to shield it from the full supercooling. A different flaw then took over the role of the nucleation centre. These nucleation centres are remarkably stable—they often survive the heating of the specimen to room temperature and subsequent recooling. Faber has interpreted them as regions of strain, in which the interphase energy is negative.

Superconductors with a negative n-s surface energy are classified as Type II (pp. 77, 115–6). Such materials are characterized by *two* critical fields H_{c1} and H_{c2}, lying respectively below and above† the usual critical field H_c. For a range of fields $H_{c1} < H_e < H_{c2}$, a Type II material will be in the mixed state described in Chapter 7, that is to say it will contain parasitic filaments of normal material (cf. § 5.1), but will remain partly superconducting up to a field H_{c2}. Hence Faber's flaws, which are examples of Type II regions, will contain superconducting inclusions. These will persist up to quite high applied fields, and are capable of acting as nuclei.

To discuss the dynamics of growth of the s-nuclei, Faber (1954) adopts Pippard's (1950*c*) suggestion that the rate of motion of a phase boundary is limited mainly by eddy currents in the n-phase. If the interphase surface area increases, the surface energy is increased. The free energy available in the supercooled phase transition is shared between this surface energy and the heat dissipated by eddy currents.

Consider an s-layer, of thickness d, growing at a velocity v (Fig. 5.8) along the surface of the rod. If $\boldsymbol{\sigma}$ is the effective electrical conductivity of the n-phase, then‡

$$\frac{1}{8\pi}(H_c^2 - B^2)d = \frac{1}{8\pi}H_c^2\delta + \tfrac{1}{2}B^2d^2\boldsymbol{\sigma}v/c^2. \qquad (5.44)^{(11)}$$

† In extreme cases H_{c2} can be very much greater than H_c.

‡ See footnote, p. 94.

Exercise. Calculate the rate of energy dissipation by eddy currents, in the configuration of Fig. 5.8. Hence derive (5.44). (Assume $\sin\theta \simeq 1$.)

When the thickness d is less than or comparable with the mean free path l of electrons, the conductivity $\boldsymbol{\sigma}$ will differ from the bulk conductivity $\boldsymbol{\sigma}_0$. Faber uses Nordheim's (1934) interpolation formula†

$$\boldsymbol{\sigma} = \frac{\boldsymbol{\sigma}_0 d}{d+\frac{4}{3}l}. \tag{5.45}$$

From (5.44) and (5.45), the velocity of propagation of the boundary (to lowest order in $(H_c-B)/H_c$) is

$$v = \frac{c^2}{4\pi}\left(\frac{d+\frac{4}{3}l}{\boldsymbol{\sigma}_0 d^3}\right)\left(\frac{H_c^2-B^2}{H_c^2}d-\delta\right). \tag{5.46}$$ (19)

Equation (5.46) has the important property that *very* thin layers will shrink ($v < 0$). Thicker layers will grow, and the rate of growth will be maximal for some finite thickness d_0. It is this maximum growth rate that is observed experimentally. Provided that $d_0 \ll l$, (5.46) gives

$$\delta = 4\times 10^{-2}\left(\frac{H_c-H}{H_c}\right)^{\frac{3}{2}} v^{\frac{1}{2}}, \tag{5.47}$$

with δ in cm, and v in cm sec^{-1}. Thus δ can be found from the measured propagation velocity v. The results for several materials, extrapolated to zero temperature, are shown in Table 5.2.

'Ideal' supercooling can be observed very near T_c (Faber 1957). In this limit, δ becomes so large that the Faber flaws are unable to serve as nuclei; the velocity v remains negative however much the field is reduced. Under these circumstances,

† In (5.44), the dependence on the angle θ (Fig. 5.8) has been suppressed; another θ-dependent factor in (5.45) has also been suppressed. A fuller discussion (Faber 1954, especially pp. 189–91) shows that these two angle-dependent factors nearly cancel.

the limiting value of the supercooling is given by the Ginzburg-Landau theory (§ 6.3):

$$H > H_{c3} = 2{\cdot}392\,\kappa H_c. \tag{5.48}$$

Equation (5.48) provides a useful method for determining the Ginzburg–Landau parameter κ (see pp. 117–8).

TABLE 5.2

Element	$\delta_0(10^{-5}$ cm)
Al	18
In	3·4
Sn	2·3

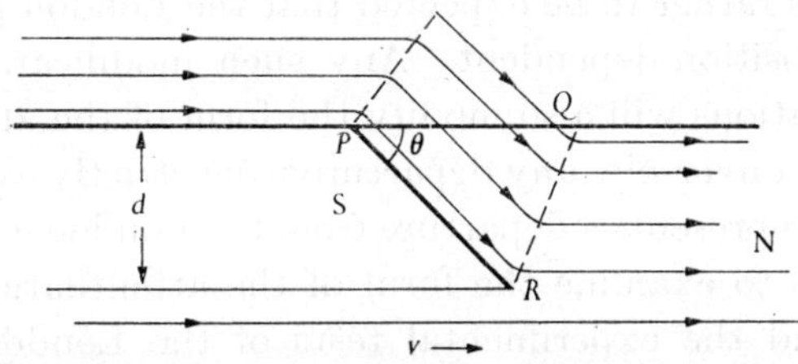

FIG. 5.8. A superconducting filament of thickness d advancing along the surface of a supercooled rod. The eddy currents are confined to the region where the magnetic field lines are distorted by the advancing boundary, i.e. roughly to the triangle PQR (Faber 1954).

CHAPTER 6

SPATIAL INHOMOGENEITY OF THE ORDER PARAMETER: IMPROVED PHENOMENOLOGICAL THEORIES

THE London equations (3.3) and (3.15) have been shown to lead to the field penetration law (3.14). The magnetic field is exponentially attenuated in the penetration layer. However, *a priori* it is hard to believe that the electromagnetic properties of a superconducting material will be unchanged near an external surface. It is rather to be expected that the London parameter Λ will be position-dependent. Any such modification of the London equations will also modify the form of the attenuation (3.14). And conversely, any departure from strictly exponential attenuation represents a departure from the London equations.† It is difficult to examine the form of the attentuation experimentally, and the experimental tests of the London electrodynamics have therefore had to be a little indirect. In § 6.1, Pippard's (1950*a*, *b*; 1953) experimental evidence for deviations from the London equations is briefly reviewed.

Pippard's results show that the London equations require modification. A new concept has to be introduced into the phenomenological theory—the concept that *the order parameter* ζ of § 2.2 *must not vary too rapidly with position* in a superconducting body. This concept forms the basis of the two improved phenomenological theories to be discussed in §§ 6.2, 6.3. Although the conceptual basis of these two theories is rather similar, the ways in which the new postulate is introduced are very different. The two theories also make rather different approximations, and therefore their ranges of validity are not identical. The theory of Ginzburg and Landau (1950) should hold for all superconductors in the immediate neighbourhood

† It will be shown in § 6.4 that the surface energy α_{ns} is a consequence of non-exponential field penetration.

of the transition temperature. The Pippard theory (1953) is not restricted to a narrow range of temperatures, but it is strictly applicable only to fairly pure Type I materials.

6.1. Experimental evidence for non-locality

Before discussing modifications of the London theory, let us look briefly at several experiments (Pippard 1950*a*, *b*; 1953) which reveal the need for an improved theory. In these experiments, Pippard measured both the inductive and the resistive components of the surface impedance of tin at microwave frequencies. From the surface reactance X and the surface resistance R, one can infer the inductive skin depth $\delta_{\mathrm{i}} = X/4\pi\omega$ and the resistive skin depth $\delta_{\mathrm{r}} = R/4\pi\omega$, where ω is the angular frequency.

In an Ohmic conductor, the ratio $\delta_{\mathrm{i}}/\delta_{\mathrm{r}}$ is unity. However at the frequencies used by Pippard ($\omega \sim 6\times 10^{10}\ \mathrm{sec}^{-1}$) the 'classical' skin depth is comparable with the mean free path of electrons; conditions are far from classical. The theory of the anomalous skin effect (Reuter and Sondheimer 1948) predicts that, in the limit of very high frequencies, $\delta_{\mathrm{i}}/\delta_{\mathrm{r}} = \surd 3$. Pippard's measurements in normal tin agree quite well with the theory.

For a superconductor, the expected behaviour is quite different. At low frequencies the resistive skin depth should vanish (since the resistance is zero), but the inductive skin depth should approach the London penetration depth. Measurements of both the inductive and resistive depths at higher frequencies should allow one to deduce the superconducting penetration depth and the Ohmic resistance of the normal fluid. In practice it is difficult to get reliable *absolute* values of the penetration depth, but the method affords an exceedingly good comparison of penetration depths under different conditions.

In the first of these experiments, Pippard (1950*a*) made measurements with a number of monocrystalline cylindrical specimens of tin. The angle between the tetragonal axis of the crystal and the axis of the cylinder differed from one specimen to another. Similar experiments were made by Dheer (1961) on indium, which is another tetragonal metal. Williams (1962) has investigated the orientation dependence in aluminium, a cubic metal. Figure 6.1 shows some typical results, illustrating the following conclusions. (a) The penetration depth does not appear to be a monotonic function of the angle between the cylinder axis and the tetragonal axis of the crystal. If this result is correct, it is in marked contrast with von Laue's

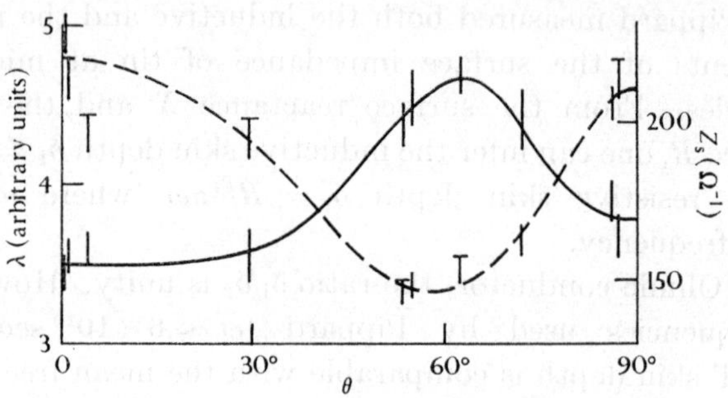

FIG. 6.1. Anisotropy of the penetration depth in superconducting tin (full curve) and of the surface conductance of normal tin (broken curve). After Pippard 1950*a*.

predictions (§ 3.5). von Laue replaces the London parameter Λ by a second-order tensor Λ_{ij} (equation (3.46)); the von Laue tensor is diagonal (equation (3.47)) relative to the crystallographic axes. If θ is the angle between the cylinder axis and the tetragonal axis, then (3.47) gives

$$(4\pi/c^2)\lambda^2 = \Lambda_2+(\Lambda_1-\Lambda_2)\cos^2\theta, \qquad (6.1)^{(60)}$$

a monotonic function of θ. (b) There is a very suggestive similarity between the anisotropy of the penetration depth λ and that of the anomalous skin depth δ_r of the *normal* phase (broken curve, Fig. 6.1).

The above experimental evidence for non-tensorial anisotropy is not very strong (Waldram 1964); in all cases the observed anisotropy is not much larger than the scatter of the individual measurements. Nevertheless it would appear more likely to be a real effect than a spurious one. Historically it was important, in that it tended to suggest the form of Pippard's non-local equation (6.11). Now, however, even if one discounts the anisotropy experiments entirely, equation (6.11) rests on firmer ground.

In a second experiment, Pippard (1950*b*) studied the dependence of the penetration depth on the magnetic field at various temperatures. These experiments were conducted using 3-cm

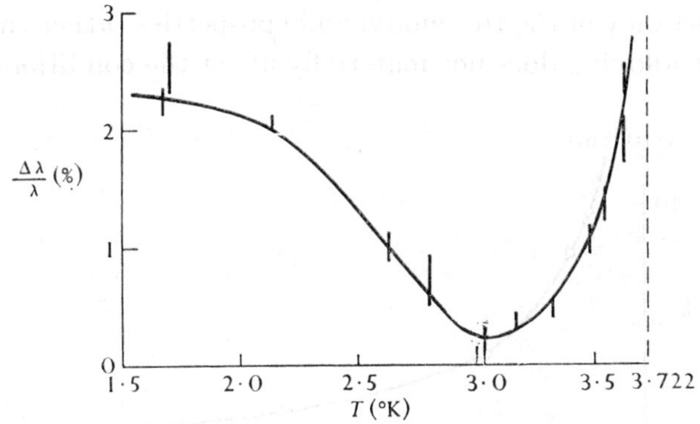

FIG. 6.2. The relative shift in the penetration depth in tin between zero field and the critical field, at temperatures between 1·5°K and the transition temperature $T_c = 3{\cdot}722$°K. The experimental conditions are a cylindrical specimen in a transverse field, at a microwave wavelength of 3 cm. (Pippard 1950*b*.)

microwaves, and with the magnetic field transverse to the cylindrical specimen. Some of the results are illustrated in Fig. 6.2. The surprising feature of this experiment is the smallness ($\sim$ 3 per cent) of the variation found.

The influence of impurities was investigated by Pippard (1953) in another series of experiments. He prepared a number of specimens of tin-indium alloys, in which the indium concentration† varied from zero to 3 per cent. The penetration depth was found to increase smoothly with increasing indium concentration, until at the composition $Sn_{0{\cdot}97}In_{0{\cdot}03}$ the penetration depth had approximately doubled (Fig. 6.3).

† More recent experiments have extended the range of compositions up to 6 per cent indium. For a detailed account, see Waldram (1964).

However, the thermodynamic properties of these alloys are very insensitive to composition. In particular, over Pippard's entire range of composition the transition temperature T_c varies by only ~ 2 per cent. Moreover, although λ itself is very greatly altered, its temperature dependence remains close to the Gorter–Casimir form

$$\lambda = \lambda_0(1-T^4/T_c^4)^{-\frac{1}{2}}. \tag{6.2}$$

The constancy of the thermodynamic properties rather suggests that the alloying does not materially affect the conditions for a

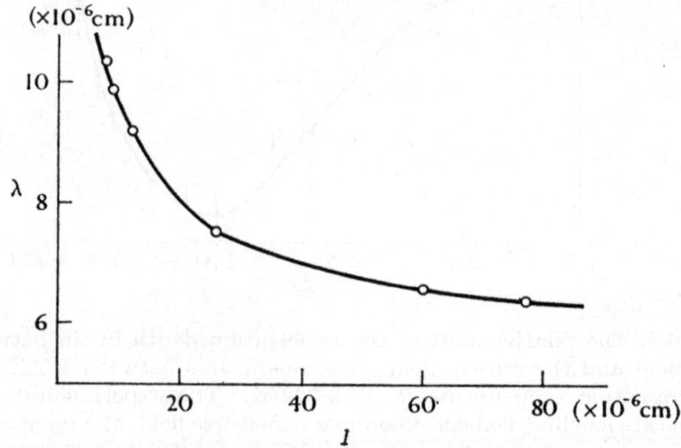

FIG. 6.3. The effect of alloying tin with indium. The abscissa is the mean free path of electrons in the normal phase, and the ordinate is the superconducting penetration depth (extrapolated to $T = 0$). (Pippard 1953.)

superconducting transition. Thus there is a marked contrast between the sensitivity of the penetration depth and the insensitivity of the thermodynamic properties to the impurity content. In Pippard's interpretation, the role of the indium atoms is merely to scatter the normal electrons. The penetration depth λ will be related to the mean free path l of electrons in the non-superconducting state.

In § 6.2, it will be shown that all three of these experiments can be explained as consequences of the existence of a minimum length ξ, the 'coherence length', within which the order parameter ζ cannot change appreciably.

6.2. Pippard's non-local equation

The most immediate evidence for some sort of spatial rigidity in the order parameter ζ (§ 2.2) comes from the second of the experiments described in § 6.1. The results of this experiment imply that an appreciable fraction of the total entropy of the system resides in a surface layer. If all this entropy were concentrated in the London penetration layer, the entropy density would have to be implausibly large.

In order to be able to discuss models in which the field penetration is not exponential, we shall require a general definition of the penetration depth. The natural definition is

$$\lambda = H_e^{-1}\int_0^\infty H(z)\,dz, \tag{6.3}$$

where the applied field $\mathbf{H}_e$ is in the x-direction, and the superconductor fills the half-space $z > 0$ (Fig. 3.1, p. 40). Note that this definition reduces to the London one if H falls exponentially with increasing z. The magnetic moment of a superconducting body with zero demagnetizing coefficient is

$$M = -\frac{1}{4\pi}H_e(\Omega - \lambda A), \qquad (6.4)^{(10)}$$

provided the body is so large that the field deep in its interior may be neglected. Here Ω is the volume and A is the surface area of the body.

In (6.4), the only quantity whose temperature dependence is significant is λ. (Volume changes may be neglected, cf. equation (2.5) et seq.) Hence

$$\left(\frac{\partial S}{\partial \mathbf{H}_e}\right)_T = \left(\frac{\partial \mathbf{M}}{\partial T}\right)_{\mathbf{H}_e} = \frac{A\mathbf{H}_e}{4\pi}\left(\frac{\partial \lambda}{\partial T}\right)_{\mathbf{H}_e}, \qquad (6.5)^{(10)}$$

where the first equality is one of the Maxwell thermodynamic relations, derived in Appendix A. From one of Pippard's experiments (Fig. 6.2), it is known that λ is nearly independent of H_e. If the slight residual dependence of λ on H_e is neglected, we may integrate (6.5) from $H_e = 0$ to $H_e = H_c$, to give

$$\Delta S = S(H_c) - S(0) = \frac{1}{8\pi} A H_c^2 \left(\frac{\partial \lambda}{\partial T}\right). \tag{6.6}$$ (11)

If it is assumed that this additional entropy associated with the applied field is contained in the London penetration layer, then the mean entropy density in the layer is

$$\Delta s = \frac{H_c^2}{8\pi}\left(\frac{1}{\lambda}\frac{\partial \lambda}{\partial T}\right). \tag{6.7}$$ (11)

For comparison, the entropy difference per unit volume at zero field between the n- and s-phases is, from (2.14),

$$\Delta s' = \frac{1}{4\pi} H_c \frac{\partial H_c}{\partial T}. \tag{6.8}$$ (10)

Both the entropy densities (6.7) and (6.8) may be evaluated for the Gorter–Casimir two-fluid model; in the neighbourhood of the transition temperature it is found that

$$\Delta s / \Delta s' = \tfrac{1}{4}. \tag{6.9}$$

Exercise. Evaluate $\lambda^{-1}\partial\lambda/\partial T$ and $H_c^{-1}\partial H_c/\partial T$ for the Gorter–Casimir two-fluid model. Hence derive (6.9).

The result (6.9) is not at all plausible. It means that, in the presence of an applied field H_c, the penetration layer can store an entropy density equal to a quarter of the difference of the entropy density between the two phases. *But in spite of this very considerable stored entropy, the depth of field penetration is to remain nearly independent of the field.*

To avoid the above paradox, Pippard suggested that the field entropy (6.6) is not stored in the London penetration layer. Rather it is spread through a much thicker layer near the surface of the superconductor. There is a characteristic length ξ, the 'range of coherence of the wave functions for superconducting electrons'. Pippard estimates that for tin, $\xi \sim 20\lambda \sim 10^{-4}$ cm. The assumption critical to the theory is that the order parameter ζ (or, equivalently, the energy gap† 2Δ) cannot change appreciably in distances $\lesssim \xi$. An additional argument which Pippard presents in support of this large coherence length is that the specific heat has a sharp discontinuity at the transition temperature. This behaviour is contrasted with the more usual λ-type singularity, such as is shown by liquid helium (Fig. 2.2, p. 25). It is believed that λ-transitions owe their logarithmic character to the possibility of short-range changes in the order parameter (cf. Onsager 1944).

The effect of impurities is to weaken the coherence of the superconducting electrons. Pippard's proposal was that in a very pure material the coherence length ξ should have a value ξ_0 characteristic of the material, but that ξ should be a function of the electronic mean free path l. In the 'dirty' limit when $l \ll \xi_0$, ξ should be comparable in magnitude with l and proportional to it. For definiteness, Pippard puts

$$\xi(l) = 1/\{\xi_0^{-1} + (\alpha l)^{-1}\}, \tag{6.10}$$

where α is a constant of order unity. Figure 6.3 shows that the penetration depth is correlated with the mean free path as expected under the above assumptions. As the free path is made shorter, the coherence length is reduced according to (6.10), and the penetration depth is able to get closer to the London value (3.16).

Thus the second and third of Pippard's experiments show departures from the London theory, and imply that the field penetration cannot be exponential (except in the limit of dirty materials). The anisotropy found by Pippard in the first of the experiments described above has even more far-reaching implications. It establishes the impossibility of a local relation between the current density and the vector potential, $\mathbf{J}(\mathbf{r}) = \mathbf{A}(\mathbf{r}) f(\mathbf{A}(\mathbf{r}))$. Pippard therefore makes a second assumption. The current density at a point $\mathbf{r}$ in the material is not uniquely related to the vector potential at that point. Instead, *the current density is proportional to some average of the vector*

† Cf. § 2.3, and especially the discussion following (2.30).

potential in the neighbourhood of the point, *taken over a region of characteristic size* ξ.

The anisotropy of the superconducting surface impedance (e.g. Fig. 6.1) is similar in character to that of a normal metal in the anomalous-skin-effect region. Pippard therefore proposes, by analogy with the theory of the anomalous skin effect,† that

$$\mathbf{J}(\mathbf{r}) = -\frac{3}{4\pi c\Lambda\xi_0}\int \frac{\mathrm{d}^3r'(\mathbf{r}-\mathbf{r}')\,(\mathbf{r}-\mathbf{r}').\mathbf{A}(\mathbf{r}')\,\mathrm{e}^{-|\mathbf{r}-\mathbf{r}'|/\xi}}{|\mathbf{r}-\mathbf{r}'|^4}. \tag{6.11}$$

(6.11)[26]

Bardeen *et al.* (1957) have shown that (6.11) can be derived from the microscopic theory (see Chapter 13). The microscopic-theoretical value for the coherence length at zero temperature is

$$\xi_0 = \hbar v_\mathrm{F}/\pi\Delta, \tag{6.12}$$

where v_F is the velocity of an electron at the Fermi surface. In a free-electron model (§ 10.1), equation (6.12) may be written

$$\xi_0 = \hbar k_\mathrm{F}/\pi\Delta m^*,$$

where k_F is the Fermi momentum and m^* is the effective mass of the electrons. In order to test (6.12) experimentally, a method is required of determining ξ_0. This is achieved by measuring the penetration depth λ in the limiting cases $\xi \ll \lambda$ (the 'London' limit) and $\xi \gg \lambda$ (the 'Pippard' limit).

When $\xi \ll \lambda$ the Pippard equation (6.11) reduces to the local form

$$\mathbf{J} = -\frac{1}{c\Lambda}\frac{\xi}{\xi_0}\mathbf{A}, \tag{6.13}$$

(6.13)[26]

† Reuter and Sondheimer (1948) have shown that under anomalous conditions, when the a.c. skin depth δ_r is comparable with the mean free path l, Ohm's law $\mathbf{J} = \boldsymbol{\sigma}\mathbf{E}$ must be replaced by the non-local expression

$$\mathbf{J}(\mathbf{r}) = \frac{3\boldsymbol{\sigma}}{4\pi l}\int \frac{\mathrm{d}^3r'(\mathbf{r}-\mathbf{r}')\,(\mathbf{r}-\mathbf{r}').\mathbf{E}(\mathbf{r}')\,\mathrm{e}^{-|\mathbf{r}-\mathbf{r}'|/l}}{|\mathbf{r}-\mathbf{r}'|^4}. \tag{6.14}$$

which differs from the London expression (3.30) only by the factor ξ/ξ_0. In other words the penetration depth in the London limit is

$$\lambda_1 = \lambda_L(\xi_0/\xi)^{\frac{1}{2}}, \tag{6.15}$$

where[19] $\lambda_L^2 = mc^2/4\pi N_e e^2$ (cf. (3.16)).

On the other hand, the penetration depth λ_2 in the Pippard limit $\xi \gg \lambda$ is given by

$$\lambda_2^3 = \frac{\sqrt{3}}{2\pi}\xi_0\lambda_L^2. \tag{6.16}$$

For very impure specimens or for thin films, ξ is to be taken equal to the mean free path (limited by impurity and/or boundary scattering). Equations (6.15) and (6.16) then suffice to determine both λ_L and ξ_0. Since it is the zero-temperature value of ξ_0 which appears in (6.12), the experimental values have to be extrapolated to zero temperature. Faber and Pippard (1955)† find $\xi_0 = 2{\cdot}1\times10^{-5}$ cm for tin, and $12{\cdot}3\times10^{-5}$ cm for aluminium. On substituting the appropriate values for v_F and Δ $(=1{\cdot}76kT_c)$ in (6.12), the theoretical values are found to be only about 20 per cent too high.

Exercise. Derive equation (6.16) from the Pippard equation (6.11).

6.3. The Ginzburg–Landau theory

Historically, the theory of Ginzburg and Landau (1950) preceded that of Pippard (1950*a*, *b*; 1953). The conceptual framework is quite similar, but in some important respects the Ginzburg–Landau theory is less general. As in Pippard's theory, the new feature is the postulate that the order parameter must not vary too abruptly with position. But in contrast with the Pippard theory, the theory of Ginzburg and Landau is *local*. This is a limitation, which leads to the failure of the theory at high frequencies. The original motivation of the theory was less clear than that of the Pippard theory; in

† For a much fuller discussion of these and more recent experimental data, and also of their theoretical significance, see Waldram (1964).

particular the new postulate is introduced in a less direct way. The main advantage of the Ginzburg–Landau theory over the Pippard form is that the equations are rather easier to manipulate. In the limit as $T \to T_c - 0$, the approximations of the theory become rigorously correct (see Chapter 14). The Ginzburg–Landau equations are particularly useful in discussing Type II superconductivity (Chapter 7).

As we saw in § 2.2, the superconducting phase is conveniently characterized by an order parameter. However, the formulation of the theory must allow the order parameter to vary spatially. Ginzburg and Landau moreover allow the order parameter to be *complex*. This is the most difficult step in the development of the theory to accept intuitively.† Ginzburg and Landau explain that their order parameter Ψ is to be thought of as some sort of averaged 'wave function of the superconducting electrons'. The expansion (2.19) of the Gibbs function in powers of ζ is retained, but with $\zeta = |\Psi|^2$. The interpretation of Ψ as a wave function implies that there is a kinetic energy density associated with regions of space where Ψ is varying, whether the variation is in amplitude or phase.

In the absence of a magnetic field, the obvious form for the kinetic contribution to the Gibbs free energy is

$$G_{\text{kin}} = (2m_s)^{-1}|-i\hbar\nabla\Psi|^2, \tag{6.17}$$

where m_s, the 'effective mass of superelectrons', is at our disposal. However, there is a difficulty about accepting (6.17): it is not gauge-invariant (§ 3.2). Even if $\mathbf{B} = 0$, the gauge can be so chosen that $\mathbf{A} \neq$ const. Under a transformation from a gauge with $\mathbf{A} = 0$ to one in which $\mathbf{A}$ is non-uniform, (6.17) cannot retain its validity. Ginzburg and Landau therefore make the natural gauge-invariant generalization

$$G_{\text{kin}} = (2m_s)^{-1}|-i\hbar\nabla\Psi - \varepsilon\mathbf{A}\Psi/c|^2. \qquad (6.18)^{(26)}$$

† Fröhlich (1966) has shown that the Ginzburg–Landau theory is a consequence of the hydrodynamics of a charged 'superfluid'. His argument, outlined on pp. 109–10, makes the need for a complex order parameter convincing.

Equation (6.18) *is valid whether there is a magnetic field or not.* Here ε is the charge on a 'superelectron'. Ginzburg and Landau assumed that $\varepsilon = e$, the electronic charge. However, in the light of the microscopic theory (see equation (14.44)), we should rather take $\varepsilon = 2e$. Moreover this value $\varepsilon = 2e$ improves the agreement of the Ginzburg–Landau theory with experiment.

Since Ψ is to be thought of as a wave function, we may normalize it in such a way that $|\Psi|^2 = n_s$, the number density of superconducting charge-carriers.† The current density is assumed to have the usual form

$$\mathbf{J} = \frac{\varepsilon\hbar}{2m_s i}\{\Psi^*\nabla\Psi-(\nabla\Psi^*)\Psi\}-\frac{\varepsilon^2}{m_s c}|\Psi|^2\mathbf{A}. \qquad (6.19)^{(26)}$$

For a spatially homogeneous system ($|\Psi|^2 = \zeta = n_s = \text{const}$), (6.19) reduces to

$$\mathbf{J} = \frac{\varepsilon\hbar}{m_s}\zeta\nabla\varphi-\frac{\varepsilon^2}{m_s c}\zeta\mathbf{A}. \qquad (6.20)^{(26)}$$

Here φ is the phase of Ψ, i.e.

$$\Psi = |\Psi| \exp \mathrm{i}\varphi. \qquad (6.21)$$

From (6.20), we note that[(16)]

$$\operatorname{curl}(c\Lambda\mathbf{J}+\mathbf{A}) = 0,$$

which is consistent with the second London equation (3.15). The significance of the Ginzburg–Landau phase angle φ is most readily seen in the London gauge. If $\mathbf{A}$ is chosen to be purely transverse, then φ is a scalar field whose gradient gives the longitudinal current (cf. (4.15)).

The magnetic Gibbs function per unit volume is obtained by augmenting the expression (2.19) with the kinetic term (6.18),

† With this normalization, (6.17) is an energy *density*.

as well as with the field energy[61] $B^2/8\pi$. Thus in the presence of a magnetic field we have

$$g_s(\mathbf{B}, T) - g_n(0, T) = \alpha|\Psi|^2 + \tfrac{1}{2}\beta|\Psi|^4 + \frac{1}{2m_s}\left|-i\hbar\nabla\Psi - \frac{\varepsilon}{c}\mathbf{A}\Psi\right|^2 + B^2/8\pi. \qquad (6.22)^{(62)}$$

In order to determine the equilibrium configuration, it is necessary to minimize (6.22) with respect to both the field and the order parameter. It is convenient to minimize with respect to **A** (remembering that **B** = curl **A**) rather than with respect to **B** itself:

$$-\nabla\wedge(\nabla\wedge\mathbf{A}) + \frac{2\pi i\varepsilon\hbar}{m_s c}\{\Psi^*\nabla\Psi - (\nabla\Psi^*)\Psi\} - \frac{4\pi\varepsilon^2}{m_s c^2}\mathbf{A}|\Psi|^2 = 0. \qquad (6.23)^{(26),(37)}$$

Similarly, minimizing with respect to Ψ^* leads to

$$\frac{\delta g_s(0, T)}{\delta\Psi^*} + \frac{1}{2m_s}\left|-i\hbar\nabla - \frac{\varepsilon}{c}\mathbf{A}\right|^2\Psi = \alpha\Psi + \beta|\Psi|^2\Psi + \frac{1}{2m_s}\left|-i\hbar\nabla - \frac{\varepsilon}{c}\mathbf{A}\right|^2\Psi = 0. \qquad (6.24)^{(26)}$$

The pair of coupled non-linear equations (6.23) and (6.24) contain the entire substance of the Ginzburg–Landau theory, apart from a boundary condition on Ψ at the surface of the superconductor. Ginzburg and Landau impose the natural condition that the velocity of a superelectron must be tangential to the boundary, i.e. the normal component must vanish:

$$\left\{-i\hbar\nabla\Psi - \frac{\varepsilon}{c}\mathbf{A}\Psi\right\}_{\perp} = 0. \qquad (6.25)^{(26)}$$

In contrast to Pippard's theory (§ 6.2), the theory of Ginzburg and Landau is local. From the very way in which the Ginzburg–Landau equations are derived, however, it is clear that they are restricted in their validity to temperatures near the transition temperature T_c. In this temperature region the penetration depth λ becomes large compared with the coherence length ξ, and the role of non-locality becomes secondary.† At very high frequencies, the non-local nature of the current response can still be seen, but by and large the two formalisms are in agreement near T_c.

The meaning of the parameters α and β is most easily found by considering a homogeneous field-free system, for which $\Psi = \Psi_0$, say. The equilibrium condition (6.24) reduces to

$$|\Psi_0^2| = -\alpha/\beta, \tag{6.26}$$

and hence the difference between the normal and superconducting free-energy densities is

$$\frac{1}{8\pi}H_c^2 = \frac{\alpha^2}{2\beta}. \qquad (6.27)^{(11)}$$

For very weak fields Ψ should be nearly constant, and under these conditions (6.20) reduces to

$$\mathbf{J} = -\frac{\varepsilon^2}{m_s c}\zeta\mathbf{A} = -\frac{\varepsilon^2}{m_s c}\Psi^*\Psi\mathbf{A}. \qquad (6.28)^{(26)}$$

This is just the second London equation (3.15), although of course there is no *a priori* reason to assume that $\zeta = N_e$, even at zero temperature.

Fröhlich (1966) has proposed a new way of looking at the Ginzburg–Landau theory, in which the physical meaning of the

† Bardeen (1954) has proposed a non-local generalization of the Ginzburg–Landau equations:[(26),(37)]

$$\nabla^2\Psi(\mathbf{r}) = \frac{m_s}{\hbar^2}\frac{\delta g_s(0,T)}{\delta\Psi^*} + \left\{\frac{\varepsilon^2}{\hbar^2c^2}A^2(\mathbf{r}) - \frac{2i\varepsilon}{\hbar c}\mathbf{A}(\mathbf{r})\,.\,\nabla_r\right\}\int d^3r' K(\mathbf{r}-\mathbf{r}')\Psi(\mathbf{r}'),$$

$$\nabla^2\mathbf{A}(\mathbf{r}) = \frac{4\pi\varepsilon^2}{m_s c^2}\Psi(\mathbf{r})\int d^3r' K(\mathbf{r}-\mathbf{r}')\Psi(\mathbf{r}')\mathbf{A}(\mathbf{r}')$$

(in a gauge with $\nabla\,.\,\mathbf{A} = 0$). However, the computational difficulties associated with these equations are formidable. Up to the present time, therefore, they have not proved useful.

complex order parameter Ψ is rather clearer. In a charged ideal hydrodynamic fluid, the velocity field may be written

$$\mathbf{J}/\nu\rho = \mathbf{v} = \mathrm{grad}(\gamma\varphi) - (\nu/c)\mathbf{A}, \qquad (6.29)^{(26)}$$

where $\gamma\varphi$ is a velocity potential, and where ν is the ratio of electric charge density to mass density ρ. The hydrodynamical equation of motion for this charged fluid is

$$\rho \frac{D\mathbf{v}}{Dt} = -\mathrm{grad}\, p + \mathbf{f}, \qquad (6.30)$$

and the equation of continuity is

$$\nu \frac{\partial \rho}{\partial t} = -\mathrm{div}\, \mathbf{J}. \qquad (6.31)$$

In the above equations, p is the pressure, $\mathbf{f}$ is the electromagnetic contribution to the density of body force, and γ is a coefficient introduced so that φ shall be dimensionless. Fröhlich notes that the magnitude of γ is determined *only* by the condition of fluxoid quantization (3.45):

$$\gamma = \hbar/m. \qquad (6.32)$$

On identifying this velocity potential with the phase φ of the Ginzburg–Landau 'wave function' Ψ, it is readily verified that (6.24) implies (6.30) and (6.31). Fröhlich's result shows that the Ginzburg–Landau equations are equivalent to ideal-fluid hydrodynamics plus London electrodynamics.

Although the structure of the Ginzburg–Landau equations (6.23) and (6.24) is much simpler than that of Pippard's equation (6.11), little is known about the general solutions. However, the equations have been studied extensively in one special configuration: the geometry of Fig. 3.1 (p. 40). Let us consider a semi-infinite slab of superconductor, occupying the half-space $z > 0$. The normal half-space $z < 0$ has a uniform magnetic field $(H_e, 0, 0)$. With this effectively one-dimensional geometry, Ψ can be chosen to be real.

It is convenient to introduce the dimensionless variables

$$\psi = \Psi/\Psi_0, \tag{6.33}$$

$$\mathscr{z} = z/\lambda_L, \tag{6.34}$$

$$\mathscr{A} = A/\lambda_L H_c\sqrt{2}, \tag{6.35}$$

$$\mathscr{B} = B/H_c\sqrt{2}, \tag{6.36}$$

where we have defined the London penetration depth λ_L (cf. (6.28)) by

$$\lambda_L^2 = m_S c^2/4\pi\varepsilon^2\Psi_0^2. \qquad (6.37)^{(19)}$$

Let us also define the Ginzburg–Landau parameter κ:

$$\kappa^2 = 2\varepsilon^2 H_c^2 \lambda_L/\hbar^2 c^2. \qquad (6.38)^{(63)}$$

The significance of κ will be discussed further, and it will be seen to be closely related to Pippard's coherence length ξ.

In terms of these new variables, the Ginzburg–Landau equations (6.23) and (6.24) simplify to

$$\mathrm{d}^2\mathscr{A}/\mathrm{d}\mathscr{z}^2 = \psi^2\mathscr{A}, \tag{6.39}$$

and

$$\mathrm{d}^2\psi/\mathrm{d}\mathscr{z}^2 = \kappa^2(\mathscr{A}^2-1)\psi+\kappa^2\psi^3. \tag{6.40}$$

Note that of the four parameters m_S, ε, α, and β, only the three combinations H_c, λ_L, and κ enter the equations. The ratio $m_S/|\Psi_0|^2$ is determined, but not the mass alone. Arbitrarily, Ginzburg and Landau choose the free-electron mass $m = 9{\cdot}1 \times 10^{-28}$ g, but this choice has neither physical significance nor consequences.†

With the particular value $\kappa = 0$ for the Ginzburg–Landau parameter, equation (6.40) has the rather trivial solution

$$\psi = 1, \tag{6.41}$$

and with (6.39) we recover the London penetration law (3.14). In order to identify (6.39) with (3.13), it is merely necessary

† The situation is analogous to that of the Bloch theory of normal metals if we exclude the possibility of measuring the Hall coefficient; neither the effect tive mass m^* of conduction electrons nor their number density N_e is measurable, but only the ratio.

to identify $|\Psi_0|^2$ with n_s and the λ_L of (6.37) with the London penetration depth. Note that λ_L need *not* be taken to have the Becker–Heller–Sauter value[19] ($\lambda^2 = mc^2/4\pi N_e e^2$, Becker *et al.* 1933) but should rather be determined empirically. Indeed, any attempt to fix λ_L *a priori* at the Becker value would immediately lead to a contradiction with the Pippard expression (6.15).

In the remaining two sections of this chapter, some of the consequences of the Ginzburg–Landau theory will be discussed, as well as the question of how the parameter κ is to be determined. When κ is known, all the coefficients of the theory are known. Both H_c and λ_L are directly measurable; λ_L is of course the penetration depth for weak magnetic fields into large specimens. The effective charge ε follows from (6.38), and is found to be $\varepsilon = 2e$.

6.4. The interphase surface energy

One of the consequences of the long-range persistence of the order parameter is that the penetration of a magnetic field into a superconducting region no longer has the exponential form (3.14). Hence the London expression[11] $-\lambda H_c^2/8\pi$ no longer gives the correct surface energy (cf. (5.6)). It will, in fact, appear that the *ad hoc* surface energy[11] $\xi H_c^2/8\pi$ of (5.9) can be accounted for in terms of the order parameter; the parameter ξ is thus to be identified with the Pippard coherence length ξ.

Exercise. With the simple geometry of Fig. 3.1, calculate the spatial dependence of the magnetic field and the order parameter from equations (6.39) and (6.40). Terms of order $\kappa^2 H_e$ may be neglected.

Qualitatively, the argument runs as follows (Pippard 1951): if ψ jumps discontinuously from zero to unity *at* the n-s surface, the field penetration is exponential and the London relation (3.14) holds. But in that case grad ψ has the form of a δ-function sheet, and the Gibbs free energy is very far removed from its minimum. To reduce the free energy let us rather assume that ψ rises gradually from the value zero (at the true n-s surface) to unity. It is not expected to attain the value unity until we are at a distance $\sim \xi$ from the surface. If $\xi > \lambda$, then the field penetration is significantly non-exponential. We can define an effective boundary for the free energy,

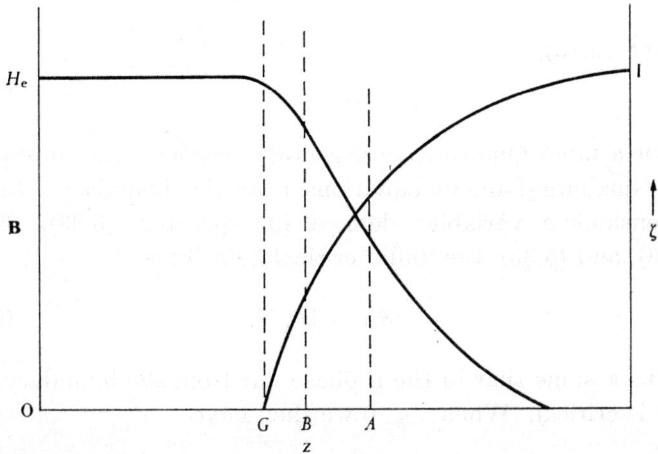

FIG. 6.4. The variation of the order parameter ζ and the magnetic field **B** in the neighbourhood of the surface of a superconductor. G is the geometrical surface of the superconductor. The order parameter ζ rises from zero at G to unity for large positive z. A represents the plane at which $\zeta = \frac{1}{2}$, i.e. the effective thermodynamic n-s boundary. B is the electromagnetic effective boundary, i.e. $GB \simeq \lambda$. From Pippard, *Proc. Camb. phil. Soc. math. phys. Sci.* **47**, 617 (1951).

i.e. a boundary such that the free energy has a value *as if* that were the position of a sharp discontinuity, but without the δ-function. Essentially this effective boundary will be the surface for which $|\psi|^2 = \frac{1}{2}$. But the effective boundary for the magnetic field is quite different. As illustrated in Fig. 6.4, it should be taken to be distant λ from the geometrical boundary. Since the penetration is not exponential, λ is to be defined as in (6.3). Provided that $GB < GA$, as drawn in Fig. 6.4, the surface energy will be positive. To see this, we need only notice that the field has been excluded from the layer BA, which is effectively normal as far as its Gibbs free energy is concerned. On the other hand when $GB > GA$, the surface energy is negative. The London model is a special case of this latter possibility, with $GA = 0$. In general GA will be very roughly equal to ξ, so that the surface energy per unit area will be

$$\alpha_{\mathrm{ns}} \simeq H_{\mathrm{c}}^2(BA)/8\pi \simeq H_{\mathrm{c}}^2(\xi-\lambda)/8\pi, \qquad (6.42)^{(11)}$$

i.e. (cf. (5.13))

$$\delta \simeq \xi-\lambda. \qquad (6.43)$$

For a more quantitative discussion, consider the solution of the Ginzburg–Landau equations near the boundary. In the dimensionless variables defined in equations (6.33), (6.34), (6.35), and (6.36), the 'bulk' critical field $\mathscr{H}_{\mathrm{c}}$ is

$$\mathscr{H}_{\mathrm{c}} = 1/\sqrt{2}. \qquad (6.44)$$

Let us assume that in the n-phase, far from the boundary, the field is critical. When $z < 0$ we then have

$$\left.\begin{aligned} \mathscr{B} = \mathscr{H}_{\mathrm{c}} = 1/\sqrt{2},\ \mathscr{A} &= z/\sqrt{2} + \text{const}; \\ \psi = 0, \qquad \psi' &= 0. \end{aligned}\right\} \qquad (6.45)$$

Far from the boundary, but in the s-phase, the limiting values are

$$\left.\begin{aligned} \mathscr{B} = \mathscr{A} &= 0; \\ \psi = 1, \quad \psi' &= 0. \end{aligned}\right\} \qquad (6.46)$$

The surface energy per unit area of the n-s interface is given by

$$\alpha_{\mathrm{ns}} = \int \mathrm{d}z \left[g_{\mathrm{n}}(B(z),\, T) - \{g_{\mathrm{n}}(0,\, T) + H_{\mathrm{c}}^2/8\pi\} - \frac{1}{4\pi} H_{\mathrm{c}}\{B(z) - H_{\mathrm{c}}\} \right]. \qquad (6.47)^{(64)}$$

The significance of the first two terms in (6.47) is obvious. The last term represents the magnetic energy $-\int \mathbf{H}_e . \mathbf{M} \, dz$; the magnetic moment M per unit volume is, of course,[65] $(1/4\pi)\{B(z) - H_c\}$. Substituting from (6.22), (6.33), (6.34), (6.36), (6.37), and (6.38) in (6.47), the surface energy is

$$\alpha_{ns} = \frac{H_c^2 \lambda_L}{2\pi} \int_{-\infty}^{\infty} dz \left\{ \frac{\psi'^2}{\kappa^2} + \mathscr{B}^2 - \frac{\mathscr{B}}{\sqrt{2}} \right\}. \tag{6.48}$$ [66]

Here $\mathscr{B}$ and ψ are the solutions of (6.39) and (6.40) satisfying the conditions (6.45) and (6.46). For general values of κ it is necessary to resort to computation. However the equations can be solved analytically in the limiting case $\kappa \ll 1$. The solution is

$$\alpha_{ns} = \xi H_c^2 / 8\pi, \tag{6.49}$$ [11]

where ξ is positive and of magnitude

$$\begin{aligned} \xi &= 8\lambda_L / 3\sqrt{2}\,\kappa \\ &= 1{\cdot}89 \lambda_L / \kappa. \end{aligned} \tag{6.50}$$

In contrast, when κ is large, the ψ'^2 term in (6.48) is small. Under these conditions α_{ns} is negative, for $\mathscr{B}^2$ will now be $< \mathscr{B}/\sqrt{2}$. Thus the rigorous solution of the Ginzburg–Landau equations confirms the suggestion that large values of κ correspond to negative n-s surface energy. In fact it can be shown that the condition for the surface energy to be positive is

$$\kappa < 1/\sqrt{2}. \tag{6.51}$$

When the condition (6.51) is not satisfied, the surface energy is negative, and parasitic solutions (§ 5.1) are possible. *Superconductors are classified as 'Type I' or 'Type II' according as they do or do not satisfy* (6.51). Type II materials have properties rather different from those described in the earlier chapters. In particular a field considerably above H_c is required in order to destroy all traces of superconductivity. This and related properties make certain Type II materials important technologically. A fuller discussion of the properties of Type II superconductors will form the contents of Chapter 7.

6.5. The limit of supercooling of the normal phase

One problem which has been deferred from § 6.3 is the determination of the Ginzburg–Landau parameter κ from empirical data. κ may be estimated from (6.50) if α_{ns} is known. The interphase surface energy α_{ns} has been determined both from measurements on the intermediate state (§ 5.2) and the rate of propagation of n-s boundaries (§ 5.3). Unfortunately, neither of these methods is capable of high precision; moreover (6.50) is itself only an approximation valid for $\kappa \ll 1/\sqrt{2}$. A better method is to study the 'ideal' supercooling of the n-phase in the neighbourbood of T_c.

Equations (6.39) and (6.40) will be studied again, but this time the superconductor will be taken to occupy the entire space. There is a totally superconducting solution, with $\psi \simeq 1$ (cf. exercise, p. 112). But there is also another (trivial) solution with $\psi \equiv 0$ everywhere. This solution corresponds to normal phase everywhere, and is stable only if $H_e \geqslant H_c$. When $H_e < H_c$, the solution $\psi = 0$ is at best metastable.

To investigate the stability of the solution $\psi = 0$ against small perturbations, let us try to find solutions which have $\psi \ll 1$ and satisfy the condition that $\psi \to 0$ as $|z| \to \infty$. The condition $\psi \ll 1$ implies that field penetration is virtually complete, so that $\mathbf{B} = \mathbf{H}_e$ everywhere. With a natural choice of gauge, we may write

$$\mathscr{A} = \mathscr{H}_e z. \tag{6.52}$$

If the non-linear term in (6.40) is neglected, then

$$\mathrm{d}^2\psi/\mathrm{d}z^2 = -\kappa^2(1-\mathscr{H}_e^2 z^2)\psi. \tag{6.53†}$$

† Note that in (6.53) the origin is arbitrary, so that the parasitic s-lamina can form anywhere in the specimen.

This equation is formally identical with the Schrödinger equation for a harmonic oscillator.

The solutions of (6.53) are not bounded everywhere unless the eigenvalue condition

$$\kappa/\mathscr{H}_e = n+\tfrac{1}{2}, \quad n = 0, 1, 2, \ldots, \tag{6.54}$$

is fulfilled. And unless the solution is bounded as $|z| \to \infty$, the condition $\psi \ll 1$ cannot hold everywhere.

When (6.54) is fulfilled, then the parastic solutions

$$\psi \propto \exp(-\tfrac{1}{2}\kappa^2 z^2) \tag{6.55}$$

of (6.54) will have lower Gibbs free energy than the normal solution $\psi = 0$. The highest field H_{c2} compatible with (6.54) is obtained by setting $n = 0$, namely $\mathscr{H}_{c2} = \kappa$ or

$$H_{c2} = \kappa\sqrt{2}\,H_c. \tag{6.56}†$$

When the external field H_e falls below H_{c2}, the Ginzburg–Landau instability sets in. The field H_{c2} therefore represents the limit of possible supercooling of the n-phase.

The situation is complicated slightly by the tendency for nucleation to occur very near the surface of a superconductor (Saint James and de Gennes 1963). A critical field H_{c3} exists:

$$H_{c3} = 2{\cdot}392\kappa H_c. \tag{6.57}$$

Below H_{c3}, it is energetically favourable for a superconducting surface sheath to form. Hence in the range $H_{c2} < H_c < H_{c3}$, supercooling of the n-phase is not possible. If $\kappa > 0{\cdot}419$, the surface sheath is stable even in a field greater than H_c.

Equation (6.57) provides the best experimental method for determining κ for a Type I superconductor (provided $\kappa < 0{\cdot}419$). Near T_c, the large penetration depth makes Faber's nucleation centres ineffective (§ 5.3). Hence at temperatures very close to T_c the supercooling tends to its ideal value, given by (6.57). Values of κ obtained by this method for a few Type I materials

† For an alternative derivation of (6.56), see Appendix B.

are shown in Table 6.1, and compared with values derived from anomalous-skin-effect measurements.

TABLE 6.1

	κ (*from supercooling*)	κ (*from anomalous skin effect*)
Al	0·0153	0·01
In	0·066	0·051
Sn	0·0968	0·149

When $\kappa > 1/\sqrt{2}$, parasitic solutions are still possible, but the interpretation of H_{c2} is different—parasitism can now occur for fields above H_c. A detailed discussion of the role of these solutions in the theory of Type II superconductors is given in Chapter 7.

CHAPTER 7

TYPE II SUPERCONDUCTIVITY

THE ideal superconducting behaviour described in Chapters 2 and 3 can occur only when the net n-s surface energy[11] $H_c^2\delta/8\pi$ is positive. It has been shown in Chapters 5 and 6 that when δ is negative parasitic solutions are not merely possible but stable; they represent a state of lower free energy than the ideal state $B = 0$. According to the Ginzburg–Landau (1950) and Pippard (1953) theories, the positive contribution to the interphase surface energy arises from the large range of coherence in a superconductor. The condition for stable parasitism is that Pippard's coherence length ξ shall be smaller than the electromagnetic penetration depth λ, or, more precisely, that the Ginzburg–Landau parameter κ shall be greater than $1\sqrt{2}$.

Abrikosov (1957) showed that when $\kappa > 1/\sqrt{2}$, the magnetization curve of a long needle-shaped specimen cannot have the form illustrated in Fig. 1.6. The ideal behaviour illustrated there consisted of (a) complete *expulsion* of magnetic fields less than H_c, and (b) complete *penetration* of fields greater than H_c. When $\kappa > 1/\sqrt{2}$, the behaviour is more complicated; the transition from pure s-phase to pure n-phase is spread over a range of field strengths $H_{c1} < H_e < H_{c2}$. In this range the field will *partially* penetrate the body. The specimen is described as being in a 'mixed state'.† Abrikosov suggested that the condition $\kappa > 1/\sqrt{2}$ might be fulfilled in some alloys. Moreover if it were indeed fulfilled, many previously puzzling observations on alloys could be interpreted in terms of the Ginzburg–Landau theory. Goodman (1961) discussed a laminar

† This mixed state should not be confused with the intermediate state of Type I materials, discussed in § 5.2. The intermediate state is a consequence of the geometry of the specimen, but the mixed state in a Type II specimen occurs even if the demagnetizing coefficient is zero.

model of the mixed state. This model does not fit the experimental data as well as Abrikosov's structure. By its greater simplicity, however, it drew attention to the existence of inhomogeneous solutions of the Ginzburg–Landau equations, and stimulated a revival of interest in Abrikosov's work.

Thus a sharp distinction is to be expected between the properties of Type I superconductors, with $\kappa < 1/\sqrt{2}$, and Type II superconductors, with $\kappa > 1/\sqrt{2}$.† In general, the mathematical description of Type II behaviour is not susceptible of development in closed expressions. For simplicity, most of the present discussion will be of *extreme* Type II superconductivity, $\kappa \gg 1/\sqrt{2}$. In this limit analytic expressions can be found. Although the discussion of the extreme Type II limit will suffice to illustrate the main features of Type II behaviour qualitatively, it will not, of course, lead to correct quantitative predictions for materials having $\kappa \sim 1/\sqrt{2}$.

In a Type II material, it is not possible to observe the 'thermodynamic' critical field H_c directly. The directly measurable critical fields are H_{c1} and H_{c2}, to be discussed in §§ 7.1, 7.2. H_c has to be determined by indirect means, e.g. by equating the area under the magnetization curve of Fig. 7.2 to $\frac{1}{2}H_c^2$.

7.1. The upper critical field: quantized flux lines

Parasitic solutions (6.55) of the Ginzburg–Landau equations are possible (cf. (6.56)) only if

$$H_e \leqslant H_{c2} \equiv \kappa\sqrt{2}\,H_c. \tag{7.1}$$

In a Type I specimen, the significance of H_{c2} is that it gives the upper limit to the possible supercooling of the n-phase. For a Type II superconductor, (7.1) implies that $H_{c2} > H_c$. Consequently the parasitic configuration is thermodynamically *stable* against conversion to pure n-phase in applied fields

† Type I and Type II superconductors are sometimes also referred to as 'Pippard' and 'London' superconductors respectively. However it is preferable (e.g. Waldram 1964) to use these terms to characterize the *electromagnetic* coherence length, compared to the penetration depth. Thus all superconductors sufficiently near to T_c are 'London' superconductors in Waldram's sense, though they need not be Type II as determined by their d.c. or low-frequency behaviour.

$<H_{c2}$; i.e. superconductivity can persist up to fields in excess of the 'bulk' critical field H_c. Superconductivity finally disappears, in an Ehrenfest second-order phase transition, at the field H_{c2}.

Following Abrikosov, the Ginzburg–Landau equations (6.23) and (6.24) will be written in the dimensionless units defined by equations (6.33), (6.34), (6.35), and (6.36), but the specialization of (6.39) and (6.40) to a one-dimensional geometry will not be made. In fact the interesting solutions will turn out to show a two-dimensional lattice structure. The dimensionless form of the Ginzburg–Landau equations is:

$$\left.\begin{aligned} -\text{curl curl}\,\mathscr{A} &= |\psi|^2\mathscr{A}+\tfrac{1}{2}\mathrm{i}\kappa^{-1}\{\psi^*\nabla\psi-(\nabla\psi^*)\psi\}, \\ (\mathrm{i}\kappa^{-1}\nabla+\mathscr{A})^2\psi &= \psi(1-|\psi|^2). \end{aligned}\right\} \tag{7.2}$$

When H_e is very near to H_{c2}, then $\psi \ll 1$. The $\psi|\psi|^2$ term may be neglected in this limit, and the second of the equations is thus linearized. Under these conditions (6.55) is a solution. Clearly, however, (6.55) is not unique; it is one of a class of solutions

$$\psi_{n,k} = \exp\{inky-\tfrac{1}{2}\kappa^2(x-nk/\kappa^2)^2\}. \tag{7.3}$$

Still more general solutions can be built by superposition of solutions (7.3):

$$\tilde{\psi} = \sum_n C_n\psi_{n,k}. \tag{7.4}$$

In (7.4) the summation is over all (positive *and* negative) integers. The parameter k is still unspecified; it will be determined variationally at a later stage.

If it were not for the degeneracy of the solutions (7.3), the linear approximation to (7.2) would be good enough for a study of the behaviour of materials in the neighbourhood of H_{c2}. But since the solutions (7.3) are degenerate, care is required. We have to choose the *right* combination of the functions (7.3) in building up a solution (7.4), in order that the neglected $\psi|\psi|^2$ term in (7.2) can legitimately be treated as a perturbation. Abrikosov (1957) finds an appropriate choice to be

$$\left.\begin{aligned} C_n &= C, \\ k &= \kappa\sqrt{(2\pi)}. \end{aligned}\right\} \tag{7.5}$$

With this choice, (7.4) reduces to†

$$\begin{aligned} \tilde{\psi} &= C\sum_n \exp\left\{inky - \frac{\pi^2}{2}(x - kn/\kappa^2)^2\right\} \\ &= C\,\mathrm{e}^{\frac{1}{2}\kappa^2x^2}\vartheta_3\left(\sqrt{(2\pi)}\,\kappa\mathrm{i}(x+\mathrm{i}y)\,\middle|\,\mathrm{i}\right). \end{aligned} \tag{7.6}$$

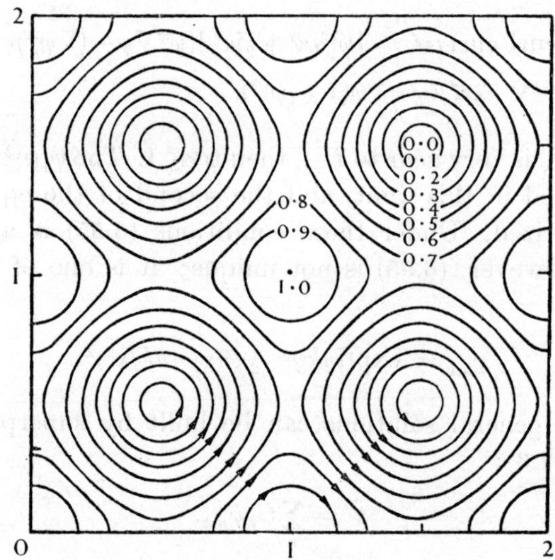

FIG. 7.1. The Abrikosov flux-line structure. The lines drawn are contours of constant $|\Psi/\Psi_0|$, as well as streamlines of current. From Abrikosov, *Zh. eksp. teor. Fiz.* **32**, 1442 (1957).

Thus $|\tilde{\psi}|^2$ is an elliptic function, with the symmetry of a square lattice (see Fig. 7.1). For fields H_e very near H_{c1} (equation (7.25)), Abrikosov finds that a triangular lattice

$$\left.\begin{aligned} C_n &= C_{n+2},\ C_1/C_0 = \pm\, i, \\ k &= \kappa\,(3\pi^2)^{1/4} \end{aligned}\right\} \tag{7·5′}$$

† The notation for the ϑ-functions is that of Whittaker and Watson (1927).

is the stable configuration, and that a 'crystallographic'' phase transition to the square lattice occurs as H_e is increased. For simplicity, this feature of the Abrikosov theory will be neglected here.†

The induction **B** is in the z-direction, and is given by[67] $\mathbf{B} = (H_e/\sqrt{2})\mathscr{B}$, where

$$\mathscr{B} = \partial\mathscr{A}/\partial x = \mathscr{H}_e - |\tilde{\psi}|^2/2\kappa. \tag{7.7}$$

From (7.6) and (7.7), it is evident that B has a maximum value H_e where $\tilde{\psi} = 0$. It falls to its minimum value,

$$\mathscr{B}_{\min} = \mathscr{H}_e - |\tilde{\psi}_{\max}|^2/2\kappa, \tag{7.8}$$

at the interstitial points (Fig. 7.1) where $\tilde{\psi}$ is maximal. The mean field $\bar{B}$ is given by

$$\begin{aligned}\bar{\mathscr{B}} &= \mathscr{H}_e - \overline{|\tilde{\psi}|^2}/2\kappa \\ &= \mathscr{H}_e - (\kappa - \mathscr{H}_e)/\beta(2\kappa^2 - 1),\end{aligned} \tag{7.9}$$

where

$$\beta = \overline{|\tilde{\psi}|^4}\Big/\Big(\overline{|\tilde{\psi}|^2}\Big)^2. \tag{7.10}$$

From (7.10) the magnetization is readily calculated. It is

$$M = (H_{c_2} - H_e)/4\pi\beta(2\kappa^2 - 1). \tag{7.11}$$ [10]

The main qualitative features of the magnetization curve are illustrated in Fig. 7.2.

Exercises. Assume (7.4), (7.5) and show that:

(a) $\mathscr{A} = \mathscr{H}_e x - (2\kappa)^{-1}\int|\tilde{\psi}|^2\,\mathrm{d}x$,

(b) $(\kappa - \mathscr{H}_e)\kappa^{-1}\overline{|\tilde{\psi}|^2} + \{(2\kappa^2)^{-1} - 1\}\overline{|\tilde{\psi}|^4} = 0$,

(c) equation (7.9) holds,

(d)[10] $g_s(H_e) - g_s(0) = (H_c^2/4\pi)\int \mathrm{d}\tau\,(\mathscr{H}^2 - \tfrac{1}{2}|\tilde{\psi}|^4 + \tfrac{1}{2})$,

(e) β is a minimum when $k = \kappa\sqrt{(2\pi)}$ and takes the value $\beta_{\min} = \vartheta_3^2(0|\mathrm{i}) = 1{\cdot}18$.

† *Added in proof:* Recent work, both experimental (Cribier *et al.* 1964) and theoretical (Kleiner *et al.* 1964, Matricon 1964, Goodman 1966) suggests that the *triangular* lattice (7·5′) is the stable structure for all fields. There is no evidence for Abrikosov's phase transition from a triangular to a square lattice. Indeed, *direct* evidence now exists for the triangular lattice structure, in the electron micrographs of Essman and Träuble (1967).

The Abrikosov solution (7.6) of the Ginzburg–Landau equations is qualitatively very similar to the Onsager–Feynman picture of vortex filaments in rotating liquid helium (see e.g. Feynman 1955). To see the similarity remember that the lines of constant **B** are *streamlines* of the current density **J**.

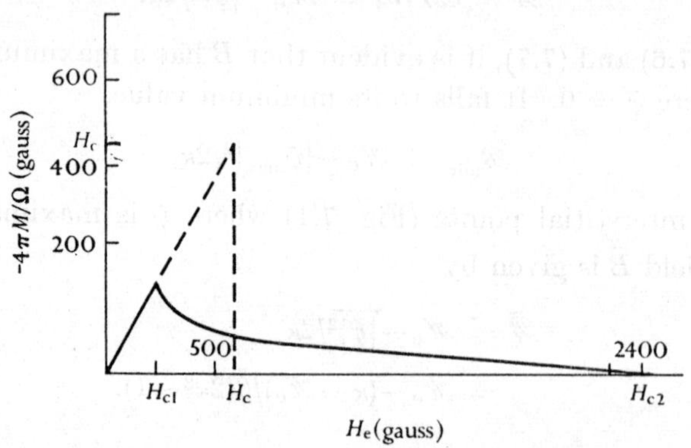

FIG. 7.2. Magnetization curve[8] for an ideal Type II superconducting needle (lead with 8·23 per cent indium). The broken curve is the magnetization curve for a Type I material (pure lead); the area under the two curves is the same (Livingston 1963).

To develop the analogy with rotating liquid helium more fully, let us examine the behaviour of (7.6) near the centre of one of the 'vortices'. We therefore change the origin of coordinates to one of the points where $\psi = 0$ (Fig. 7.1). The ϑ_3-function has the property (e.g. Whittaker and Watson 1927)

$$\vartheta_3(z+\tfrac{1}{2}\pi-\tfrac{1}{2}\pi\tau \mid \tau) = e^{i(z-\frac{1}{4}\pi\tau)}\vartheta_1(z|\tau). \tag{7.12}$$

If (cf. (7.6)) the argument z and the parameter τ of the ϑ-functions are taken as

$$\left.\begin{aligned} z &= -\kappa i\sqrt{(2\pi)}(x+iy), \\ \tau &= i, \end{aligned}\right\} \tag{7.13}$$

equation (7.12) implies that

$$\tilde{\psi}(x+\sqrt{\pi}/2\kappa\sqrt{2},\, y+\sqrt{\pi}/2\kappa\sqrt{2})$$
$$\propto \vartheta_1\Big(-\mathrm{i}\kappa\sqrt{(2\pi)}\,(x+\mathrm{i}y)\Big|\,\mathrm{i}\Big)\exp\{-\kappa\sqrt{(2\pi)}\,(x+\mathrm{i}y)\}$$
$$= \vartheta_1(z|\mathrm{i})\mathrm{e}^{-iz}, \qquad (7.14)$$

or, for small z,

$$\tilde{\psi} = Cz\vartheta_1'(0). \qquad (7.15)$$

In other words, putting $z = z_0\,\mathrm{e}^{i\theta}$, the phase angle φ of the Ginzburg–Landau order parameter ψ is the polar angle θ. Not only is ψ single-valued, but its phase increases by 2π every time we go around the core of the vortex.

The arguments of § 3.4.1 are not strictly applicable, since there are no integration contours far enough away from all vortex cores to justify the neglect of $\partial\psi/\partial|z|$. However, provided Φ is taken to be the *fluxoid*, rather than the flux, the quantization condition (3.45) holds rigorously for *any* contour (cf. footnote, p. 52). The Abrikosov vortices are thus to be interpreted as flux lines, each carrying a single fluxoid quantum[16] $2\pi\hbar c/\varepsilon = \pi\hbar c/e$.

7.2. Energy of a single flux line: the lower critical field

In a Type II material, the London theory would predict the existence of parasitic solutions for *all* values of the applied field satisfying $H_e < H_{c2}$. In particular there should never be a Meissner effect, *however small* the applied field. In the Ginzburg–Landau theory this conclusion is modified. Abrikosov shows that the flux lines themselves make a contribution to the Gibbs free energy. For small enough fields, there *is* a Meissner effect; a 'lower critical field' H_{c1} exists, as a threshold for the appearance of parasitic solutions.

For arbitrary values of κ, the Ginzburg–Landau equations are intractible and require numerical solution. In the extreme Type II limit $\kappa \gg 1$, however, the solution becomes simple. If the flux is very small, there will be few flux lines, i.e. the flux lines will be well separated. In the first instance the interaction between the flux lines will be neglected. This approximation reduces the problem to a study of the properties of a single flux line.

From Fig. 7.1 it may be seen that each flux line has a central core in which $|\psi|$ is small. Abrikosov shows that the core radius is κ^{-1} in Ginzburg–Landau units, i.e., from (6.34), λ_L/κ in laboratory units. Outside the core region ψ rapidly approaches unity.

On making the substitutions

$$\left.\begin{aligned} Q &= |\mathscr{A} - \kappa^{-1}\nabla\varphi|, \\ \psi &= f \exp i\varphi, \end{aligned}\right\} \tag{7.16}$$

the Ginzburg–Landau equations (7.2) become

$$-(\kappa^2 r)^{-1}\frac{\mathrm{d}}{\mathrm{d}r}\left(r\frac{\mathrm{d}f}{\mathrm{d}r}\right) + Q^2 f = f - f^3, \tag{7.17}$$

$$\frac{\mathrm{d}}{\mathrm{d}r}\left(\frac{1}{r}\frac{\mathrm{d}}{\mathrm{d}r}(rQ)\right) = Qf^2. \tag{7.18}$$

It is important to note that κ^{-2} appears as a factor in the '∇^2' term in (7.17), but is absent from (7.18). Thus we may conclude from (7.17) that most of the variation of ψ occurs in the region $r \lesssim \kappa^{-1} \ll 1$. In contrast, (7.18) implies that most of the variation of Q (and hence of A) occurs in the region $r \lesssim 1$.

Since by hypothesis $\kappa \gg 1$, it is a good approximation (de Gennes and Matricon 1964) to regard the core as a line singularity. Outside the core, where the field is varying rapidly, ψ is effectively unity and the London theory is adequate. From the London equation (3.15) it follows that

$$\mathbf{B} + \lambda^2 \operatorname{curl}\operatorname{curl}\mathbf{B} = \mathbf{\Phi}_0\delta(\mathbf{r}). \tag{7.19}$$

The δ-function on the right-hand side represents the effect of the core, regarded as a line source. The coefficient Φ_0 is just the fluxoid quantum (3.45).

Exercise. From (7.19), calculate the fluxoid through a contour containing the core, and hence verify that the coefficient Φ_0 in (7.19) has been chosen correctly.

The two-dimensional equation (7.19) has the asymptotic solution

$$B = \frac{\Phi_0}{2\pi\lambda_L^2}\ln(\lambda_L/r) \tag{7.20}$$

for $r \ll \lambda$. In the region where $r \gg \lambda$, B is negligible. Hence the energy of a flux line, per unit length, is[68]

$$\mathscr{E} = \int_0^{\lambda_L} \mathrm{d}x\,\mathrm{d}y\left\{\frac{1}{8\pi}B^2 - \tfrac{1}{2}n_s m_s v_s^2\right\}$$

$$= \int_0^{\lambda_L} \mathrm{d}x\,\mathrm{d}y\left\{\frac{1}{8\pi}B^2 + \lambda_L^2(\mathrm{curl}\,\mathbf{B})^2\right\} \tag{7.21}$$

$$= (\Phi_0/4\pi\lambda_L)^2 \ln\kappa. \tag{7.22}$$

By numerical integration of the Ginzburg–Landau equations (7.17) and (7.18), Abrikosov has derived a more accurate expression for the energy of a flux line: (7.22) is corrected to

$$\mathscr{E} = (\Phi_0/4\pi\lambda_L)^2(0{\cdot}081 + \ln\kappa). \tag{7.23}$$ [69]

With this knowledge of the energy of a flux line, we are now in a position to find the lower critical field H_{c1}. H_{c1} is defined as the lowest field for which it is energetically favourable for a flux line to penetrate the specimen. When there is a single flux line piercing the body, the reduction in the magnetic energy is

$$\mathbf{H}_e.\Delta\mathbf{M} = \frac{1}{4\pi}H_e\Phi_0 \tag{7.24}$$ [9]

per unit length. Provided that (7.24) is greater than (7.23), it will be energetically favourable to create at least one line. Thus

$$H_{c1} = \frac{\Phi_0}{4\pi\lambda_L^2}(0{\cdot}081 + \ln\kappa). \tag{7.25}$$ [82]

However, there is no reason why the process should end with the creation of only *one* line. The condition for the creation of a second flux line remains $H_e > H_{c1}$, and so on. The process will terminate only when the interaction between flux lines—an interaction which has been neglected until now—is sufficient to change the energy balance.

To calculate the interaction energy between flux lines, let us note that the free energy density due to the magnetic field is the integrand of (7.21), i.e. after integration by parts,

$$f = \frac{1}{8\pi}(B^2 - \lambda_L^2\,\mathbf{B}\,.\,\nabla^2\mathbf{B}). \qquad (7.26)^{(11)}$$

In the extreme Type II limit $\kappa \gg 1$, Abrikosov (1957) has shown that the field due to a single flux line is

$$B = \frac{H_c}{\kappa\sqrt{2}}K_0(r/\lambda_L), \qquad (7.27)^{(70)}$$

where r is the distance from the axis of the flux line, and K_0 denotes a modified Bessel function of the second kind. For large arguments the K_0 function vanishes exponentially ($K_0(x) \sim e^{-x}$), while for small arguments $K_0(x) \sim -\ln x$; hence we recover the approximation (7.20).

From (7.26) and (7.27), the interaction energy is

$$f = \frac{H_c^2}{4\kappa^2}\sum_{i,j} K_0(|\mathbf{r}_i - \mathbf{r}_j|/\lambda_L), \qquad (7.28)^{(71)}$$

where the summation is over the (two-dimensional) coordinates of the flux lines. Anderson and Kim (1964) make the illuminating comment that the interactions (7.28) are formally identical with the screened electrostatic interactions between line charges, if λ_L is identified as the screening length.

From (7.28) the interaction between two flux lines is seen to be appreciable only when their separation is $\lesssim \lambda_L$. Hence it is clear that the spacing of the Abrikosov flux lines will be $\sim\lambda_L$, even for fields only very slightly in excess of H_{c1}. Abrikosov has shown that at the lower critical field H_{c1} the body enters the mixed state through an Ehrenfest second-order transition, and that the magnetization curve has a vertical tangent at H_{c1} (Fig. 7.2).

In a Type I material, the n-s phase transition is of first order in the presence of an applied magnetic field, and is only of second order in *zero* field. In contrast, the transitions of a Type II superconductor from the pure superconducting state to the mixed state at H_{c1}, and from the mixed state to the normal state at H_{c2}, are *always* of second order. However, κ is insensitive to temperature. Thus, since $H_{c2} = \sqrt{2}\,\kappa H_c$, the temperature dependence of H_{c2} is qualitatively similar to that of H_c (cf. Fig. 1.2, p. 7).

7.3. The critical current: flux pinning

In the Abrikosov flux-line model, as described so far, a Type II superconductor can withstand static magnetic fields greatly in excess of the bulk critical field H_c, but cannot carry an electric current. To see this, note that if the distribution of flux lines is uniform, the current density[(72)] $\mathbf{J} = c \operatorname{curl} \mathbf{H}/4\pi$ will be, on the average, zero. If the density of flux lines is not uniform, the magnetic interaction between lines acts like a pressure. The lines will move under the influence of this pressure, until their density is once again uniform and $\mathbf{J} = 0$. This theoretical behaviour is in marked contrast with experiment, and the only way in which theory and experiment can be reconciled is to find some mechanism which impedes the motion of flux lines.

Much of the present technological importance of superconductors arises from the possibility of constructing solenoids as sources of very high magnetic fields. These devices would not be practicable if the current-carrying capacity of Type II superconductors were indeed zero or nearly zero.

Typically, the materials used in the construction of high-fields solenoids (e.g. Nb_3Sn) are sintered alloys. They are very brittle materials and are far from being homogeneous monocrystalline bodies. Moreover, the Abrikosov theory would predict that the magnetization is uniquely defined by the applied field (Fig. 7.2). But in real hard superconductors there is usually a great deal of hysteresis. Many years ago Mendelssohn (1935) proposed a 'sponge' model for the superconductivity of alloys. He postulated that in the alloy the superconducting material forms a complicated filamentary structure with inclusions of non-superconducting material. Such a Mendelssohn sponge has indeed been constructed (Charles and Harrison 1963) by forcing lead into porous Vycor glass at high pressure.

Most of the Type II superconducting materials used for high-field magnets are rather 'dirty' metallurgically, and probably *do* contain normal inclusions. But it is unlikely that many of them are really quite as bad as the Charles–Harrison lead-in-Vycor specimen. The flux-pinning model of Anderson and Kim (1964) is probably wider in its applicability than the Mendelssohn sponge. In the flux-pinning model the material is assumed to be an Abrikosov Type II specimen, in which the motion of quantized flux lines is impeded by imperfections of some sort in the structure. Flux lines (or rather bundles of lines) are pinned on these imperfections, and a finite activation energy is required to shake them loose again. There is still some controversy regarding the detailed mechanism: Anderson and Kim treat the pinning centres as point defects, but Friedel *et al.* (1963) consider that cavities of a size comparable to the line spacing dominate. Campbell *et al.* (1964) have discussed the form of magnetization curve to be expected in both cases, i.e. both for many weak pinning centres and for well-separated strong centres. For simplicity the present discussion will be confined to the Anderson–Kim model.

The Lorentz force experienced by unit length of a flux line is

$$\mathscr{F} = \frac{1}{c}\mathbf{J}\wedge\mathbf{\Phi}_0, \qquad (7.29)^{(26)}$$

where Φ_0 is the fluxoid quantum (3.45), and the direction of the vector $\mathbf{\Phi}_0$ is along the flux line. Per unit volume the Lorentz force is

$$\boldsymbol{\alpha} = \frac{1}{c}\mathbf{J}\wedge\mathbf{B}. \qquad (7.30)^{(26)}$$

In a steady state, this force $\boldsymbol{\alpha}$ has to be balanced by the quasi-viscous pinning force.

However, as was emphasized by Anderson (1962), the short-range interaction between flux lines acts to keep their density very uniform. A single flux line cannot easily jump a pinning barrier, while remaining at a nearly constant distance from all its neighbours. Anderson therefore introduced the concept of a pinned bundle of lines, the radius of the bundle being $\sim\lambda$. From (7.28), the range of repulsion between flux lines is λ, so that this estimate of the size of the bundle is reasonable. The activation energy required for the hopping of a bundle over a pinning centre is estimated to be

$$E_b \simeq pH_c^2\,\xi_0^3/8\pi - JB\xi_0^3\,l/c. \qquad (7.31)^{(11),(26)}$$

In (7.31) the first term represents the contribution to the free energy made by the inhomogeneity. The assumption is that the volume of the flux bundle is $\sim\xi_0^3$, so that the *maximum* possible energy associated with the bundle is[11] $\xi_0^3 H_c^2/8\pi$. In practice the pinning will be much less effective; the small parameter p represents the fractional efficacy of the pinning centres. The second term of (7.31) is the Lorentz-force contribution to the free energy. The assumption here is that the force acts over a distance l. The meaning of l is presumably the average distance between pinning centres.

If the pinning mechanism were totally effective, and the flux lines were quite unable to move, there would be no limit to the current that could be carried. In 1962 Bean introduced the idea of a 'critical state', characterized by a critical value of the current density J. By the critical state is meant the condition in which the trapped flux is *just* able to move (and hence to escape). In other words it represents the threshold for break-down of the pinning mechanism.

The experiments of Kim *et al.* (1962, 1963) confirm that there is a critical state. But the critical parameter is found not to be J but rather

$$J(B+B_0). \qquad (7.32)$$

In the theory of the critical parameter (Anderson 1962, Anderson and Kim 1964), it is assumed that E_b is the activation energy for a flux bundle to hop over a pinning centre. The rate $\mathscr{R}$ at which flux lines creep past the pinning centre will be

$$\mathscr{R} = \omega \exp(-E_b/kT), \tag{7.33}$$

where ω is some frequency characteristic of the bundle. Anderson estimates ω to lie between 10^5 and 10^{10} sec^{-1}; closer limits are not needed since ω will appear finally only in the argument of a logarithm. Let us define $\mathscr{R}_c$ as the value of $\mathscr{R}$ which is *just* too slow for flux creep to be observable. Then α will take the critical value (from (7.30), (7.31), and (7.33))

$$c^{-1}(JB)_{\text{crit}} \equiv \alpha_{\text{crit}} = \frac{pH_c^2}{8\pi l} - \frac{kT}{l\xi_0^3}\ln\left(\frac{\mathscr{R}_c}{\omega}\right). \tag{7.34}$$

(7.34)[11],[26]

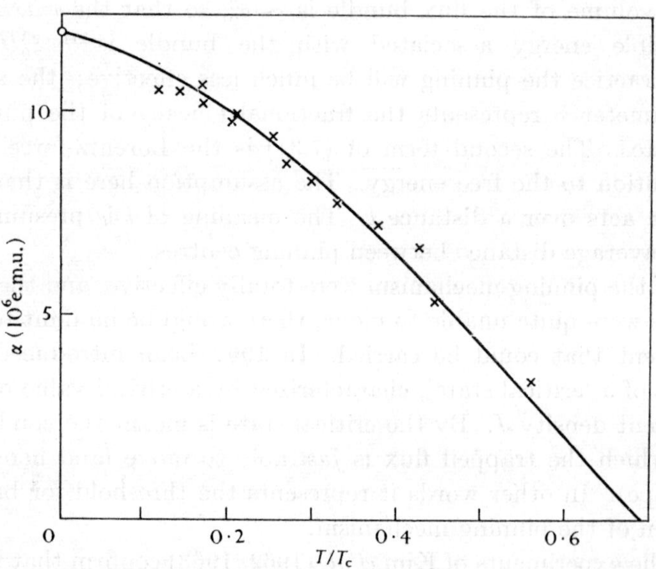

FIG. 7.3. Temperature dependence of the critical parameter α. The full curve is the theoretical curve of Anderson (1962); the points are the experimental values of Kim *et al.* (1962). The values $p = 7\times10^{-3}$, $l = 1{\cdot}0\times10^{-5}$ cm, have been chosen to fit the data.

Equation (7.34) agrees with (7.32), apart from the constant B_0. This constant is interpreted by Anderson as being a consequence of the fact that *there must be at least one flux line in a bundle.* B cannot therefore vanish, but has a minimum value $B_0 \sim \Phi_0/\xi_0^2$. For fields much larger than B_0, (7.32) and (7.34) are in agreement. On selecting reasonable values for the parameters p and l ($p = 7 \times 10^{-3}$, $l = 1{\cdot}0 \times 10^{-5}$ cm), the temperature dependence of (7.34) is also found to be in good agreement with experiment (see Fig. 7.3).

Thus the Anderson–Kim model accounts, at least qualitatively, for the existence of a critical value of JB. But it is clear from the derivation that this critical value is not very sharply defined; the definition of $\mathscr{R}_c$ is somewhat arbitrary. Even apart from this uncertainty, it is far from being the whole story. Much of the most recent work has been concerned with problems of the motion of flux lines, and the consequent Ohmic resistance.† It is also found experimentally that the critical parameter α depends on the previous history of the specimen. For example, by a 'training' sequence of magnetizations and demagnetizations, α can be considerably increased. For a discussion of the way in which α can depend on the configuration of flux lines and pinning centres, the reader is referred to Evetts *et al.* (1964).

Another factor which may influence the current-carrying capacity of Type II superconductors is the formation of a surface sheath (cf. § 6.5). The surface critical field H_{c3} is $\sim$ 70 per cent greater than H_{c2}, and the surface s-layer may be capable of carrying a considerable current. The theory of this effect is still incomplete, and it is not possible yet to give quantitative estimates of its importance.

† See, e.g. Borcherds *et al.* (1964), Volger *et al.* (1964), Bardeen and Stephen (1965), Kim *et al.* (1965).

CHAPTER 8

TUNNELLING

THE present chapter is a brief survey of the phenomenon which presents the clearest experimental evidence for the existence of an energy gap: the tunnel effect through an insulating layer.

The order of magnitude of the energy gap in a superconductor is given by the transition temperature, $T_c \sim 1°K$, i.e. $\Delta \sim 10^{-3}$ eV. If a d.c. potential difference of the order of a millivolt is applied across an insulating layer between two superconductors, or between a normal metal and a superconductor, the current-voltage characteristic will show strong departures from linearity. In § 8.1 the shape of this characteristic will be calculated from the 'BCS model' (§ 2.3). On the whole the results agree quite well with the experiments of Giaever *et al.* (1960, 1961, 1962), but there are some minor discrepancies.

Besides the tunnelling of quasiparticles, as studied in the above experiments of Giaever, another type of tunnel current exists, predicted theoretically by Josephson (1962) and subsequently found experimentally (Anderson and Rowell 1963). In this case it is not the quasiparticles which tunnel through the barrier, but the Cooper pairs (see § 11.1). The consequence is a current which does not vanish even though the potential difference is zero. A non-zero potential difference produces an alternating current! In § 8.2 the Josephson effect will be studied in terms of the Ginzburg–Landau theory.

If the insulting layer between the two superconductors is replaced by a layer of a normal metal such as copper (or aluminium above 1·2°K), there is another interesting phenomenon, in which the normal layer acquires superconducting properties. This is one of the 'proximity' effects, in which the properties of both normal and superconducting metals are modified near a boundary. However, the phenomenon does not admit of a satisfactory explanation in terms of the Ginzburg–Landau theory. Its discussion is accordingly deferred to § 14.5.

8.1. Quasiparticle tunnelling

According to the 'BCS model' (§ 2.3), the density of quasiparticle states of energy E_k is given by (2.28):

$$\begin{aligned} \Xi_s &= \Xi_0 \partial\eta_k/\partial E_k = \Xi_0 E_k/\eta_k \\ &= \Xi_0 E_k/\sqrt{(E_k^2-\Delta^2)}, \end{aligned} \tag{8.1}$$

where η_k and E_k are measured from the Fermi surface. We note that Ξ_s is singular at $E_k = \pm\Delta$ (and, of course, that it has to be zero for $|E_k|<\Delta$). Nevertheless there is a one-to-one correspondence between the BCS quasiparticle states and the Bloch electronic states of the normal metal. The effect of the energy gap is to displace the states in the forbidden region $|\eta_k|<\Delta$ into the energy range outside the gap (Fig. 8.1).

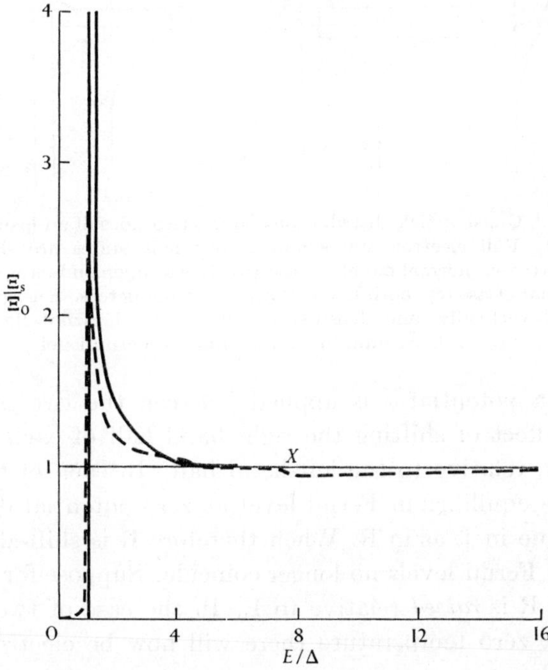

FIG. 8.1. Density of states. The full curve is the BCS theoretical density (equation (8.1)); the broken curve shows the experimental results of Giaever and Megerle (1961) for lead at 0·33°K. An anomaly X occurs at $E \simeq k\Theta$, the Debye energy.

In Fig. 8.2(a), (b), and (c), the quasiparticle distribution is illustrated schematically for two metals (at a finite temperature) separated by an insulating layer. In Case (a), both metals are normal. In Case (b), the metal on the left of the barrier (L) is a superconductor but the one on the right (R) is normal. Finally, in Case (c), both metals are superconductors, but R has a larger energy gap than L. The shaded regions in these figures indicate occupied 'electron' states or unoccupied 'hole' states. The absence of available states within the energy gap is reflected by the vanishing of $\Xi_s(E_k)$ for $|E_k| < \Delta$.

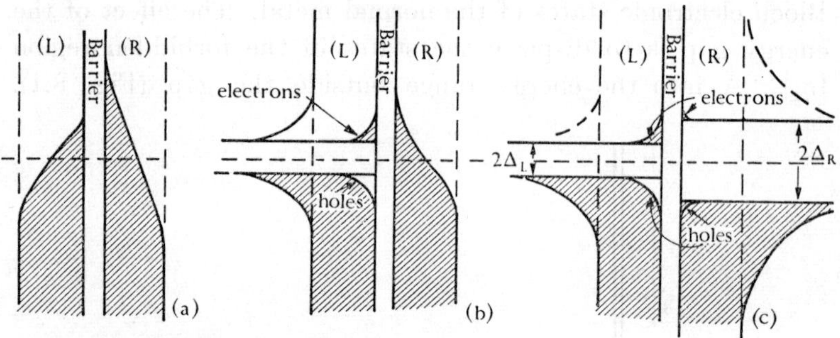

FIG. 8.2. Quasiparticle distributions on the two sides of an insulating barrier. Full electron states and empty hole states are shaded. Case (a): two normal metals. Case (b): L is a superconductor and R is normal. Case (c): both L and R are superconductors. Energies are plotted vertically, and densities of *filled* states horizontally. The broken horizontal line represents the Fermi level.

When a potential ϕ is applied between the two metals, it has the effect of shifting the right-hand half of each diagram vertically, relative to the left-hand half. In each of the three cases, the equilibrium Fermi level at zero potential difference is the same in L as in R. When therefore R is shifted relative to L, the Fermi levels no longer coincide. Suppose for definiteness that R is *raised* relative to L. In the case of two normal metals at zero temperature there will now be electrons in R having the same energy as unoccupied states† in L; their

† The corresponding statement at non-zero temperature is that for a given quasiparticle energy E, the statistical probability of occupation in R exceeds that in L.

number will be proportional to ϕ. These electrons are capable of tunnelling, and the tunnel current will obey Ohm's law (Fig. 8.3(a)).

In case (b), the zero-temperature situation is that there are no electrons above the Fermi level, nor holes below it. Hence there are *no* available states in the superconductor L for the electrons in R to tunnel into, unless the displacement of R is at least Δ, i.e. unless $e\phi \geqslant \Delta$. Thus for $\phi > \Delta/e$ the tunnel current is zero, and it rises at $\phi = \Delta/e$ with infinite slope. At finite temperatures the picture is modified slightly—the rising characteristic is still quite abrupt, but its slope is finite (Fig. 8.3(b)).

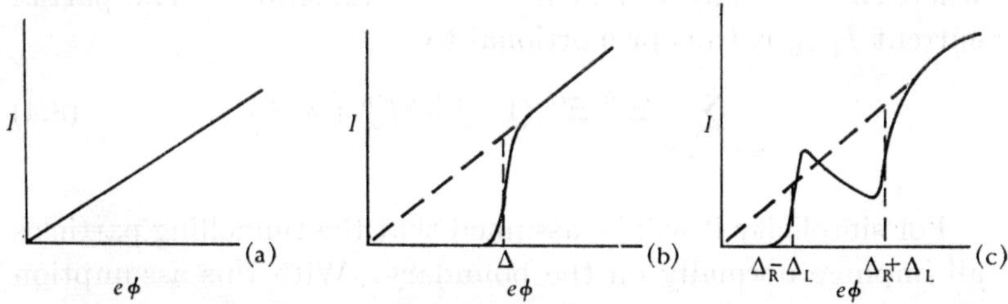

FIG. 8.3. Current–voltage characteristics for the tunnel current. The three cases correspond to the three cases of Fig. 8.2.

Case (c) leads to the characteristic illustrated in Fig. 8.3(c). The current is practically zero until $e\phi > \Delta_R - \Delta_L$, the half difference between the two energy gaps. On account of the high density of states near the edge of the energy gap, the current overshoots the Ohmic value. It then falls, to reach a minimum near the half sum of the energy gaps, $\Delta_R + \Delta_L$. This minimum occurs when the energy gap of R lies opposite the electrons of L available for tunnelling, so that only the *holes* of L can tunnel.

In a quantitative theory of the tunnelling current, we have to calculate the net current, i.e. the difference between the electron current from L to R and the current from R to L:

$$I_{\text{nett}} = I_{L\to R} - I_{R\to L}. \tag{8.2}$$

Let us use $f_{\bar{k}}$ to denote the thermal average occupation number of the quasiparticle state $\bar{k}$; Ξ_S will denote the density of quasiparticle states. The symbol $\bar{k}$ is used as a shorthand for the momentum and spin suffixes $(\mathbf{k}, \sigma)$. The probability that an electron of wave number $\mathbf{k}$ will tunnel from left to right is evidently proportional to the probability $f_{\bar{k}}^{(L)}$ that there was initially an electron in L in the state $\bar{k}$. Moreover this electron can tunnel into the right-hand state $\bar{k}'$ only if $\bar{k}'$ is initially empty in R. By Fermi's golden rule (see, e.g. Schiff 1955), the transition probability is

$$P_{L\to R}(\bar{k};\bar{k}') = \frac{2\pi}{\hbar}|\mathscr{M}|^2\,\Xi^{(R)}\Xi^{(L)}(1-f_{\bar{k}'}^{(R)})f_{\bar{k}}^{(L)}, \tag{8.3}$$

where $\mathscr{M}$ is a matrix element for the transition.† The partial current $I_{L\to R}$ is thus proportional to

$$\sum_{\bar{k}^{(L)},\,\bar{k}'^{(R)}} \Xi^{(R)}\Xi^{(L)}(1-f_{\bar{k}'}^{(R)})f_{\bar{k}}^{(L)}\,|\mathscr{M}|^2\,\delta_{\sigma\sigma'}. \tag{8.4}$$

For simplicity it will be assumed that the tunnelling particles all impinge normally on the boundary. With this assumption there is a very simple one-to-one correspondence between the specification of states in L and R: $\bar{k}^{(L)}$ 'belongs' to $\bar{k}'^{(R)}$ when the corresponding energies are equal (in the presence of the applied potential difference). If E' is the level in R which is shifted by the field into coincidence with E in L, then (8.4) is equivalent to

$$I_{L\to R} \propto \int \mathrm{d}E\,|\mathscr{M}|^2\,\Xi^{(R)}(E')\,\Xi^{(L)}(E)\,f^{(L)}(E)\,(1-f^{(R)}(E')). \tag{8.5}$$

In (8.5) the sum over states $\bar{k}, \bar{k}'$ has been replaced by an integral over E in the usual way.

In fact it can be shown that the above postulate of normal incidence is unnecessarily restrictive. Equation (8.5) holds under the much more general assumption of specular transmission. The assumption is that the matrix element $\mathscr{M}$ vanishes

† In tunnelling processes, the spin never flips; i.e. $|\mathscr{M}|^2$ contains a factor $\delta_{\sigma\sigma'}$. This factor is written explicitly in (8.4).

unless the component of **k** parallel to the surface is equal to the corresponding component of **k**′ both in magnitude and direction.

An expression similar to (8.5) holds for the reverse current $I_{R\to L}$. Hence the net current from L to R is

$$I_{\text{nett}} = I_{L\to R}-I_{R\to L}$$
$$\propto \int dE\,|\mathcal{M}|^2\,\Xi^{(R)}(E')\,\Xi^{(L)}(E)\,\{f^{(L)}(E)-f^{(R)}(E')\}. \qquad (8.6)$$

When the explicit correspondence between L and R states,

$$E' = E+e\phi, \qquad (8.7)$$

is substituted into (8.6), the tunnelling current is finally

$$I \propto \int dE\,\Xi^{(R)}(E+e\phi)\,\Xi^{(L)}(E)\,\{f(E)-f(E+e\phi)\}. \qquad (8.8)$$

If the density of states $\Xi^{(R)}$ is known then $\Xi^{(L)}$ can be calculated from the experimental I—ϕ characteristic. In particular, if R is a normal metal, as in case (b), $\Xi^{(R)}$ is constant and $\Xi^{(L)}$ can be found readily. The broken curve in Fig. 8.1 illustrates such an 'experimental' density of states in lead. Qualitatively it agrees quite well with the BCS density-of-states expression (8.1); in particular the energy gap is clearly exhibited. But the singularities of equation (8.1) are smeared out, and there is some fine structure which is not accounted for by the simple theory. In particular there is a small anomaly (X, Fig. 8.1) at $E \simeq k\Theta$, the Debye temperature. An explanation of the observed anomalies has been proposed by Schrieffer *et al.* (1963).

In this section the semiconductor picture of quasiparticle excitations has been employed. As was stressed in the footnote on p. 32, this picture can be very misleading; quasi-electrons and quasi-holes are not clearly distinguishable entities. In a sense the correctness of the above treatment is fortuitous, and arises out of the fortunate cancellation of two of the coherence factors discussed in § 12.3. The situation is different when the simultaneous tunnelling of two or more quasi-particles is calculated (Schrieffer and Wilkins 1963): the coherence factors do *not* cancel, and the naïve treatment above is inadequate.

8.2. The Josephson effect

In 1962 Josephson predicted that the tunnelling current should not vanish in the limit as the potential difference across an insulating barrier becomes zero, i.e. 'tunnelling supercurrents' should occur. The prediction was based on the microscopic BCS theory. However, it is now clear (de Gennes 1963, Josephson 1964, 1965*a*) that the effect should be expected within the framework of the Ginzburg–Landau theory, although its magnitude is not determined.

The relative phase φ of the Ginzburg–Landau function Ψ between two completely isolated superconductors is arbitrary. If however the two superconductors are separated by an insulating barrier, there is a small coupling energy between the two bodies. As the insulating barrier is reduced in thickness, the coupling between the two bodies is increased, and the relative phase of the Ginzburg–Landau function becomes locked. There are two distinct possibilities: the phases can be locked together in the true equilibrium state with no current, or in one of many metastable configurations carrying a current.

The picture is perhaps clearer for a superconducting *ring* of material interrupted by a narrow insulating barrier. The condition of single-valuedness of Ψ requires that the phase φ of Ψ must change by an integral multiple of 2π every time we traverse the ring. The ground state, with no supercurrent, has zero phase change around the ring, but the Ginzburg–Landau equations admit of other solutions with a phase shift. These are current-carrying states.

In the limit of a very thin barrier, the Ginzburg–Landau equations are applicable everywhere except *on* the barrier, which may be regarded as a surface of separation. To link the Ginzburg–Landau function in R with that in L, we need a boundary condition. This boundary condition cannot be derived from the phenomenological theory, but must be introduced as a postulate. From the microscopic theory the correct boundary conditions are (de Gennes 1963):

$$\left(\frac{\partial\Psi}{\partial x}+\frac{\mathrm{i}\varepsilon}{\hbar c}\Psi A_x\right)_{\mathrm{R}} = \left(\frac{\partial\Psi}{\partial x}+\frac{\mathrm{i}\varepsilon}{\hbar c}\Psi A_x\right)_{\mathrm{L}} = \frac{\alpha\mathscr{T}}{2\xi_0}(\Psi_{\mathrm{R}}-\Psi_{\mathrm{L}}). \tag{8.9}$$

(8.9)[26]

Here α is a numerical factor ~ 1, and $\mathscr{T}$ is the single-quasi-particle transmission coefficient of § 8.1.

In particular, if the magnetic field does not vary appreciably across the barrier, (8.9) gives

$$\Psi_{\mathrm{R}}-\Psi_{\mathrm{L}} = \frac{2\xi_0}{\alpha\mathscr{T}}\frac{\partial\Psi}{\partial x}, \tag{8.10}$$

and $\partial\Psi/\partial x$ is continuous. From (6.19), the current crossing the barrier is

$$\begin{aligned} I &= -\frac{\mathrm{i}\hbar\varepsilon S}{2m}\{\Psi^*\nabla\Psi-(\nabla\Psi^*)\Psi\}_{\mathrm{R}} \\ &= -\frac{\mathrm{i}\hbar\varepsilon S}{2m}\{\Psi^*\nabla\Psi-(\nabla\Psi^*)\Psi\}_{\mathrm{L}}, \end{aligned} \tag{8.11}$$

or, using (8.10),

$$\begin{aligned} I &= \frac{\mathrm{i}\hbar\varepsilon S}{2m}\frac{\alpha\mathscr{T}}{2\xi_0}(\Psi_{\mathrm{R}}^*\Psi_{\mathrm{L}}+\Psi_{\mathrm{L}}^*\Psi_{\mathrm{R}}) \\ &= \frac{\mathrm{i}\hbar\varepsilon S}{2m}\frac{\alpha\mathscr{T}}{\xi_0}|\Psi_{\mathrm{R}}|^2\sin\delta \\ &= I_{\mathrm{C}}\sin\delta, \end{aligned} \tag{8.12}$$

say. Here δ is the difference in phase between Ψ_{R} and Ψ_{L}; the equality of $|\Psi_{\mathrm{R}}|$ and $|\Psi_{\mathrm{L}}|$ follows from the geometry. S is the area of the junction.

In the absence of an applied magnetic field, (8.12) allows a steady current of magnitude $\leqslant I_{\mathrm{C}}$ to flow. The maximum Josephson current I_{C} has been given above in terms of the coefficient α, whose evaluation is outside the scope of the phenomonological theory. It can be shown, by microscopic theory, that I_C is in fact equal to the quasiparticle tunnelling current produced by an applied potential difference $\pi\Delta/2e$ across the barrier.

The remarkable feature of (8.12) is that the Josephson current is proportional to $\mathscr{T}$, the *single*-quasiparticle tunnelling transmission coefficient. (Since the superconducting charge carriers are pairs of electrons, we might naïvely have expected a current proportional to $\mathscr{T}^2$.)

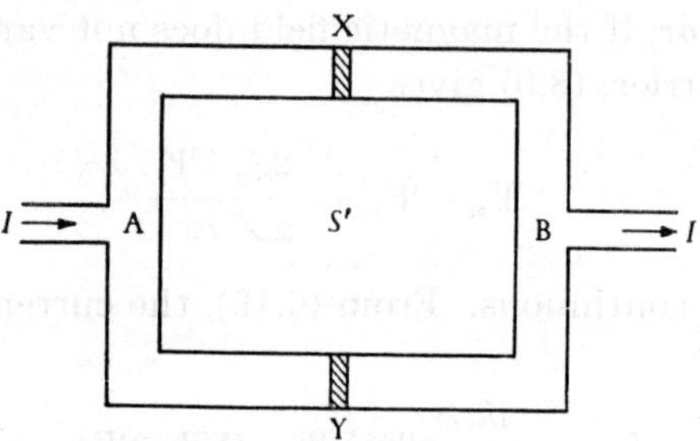

FIG. 8.4. Two Josephson junctions in parallel.

If an electrostatic potential difference ϕ is applied across the barrier, the difference between the phases of the Ginzburg–Landau functions in L and R will vary with time,

$$\frac{\partial\delta}{\partial t} = \varepsilon\phi/\hbar,$$

$$=\omega,$$

say. Then, from (8.12), the Josephson current I will be oscillatory with angular frequency ω. *A steady voltage produces an alternating current*. This a.c. Josephson effect has been observed experimentally (Shapiro *et al.* 1964).

In the presence of an external magnetic field, a Josephson junction responds in a very interesting manner. To illustrate the principles, let us look not at a single junction of finite spatial extension but rather at two point junctions X and Y in parallel (Fig. 8.4). More precisely, the junctions are to be assumed so small that the magnetic flux *in* either junction is negligible but the superconducting wires are thick enough to allow us to neglect the thickness λ of the penetration layer.

Well inside the wire the supercurrent density $J_{\mathrm{S}} = 0$. Hence, from (6.20), the phase φ of the Ginzburg–Landau wave function is related to the vector potential:

$$\nabla\varphi = \frac{\varepsilon}{\hbar c}\mathbf{A}. \qquad (8.13)^{(26)}$$

Hence, integrating along the contour XBY,

$$\varphi_{\mathrm{R}}(\mathrm{X}) - \varphi_{\mathrm{R}}(\mathrm{Y}) = \frac{\varepsilon}{\hbar c}\int_{\mathrm{XBY}} \mathbf{A}.\mathrm{dl}, \qquad (8.14\mathrm{a})^{(26)}$$

and integrating along the contour XAY,

$$\varphi_{\mathrm{L}}(\mathrm{X}) - \varphi_{\mathrm{L}}(\mathrm{Y}) = \frac{\varepsilon}{\hbar c}\int_{\mathrm{XAY}} \mathbf{A}.\mathrm{dl}. \qquad (8.14\mathrm{b})^{(26)}$$

Subtracting these equations gives

$$\begin{aligned}\varphi_{\mathrm{R}}(\mathrm{X}) - \varphi_{\mathrm{L}}(\mathrm{X}) + \varphi_{\mathrm{L}}(\mathrm{Y}) - \varphi_{\mathrm{R}}(\mathrm{Y}) &= \frac{\varepsilon}{\hbar c}\oint \mathbf{A}.\mathrm{dl} \\ &= \frac{\varepsilon}{\hbar c}\Phi \\ &= 2\pi\Phi/\Phi_0. \qquad (8.15)^{(26)}\end{aligned}$$

It is convenient to write

$$\left.\begin{aligned}\varphi_{\mathrm{R}}(\mathrm{X}) - \varphi_{\mathrm{L}}(\mathrm{X}) &= \delta + \pi\Phi/\Phi_0, \\ \varphi_{\mathrm{R}}(\mathrm{Y}) - \varphi_{\mathrm{L}}(\mathrm{Y}) &= \delta - \pi\Phi/\Phi_0.\end{aligned}\right\} \qquad (8.16)$$

The supercurrent from A to B is the sum of the Josephson currents through the two junctions X and Y. From (8.12), it is

$$\begin{aligned} I &= I_{\mathrm{C}}\{\sin(\varphi_{\mathrm{R}}(\mathrm{X}) - \varphi_{\mathrm{L}}(\mathrm{X})) + \sin(\varphi_{\mathrm{R}}(\mathrm{Y}) - \varphi_{\mathrm{L}}(\mathrm{Y}))\} \\ &= I_{\mathrm{C}}\{\sin(\delta + \pi\Phi/\Phi_0) + \sin(\delta - \pi\Phi/\Phi_0)\} \\ &= 2I_{\mathrm{C}} \sin\delta \cos\pi\Phi/\Phi_0. \qquad (8.17)\end{aligned}$$

For a given flux Φ, the current is maximized by choosing the phases so that $\sin\delta = 1$. This current is seen to be a periodic function of the flux Φ through the loop, and is a maximum when the flux is an integer multiple of the flux quantum Φ_0. The double junction is acting as an interferometer, which detects the interference between the coherent superconducting electrons propagating along the two alternative paths AXB and AYB. These predictions have been verified experimentally by Jaklevic *et al.* (1964).

This interferometer provides an exceedingly sensitive magnetometer; if the area enclosed by the double junction is S', successive current maxima occur at fields differing by Φ_0/S'. But S' can easily be made as large as 10^{-2} cm, so that maxima are only $\sim$ 20 microgauss apart!

The properties of a single extended junction have been studied by Ferrell and Prange (1963); see also de Gennes (1966). Ferrell and Prange show that the insulating barrier behaves rather like a Type II superconductor with an extremely large effective penetration depth

$$\lambda_{\text{eff}} \sim (\lambda\xi_0/\mathscr{T})^{\frac{1}{2}}, \tag{8.18}$$

where λ and ξ_0 are the penetration depth and Pippard length of the superconducting material, and $\mathscr{T}$ is the tunnelling transmission coefficient for single quasiparticles. Typically, $\lambda \sim \xi_0 \sim 10^{-5}$ cm and, for a barrier 10 to 15 atoms thick, $\mathscr{T} \sim 10^{-8}$. Thus $\lambda_{\text{eff}} \sim 10^{-1}$ cm.

PART II

MICROSCOPIC THEORY

CHAPTER 9

SECOND QUANTIZATION

ONE of the difficulties in developing the microscopic theory of superconductivity was the extreme smallness of the condensation energy ($H_c^2/8\pi \sim 10^4$ erg cm^{-3} for $H_c \sim 500$ gauss) compared with the Fermi energy $\mu \sim 10^{11}$ erg cm^{-3}. But in spite of the smallness of this energy difference, many of the properties of the superconducting phase are very different from those of the normal phase.

Bloch's (1928) theory of conductivity gives an adequate account of the properties of normal metals, but for many years it was believed that a *different* mechanism would be needed for superconductivity. One of the limitations of the Bloch theory is its neglect of the Coulomb correlations between the electrons in a metal. Heisenberg (1947, 1948) and Koppe (1947) very reasonably sought the mechanism of superconductivity in this Coulomb interaction. Born and Cheng (1948) suggested that the Brillouin zone structure was also important. These early theories postulated that the ground state carried a current. However, this postulate violates a theorem of Bloch (see Bohm 1949) that in the ground state of any quantum-mechanical system the current is zero.

Fröhlich (1950, 1952) was the first to observe that the Bloch theory can be formulated as a field theory, with an explicit electron–phonon interaction Hamiltonian. This interaction scatters electrons, and leads to the usual Bloch expression for the resistance. But the interaction has other consequences as well: (a) it leads to a 'self-energy' correction to the single-electron states,† because the electron interacts with the lattice which in turn reacts back on the electron; (b) the lattice, polarized by one electron, may interact with a second electron, leading to an effective electron–electron interaction.‡ Fröhlich attributed superconductivity to this electron–electron interaction.

In order to describe systems of particles interacting through the mediation of a field, the formalism of second quantization is almost indispensable. This formalism is not essentially difficult, but it is still unfamiliar to many pure and applied physicists. A brief account will follow.

9.1. Normal modes of a lattice

The formalism of second quantization was constructed to discuss the quantization of a field (e.g. the electromagnetic field) defined at all points of a continuum. However, it will be sufficient here to consider the simpler problem of the elastic vibrations of a lattice of discrete point masses, coupled by Hookean springs. Such a lattice, in two dimensions, is illustrated in Fig. 9.1. To simplify the discussion still further, we shall restrict ourselves to longitudinal displacements in a one-dimensional lattice containing an odd number N of identical masses M. Born–Karman cyclic boundary conditions will be imposed.

If x_j is the displacement of the jth atom from its equilibrium position and K is the spring constant, the equations of motion are

$$M\ddot{x}_j = K^2(x_{j+1}-2x_j+x_{j-1}), \tag{9.1}$$

with end conditions

$$x_1 = x_{N+1}.$$

† The self energy of a particle is the change in its energy resulting from the interaction with the field. The concept was first used in quantum electrodynamics.

‡ From the standpoint of field theory, this electron–electron interaction is described as arising from the exchange of virtual phonons. This description is analogous to the rather more familiar field-theoretical description of nuclear forces in terms of exchange of mesons, and Coulomb forces in terms of exchange of photons.

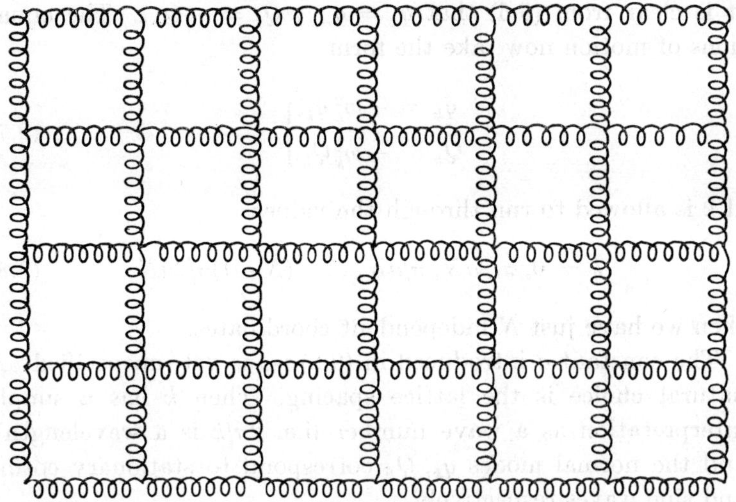

Fig. 9.1. Coupled-harmonic-oscillator model for a lattice in two dimensions.

The transformation to normal coordinates

$$\xi_k = N^{-\frac{1}{2}} \sum_j x_j \, e^{ijka} \tag{9.2}$$

yields the simple equations of motion

$$\ddot{\xi}_k = -\omega_k^2 \xi_k, \tag{9.3}$$

where

$$\omega_k^2 = 2(K^2/M)(1-\cos ka). \tag{9.4}$$

The equations of motion (9.3) have the form of harmonic oscillator equations. However the ξ_k, as defined by (9.2), are complex quantities. They have to satisfy the relations

$$\xi_k^* = \xi_{-k} \tag{9.5}$$

in order that the x_j shall be real. In other words the ξ_k, ξ_{-k} are not mutually independent. We therefore introduce the *real* normal coordinates ($k \neq 0$)

$$\left.\begin{aligned} q_k &= (\xi_k + \xi_{-k})/\sqrt{2}, \\ Q_k &= -i(\xi_k - \xi_{-k})/\sqrt{2}. \end{aligned}\right\} \tag{9.6}$$

It is clear from (9.6) that $q_k = q_{-k}$; $Q_k = -Q_{-k}$. The equations of motion now take the form

$$\left.\begin{aligned}\ddot{q}_k &= -\omega_k^2\, q_k,\\ \ddot{Q}_k &= -\omega_k^2\, Q_k.\end{aligned}\right\} \tag{9.7}$$

If k is allowed to run through the values

$$k = 0,\ \pi/2aN,\ \pi/aN,\ \ldots,\ (N-1)\pi/2aN, \tag{9.8}$$

then we have just N independent coordinates.

The constant a introduced in (9.2) is, as yet, unspecified. A natural choice is the lattice spacing. Then k has a simple interpretation as a wave number (i.e. $2\pi/k$ is a wavelength), and the normal modes q_k, Q_k correspond to stationary cosine and sine waves respectively.

From the equations of motion (9.7), it is straightforward to infer a Hamiltonian

$$\hat{H}_L = \tfrac{1}{2}\sum_k (\Pi_k^2/M + M\omega_k^2 Q_k^2 + \pi_k^2/M + M\omega_k^2 q_k^2), \tag{9.9}$$

where the canonical momenta π_k, Π_k are $M\dot{q}_k$, $M\dot{Q}_k$ respectively.

Exercise. Construct a Lagrangian $\mathscr{L} = \hat{H}_{\text{kin}} - V$ which yields the equations of motion (9.7). Calculate the canonical momenta $\pi_k = \partial\mathscr{L}/\partial\dot{q}_k$, $\Pi_k = \partial\mathscr{L}/\partial\dot{Q}_k$, and hence show that the Hamiltonian $\hat{H}_L = \Sigma_k(\pi_k\dot{q}_k + \Pi_k\dot{Q}_k) - \mathscr{L}$ has the form (9.9). Show how the mass M in (9.9) can be suppressed by a canonical transformation.

It has been assumed above that the displacements were longitudinal; the calculation for transverse displacements proceeds similarly. In a three-dimensional lattice normal modes can be introduced in the same way, but the secular equation is much more complicated than (9.4). In general there are three polarizations for a given wave vector **k**. But, except in special crystallographic directions, these polarizations are neither transverse nor longitudinal. Following Debye, we assume for simplicity that the polarization directions *are* longitudinal and transverse, and that the 'acoustic' approximation holds, i.e.

$$\omega_{\mathbf{k},\mathbf{e}} = ks_{\mathbf{e}}, \quad k \leqslant k_0. \tag{9.10}$$

Here $\mathbf{e}$ is a polarization vector ($\|$ or $\perp$), $s_{\mathbf{e}}$ is the velocity of acoustic waves of polarization $\mathbf{e}$, and k_0 is the Debye cut-off wave number, chosen so as to give the correct number of normal modes.

9.2. Quantum mechanics of a harmonic oscillator

In § 9.1 the vibrations of a lattice were described in terms of a set of independent harmonic oscillators. But that account was entirely classical; we now have to quantize the system. Consider first a single oscillator, with Hamiltonian

$$\hat{H} = \tfrac{1}{2}(\pi^2+\omega^2\xi^2). \tag{9.11}$$

Since π and ξ are canonically conjugate variables, the system is quantized by requiring

$$[\xi, \pi] \equiv \xi\pi - \pi\xi = \mathrm{i}\hbar. \tag{9.12}$$

In a Schrödinger representation, (9.12) implies

$$\pi = -\mathrm{i}\hbar\frac{\partial}{\partial\xi}. \tag{9.13}$$

By direct substitution into the Schrödinger equation

$$\hat{H}\phi = E\phi, \tag{9.14}$$

it can be verified that

$$\phi_0 = C\mathrm{e}^{-\alpha\xi^2} \tag{9.15}$$

is an eigenfunction of $\hat{H}$, with eigenvalue

$$E_0 = \tfrac{1}{2}\hbar\omega. \tag{9.16}$$

The coefficient α must be chosen to be

$$\alpha = \omega/2\hbar. \tag{9.17}$$

Moreover, since ϕ_0 has no nodes, it is the ground state.

It is convenient to define the non-Hermitian operators

$$\left.\begin{aligned}\hat{b} &= (\pi - \mathrm{i}\omega\xi)/\sqrt{(2\hbar\omega)},\\ \hat{b}^{+} &= (\pi + \mathrm{i}\omega\xi)/\sqrt{(2\hbar\omega)}.\end{aligned}\right\} \tag{9.18}$$

It may be verified (by direct substitution and the use of (9.12)) that

$$[\hat{b}, \hat{b}^{+}] = 1. \tag{9.19}$$

Moreover, in terms of $\hat{b}$, $\hat{b}^{+}$, the Hamiltonian (9.11) can be written in any of the alternative forms:†

$$\begin{aligned}\hat{H} &= \tfrac{1}{2}\hbar\omega(\hat{b}^{+}\hat{b} + \hat{b}\hat{b}^{+})\\ &= \hbar\omega(\hat{b}^{+}\hat{b} + \tfrac{1}{2})\\ &= \hbar\omega(\hat{b}\hat{b}^{+} - \tfrac{1}{2}).\end{aligned} \tag{9.20}$$

The property of the operators $\hat{b}$ and $\hat{b}^{+}$ which makes them particularly useful is that they are respectively lowering (or 'annihilation') and raising (or 'creation') operators. By this we mean that if ϕ_n is an eigenstate of $\hat{H}$ with energy E_n, then $\hat{b}^{+}\phi_n$ is an eigenstate with energy $E_n + \hbar\omega$. Likewise, $\hat{b}\phi_n$ is either zero or an eigenstate with energy $E_n - \hbar\omega$. The proof follows: Define

$$\hat{\mathscr{H}} = \hat{H}/\hbar\omega = \hat{b}^{+}\hat{b} + \tfrac{1}{2}. \tag{9.21}$$

Then

$$\hat{\mathscr{H}}\phi_n = (E_n/\hbar\omega)\phi_n. \tag{9.22}$$

Hence

$$\begin{aligned}\hat{\mathscr{H}}\hat{b}^{+}\phi_n &= (\hat{b}^{+}\hat{b} + \tfrac{1}{2})\hat{b}^{+}\phi_n\\ &= (\hat{b}^{+}\hat{b}\hat{b}^{+} + \tfrac{1}{2}\hat{b}^{+})\phi_n\\ &= \hat{b}^{+}(\hat{\mathscr{H}} + 1)\phi_n\end{aligned}$$

(from (9.20))

$$= \hat{b}^{+}(E_n/\hbar\omega + 1)\phi_n$$

(from (9.22))

$$= (E_n/\hbar\omega + 1)\hat{b}^{+}\phi_n.$$

That is

$$\hat{H}\hat{b}^{+}\phi_n = (\mathrm{E}_n + \hbar\omega)\hat{b}^{+}\phi_n, \tag{9.23}$$

† To supply a partial motivation for the use of the $\hat{b}$, $\hat{b}^{+}$ operators, note that, if $\hat{b}$ and $\hat{b}^{+}$ were allowed to commute, $\hat{H}$ would be factorized: $\hat{H} = \hbar\omega\hat{b}^{+}\hat{b}$.

and the first part of the theorem is proved. The second part is proved similarly, with the lowering operator $\hat{b}$ in place of the raising operator $\hat{b}^+$.

Thus the $\hat{b}$, $\hat{b}^+$ operators generate a 'ladder' of states, starting from any eigenstate ϕ_n. It can be shown (see, e.g. Dirac 1958 pp. 137–9) that the ladder is complete, i.e. starting from *any* eigenstate ϕ_n, we generate *all* the eigenstates by repeated application of $\hat{b}$ or $\hat{b}^+$. Furthermore the Hamiltonian (9.11) is manifestly positive definite and hence the ladder must possess a lowest rung (with non-negative eigenvalue). In fact ϕ_0 is this lowest state; $\hat{b}\phi_0$ is not a new eigenstate, but is identically zero:

$$\begin{aligned}\hat{b}\phi_0 &= \surd(2\hbar\omega)\,(\pi - \mathrm{i}\omega\xi)\exp(-\omega\xi^2/2\hbar^2)\\ &= \surd(2\hbar\omega)\,(-\mathrm{i}\hbar\partial/\partial\xi - \mathrm{i}\omega\xi)\exp(-\omega\xi^2/2\hbar^2)\\ &\equiv 0.\end{aligned} \tag{9.24}$$

Hence the energy eigenvalues are given by

$$E_n = (n+\tfrac{1}{2})\hbar\omega. \tag{9.25}$$

The *number operator* $\hat{n}$ is defined by

$$\hat{n} \equiv \hat{b}^+\hat{b}. \tag{9.26}$$

It is appropriately named, since

$$\begin{aligned}\hat{n}\phi_n &= \hat{b}^+\hat{b}\phi_n\\ &= (\hat{\mathscr{H}} - \tfrac{1}{2})\phi_n\\ &= n\phi_n.\end{aligned} \tag{9.27}$$

By a similar proof, it can be shown that

$$\hat{b}\hat{b}^+\phi_n \equiv (\hat{n}-1)\phi_n = (n-1)\phi_n.$$

Thus the Schrödinger equation (9.14) has been solved by the use of the ladder operators $\hat{b}$, $\hat{b}^+$; the eigenstates are

$$\phi_n = \mathscr{A}_n(\hat{b}^+)^n\phi_0, \tag{9.28}$$

with eigen-energies $(n+\frac{1}{2})\hbar\omega$. All that remains is to choose the coefficients $\mathscr{A}_n$ to normalize the wave functions.

Let ϕ_n be normalized, i.e.

$$\int |\phi_n|^2\, \mathrm{d}\xi = 1. \tag{9.29}$$

Then let

$$\hat{b}^+\phi_n = c_n\phi_{n+1}.$$

The norm of $\hat{b}^+\phi_n$ is

$$\begin{aligned}\int \mathrm{d}\xi\, (\phi_n^*\hat{b})\, (\hat{b}^+\phi_n) &= \int \mathrm{d}\xi\, \phi_n^*(\hat{b}\hat{b}^+)\phi_n \\ &= \int \mathrm{d}\xi\, \phi_n^*(\hat{\mathscr{H}}+\tfrac{1}{2})\phi_n \\ &= \int \mathrm{d}\xi\, \phi_n^*(\hat{n}+1)\phi_n \\ &= (n+1)\int \mathrm{d}\xi\, \phi_n^*\phi_n \\ &= n+1. \end{aligned} \tag{9.30}$$

Hence $c_n = \sqrt{(n+1)}$, i.e.

$$\hat{b}^+\phi_n = \sqrt{(n+1)}\phi_{n+1}. \tag{9.31}$$

Hence, applying (9.31) repeatedly, we find $\mathscr{A}_n$:

$$(\hat{b}^+)^n\phi_0 = \sqrt{1}.\sqrt{2}.\sqrt{3}\ldots\sqrt{n}\,\phi_n, \tag{9.32}$$

or

$$\mathscr{A}_n = (n!)^{-\frac{1}{2}}. \tag{9.33}$$

By a proof similar to that of (9.31), it can be shown that

$$\hat{b}\phi_n = \sqrt{n}\,\phi_{n-1}. \tag{9.34}$$

Before leaving the problem of a single oscillator, it will be convenient to calculate the matrix elements of $\hat{b}$ and $\hat{b}^+$; these will be required later. We define the matrix elements, as usual, by

$$\left.\begin{aligned}(\hat{b}^+)_{n'n} &= \int \mathrm{d}\xi\, \phi_{n'}^*\hat{b}^+\phi_n\,; \\ (\hat{b})_{n'n} &= \int \mathrm{d}\xi\, \phi_{n'}^*\hat{b}\phi_n. \end{aligned}\right\} \tag{9.35}$$

From (9.31) and (9.34) it follows that

$$\left.\begin{aligned}(\hat{b}^{+})_{n'n} &= \sqrt{(n+1)}\int \mathrm{d}\xi\, \phi^{*}_{n'}\phi_{n+1}\\ &= \sqrt{(n+1)}\ \delta_{n',n+1};\\ (\hat{b})_{n'n} &= \sqrt{n}\int \mathrm{d}\xi\, \phi^{*}_{n'}\phi_{n-1}\\ &= \sqrt{n}\ \delta_{n',n-1},\end{aligned}\right\} \tag{9.36}$$

where δ is the usual Kronecker symbol.

These matrix elements have a simple physical interpretation. In processes in which quanta of a field are emitted or absorbed, the transition matrix element contains (9.36) as a factor. Thus emission and absorption probabilities are respectively proportional to $|(\hat{b}^{+})_{n,n+1}|^2$ and $|(\hat{b})_{n,n-1}|^2$, i.e. to $n+1$ and n. Since the spontaneous emission probability means the excess of emission probability over absorption probability, it follows that the spontaneous emission rate is independent of the number of quanta present (cf. Einstein's (1917) A and B coefficients).

9.3. Quantization of a lattice-displacement field

The classical normal-mode analysis of § 9.1 led to the Hamiltonian (9.9)—a set of independent harmonic oscillators. These can be quantized by the procedure of § 9.2. First, however, it is convenient to introduce a new set of variables describing progressive waves, instead of using the standing-wave variables q_k, Q_k and their conjugates. From (9.6),

$$\left.\begin{aligned}\xi_k &= (1/\sqrt{2})(q_k+\mathrm{i}Q_k),\\ \xi_{-k} &= (1/\sqrt{2})(q_k-\mathrm{i}Q_k).\end{aligned}\right\} \tag{9.37}$$

From (9.37), it follows at once that

$$\left.\begin{aligned}\partial/\partial\xi_k &= (1/\sqrt{2})(\partial/\partial q_k - \mathrm{i}\partial/\partial Q_k),\\ \partial/\partial\xi_{-k} &= (1/\sqrt{2})(\partial/\partial q_k + \mathrm{i}\partial/\partial Q_k),\end{aligned}\right\} \tag{9.38}$$

whence

$$\partial\xi_{-k}/\partial\xi_k = \partial\xi_k/\partial\xi_{-k} = 0. \tag{9.39}$$

The commutation rules for the Hermitian variables q_k, Q_k, π_k, Π_k are

$$\left.\begin{aligned}[q_k, \pi_l] &= [Q_k, \Pi_l] = i\hbar\delta_{kl},\\ [q_k, q_l] &= [\pi_k, \pi_l] = [Q_k, Q_l] = [\Pi_k, \Pi_l] = 0,\\ [q_k, Q_l] &= [\pi_k, \Pi_l] = [q_k, \Pi_l] = [Q_k, \pi_l] = 0,\end{aligned}\right\} \tag{9.40}$$

since, by (9.9), the different degrees of freedom are independent. From (9.39) and (9.40), it follows that

$$\left.\begin{aligned}[\xi_k, \partial/\partial\xi_l] &= [\xi_{-k}, \partial/\partial\xi_{-l}] = -\delta_{kl},\\ [\xi_k, \xi_l] &= \quad [\xi_k, \xi_{-l}] = 0\\ [\partial/\partial\xi_k, \partial/\partial\xi_l] &= [\partial/\partial\xi_k, \partial/\partial\xi_{-l}] = 0.\end{aligned}\right\} \tag{9.41}$$

Exercise. Using (9.39) and (9.40), verify that

$$[\xi_k, \xi_{-k}] = [\partial/\partial\xi_k, \xi_{-k}] = [\partial/\partial\xi_k, \partial/\partial\xi_{-k}] = 0,$$

and hence that (9.41) is correct for all choices of k and l.

Let us now define the set of variables

$$\left.\begin{aligned}\hat{b}_k &= \sqrt{(\omega M/2\hbar)}\, \xi_k + \sqrt{(\hbar/2M\omega)}\, \partial/\partial\xi_{-k},\\ \hat{b}_k^+ &= \sqrt{(\omega M/2\hbar)}\, \xi_k - \sqrt{(\hbar/2M\omega)}\, \partial/\partial\xi_{-k}.\end{aligned}\right\} \tag{9.42}$$

In contrast to (9.8), $2aNk/\pi$ is now allowed to assume *all* integer values between $-N+1$ and $N-1$:

$$\begin{aligned}k = -(N-1)\pi/2Na, \ldots, -\pi/2Na, 0,\\ \pi/2Na, \ldots, (N-1)\pi/2Na.\end{aligned} \tag{9.43}$$

The $\hat{b}$'s satisfy the commutation rules

$$\left.\begin{aligned}[\hat{b}_k, \hat{b}_l] &= [\hat{b}_k^+, \hat{b}_l^+] = 0,\\ [\hat{b}_k, \hat{b}^+] &= \delta_{kl}.\end{aligned}\right\} \tag{9.44}$$

These are the natural generalization of (9.19) to a system of many degrees of freedom; (9.19) holds for any *one* normal mode, while variables belonging to different modes commute. In the new variables, the Hamiltonian (9.9) takes the form

$$\hat{H}_{\mathrm{L}} = \sum_k \hat{H}_k = \sum_k \hbar\omega_k(\hat{b}_k^+\hat{b}_k + \tfrac{1}{2}), \tag{9.45}$$

a sum of oscillator Hamiltonians (9.20). Number operators $\hat{n}_k$ can be introduced for each normal mode by (9.26), and the total number operator is

$$\hat{N} = \sum_k \hat{n}_k = \sum_k \hat{b}_k^+ \hat{b}_k. \tag{9.46}$$

The ground state of the system, or 'Boson vacuum', is the state in which *all* the oscillators are in their individual ground states; its wave function is

$$\prod_k \exp(-\omega_k \xi_k^2 / 2\hbar^2). \tag{9.47}$$

Evidently (9.47) is an eigenstate of $\hat{N}$ with the eigenvalue zero. Excited states are generated from (9.47) by the various creation operators; their normalization is given by a trivial extension of (9.33).

The generalization to three dimensions is immediate—the wave number k in (9.47) is replaced by a wave vector $\mathbf{k}$ and a polarization vector $\mathbf{e}_\mathbf{k}$ (cf. the account leading to (9.10)). In the limit of a continuum the formalism is unchanged except insofar as the *number* of normal modes increases without limit; the Debye cut-off wave number $k_0 \to \infty$. For a given eigenvalue N of the total number operator $\hat{N}$, the states of the system are in one-to-one correspondence with those of a system of N non-interacting Bosons with an energy spectrum $E_k = \hbar\omega_k$ (e.g. Dirac 1958 p. 227). If we regard $\hat{b}^+_{\mathbf{k},\mathbf{e}_\mathbf{k}}$ as creating a 'phonon' of wave vector $\mathbf{k}$, then these phonons have the properties of particles obeying Bose statistics.

We may regard the operators $\hat{b}^+_\mathbf{k}$, $\hat{b}_\mathbf{k}$ as the Fourier transforms of operators $\hat{\Psi}^+(\mathbf{r})$, $\hat{\Psi}(\mathbf{r})$, which are functions of the position variable $\mathbf{r}$:

$$\left.\begin{aligned} \hat{\Psi}^+(\mathbf{r}) &= \Omega^{-\frac{1}{2}} \sum_{\mathbf{k},\mathbf{e}_\mathbf{k}} \hat{b}^+_{\mathbf{k},\mathbf{e}_\mathbf{k}}\, \mathrm{e}^{-i\mathbf{k}.\mathbf{r}} \\ \hat{\Psi}(\mathbf{r}) &= \Omega^{-\frac{1}{2}} \sum_{\mathbf{k},\mathbf{e}_\mathbf{k}} \hat{b}_{\mathbf{k},\mathbf{e}_\mathbf{k}}\, \mathrm{e}^{i\mathbf{k}.\mathbf{r}}. \end{aligned}\right\} \tag{9.48}$$

Here Ω is the volume of the Born–Karman box. $\hat{\Psi}(\mathbf{r})$, $\hat{\Psi}^+(\mathbf{r})$ obey the commutation relations

$$\left.\begin{aligned} [\hat{\Psi}^+(\mathbf{r}), \hat{\Psi}^+(\mathbf{r}')] &= [\hat{\Psi}(\mathbf{r}), \hat{\Psi}(\mathbf{r}')] = 0, \\ [\hat{\Psi}(\mathbf{r}), \hat{\Psi}^+(\mathbf{r}')] &= \delta(\mathbf{r}-\mathbf{r}'), \end{aligned}\right\} \tag{9.49}$$

where $\delta(\mathbf{r}-\mathbf{r}')$ is a Dirac δ-function. As $\hat{b}_k^+$ creates a phonon of wave vector $\mathbf{k}$, we may think of $\hat{\Psi}^+$ as the operator creating a phonon at the point $\mathbf{r}$. In the same way as $\hat{b}_k^+\hat{b}_k$ gives the number of phonons with wave vector $\mathbf{k}$, so $\hat{\Psi}^+(\mathbf{r})\hat{\Psi}(\mathbf{r})$ gives the number density of phonons at the point $\mathbf{r}$. This should be compared with the interpretation of $\psi^*(\mathbf{r})\psi(\mathbf{r})$ as the probability density in ordinary quantum mechanics. The formalism of quantized fields may be derived rigorously by a process of quantization of classical fields (see e.g. Mandl 1959). Applying this method to the wave function ψ of a single-particle system reproduces the above formalism. Thus this formalism is a generalization of the quantum mechanics of a single particle to a system of many non-interacting particles obeying Bose statistics.

The formalism described here is often called second quantization, since we may regard the passage from classical mechanics to the Schrödinger equation as a 'first' quantization, in which p and q do not commute although the wave function ψ and its conjugate ψ^* do. In 'second' quantization ψ itself becomes an operator, obeying the commutation rules (9.49).

Following Dirac (1958 p. 138), we shall denote the 'vacuum' state (9.47) by the ket symbol $|0\rangle$ and its conjugate $\langle 0|$. This state is a product of factors, one for each normal mode. These individual factors can conveniently be written $|0_{\mathbf{k},\mathbf{e_k}}\rangle$. The nth excited state of the $(\mathbf{k},\mathbf{e_k})$th mode can be written explicitly in terms of creation operators. A convenient notation replaces the zero in the ket symbol by $n_{\mathbf{k},\mathbf{e_k}}$. Thus

$$|n_{\mathbf{k},\mathbf{e_k}}\rangle = (n_{\mathbf{k},\mathbf{e_k}}!)^{-\frac{1}{2}}(\hat{b}^+_{\mathbf{k},\mathbf{e_k}})^{n_{\mathbf{k},\mathbf{e_k}}}|0\rangle. \tag{9.50}$$

9.4. Second quantization for Fermions

A collection of particles obeying Bose–Einstein statistics has a wave function which is symmetrical under the interchange of any two particles. In second-quantized notation, this symmetry is reflected in, for example, the identity of the states $\hat{b}^+_{\mathbf{k},\mathbf{e_k}}\hat{b}^+_{\mathbf{l},\mathbf{e_l}}|0\rangle$ and $\hat{b}^+_{\mathbf{l},\mathbf{e_l}}\hat{b}^+_{\mathbf{k},\mathbf{e_k}}|0\rangle$ (where $|0\rangle$ is the Boson vacuum).

It is well known that the wave function for a system of N Fermions can be written† in Slater's (1929) determinantal form

$$\psi(\mathbf{r}_1, \mathbf{r}_2, \ldots \mathbf{r}_N;\ \sigma_1, \ldots \sigma_N) = \det \psi_{\bar{k}_1}(\mathbf{r}_1)\, \psi_{\bar{k}_2}(\mathbf{r}_2) \ldots \psi_{\bar{k}_N}(\mathbf{r}_N). \tag{9.51}$$

Here $\psi_{\bar{k}}(\mathbf{r}) = e^{i\mathbf{k}.\mathbf{r}}\chi(\sigma)$ is the wave function of a single free Fermion with energy $\hbar^2k^2/2m$. The wave function (9.51) is antisymmetric against interchange of particles, and therefore automatically satisfies the Pauli exclusion principle. In the presence of an external potential $V(\mathbf{r})$, the wave function is still of Slater form. However, the one-particle functions $\psi_{\bar{k}}(\mathbf{r})$ are now the eigenstates of the Schrödinger equation

$$-\hbar^2\nabla^2\psi/2m + V(\mathbf{r})\psi = \varepsilon(\bar{k})\psi.$$

When interactions are present, (9.51) is no longer an eigenstate of the N-body Hamiltonian. Nevertheless the states (9.51) still form a complete orthogonal set, i.e. *any* N-body state can be written as a linear combination of states (9.51). In § 9.3, we saw how to describe a system of non-interacting Bosons by means of creation and annihilation operators. In this section an analogous formalism will be developed for Fermions. As in the previous section, the new description regards the total number of particles as an operator; the number of Fermions is not specified *ab initio*. In other words all states of the form (9.51), for *all possible* numbers of particles, are required as a basis. In more mathematical language, we require to span not merely the Hilbert space of N Fermions, but rather the set of *all* N-particle Hilbert spaces, for all values of N. This product space is often called the Fock space.

Let us denote the Fermion vacuum (or state of no particles) $|0\rangle$, and suppose that it can be decomposed into a product of empty one-particle states $|0_{\bar{k}}\rangle$ by analogy with the Boson

† For Fermions, the one-particle states require both a momentum label $\mathbf{k}$ and a spin label σ representing the z-component of the spin angular momentum. For brevity, the symbol $\bar{k}$ will be used for the two labels $(\mathbf{k}, \sigma)$.

vacuum. It is necessary to take cognizance of the order of factors in this decomposition of $|0\rangle$. Let us therefore define some arbitrary order

$$\bar{k}_1, \bar{k}_2, \ldots, \bar{k}_n, \ldots \tag{9.52}$$

to be the 'standard' order, and write

$$|0\rangle = |0_{\bar{k}_1}\rangle|0_{\bar{k}_2}\rangle \ldots |0_{\bar{k}_n}\rangle \ldots \tag{9.53}$$

States containing particles will have some of the factors $|0_{\bar{k}}\rangle$ replaced by factors $|1_{\bar{k}}\rangle$.

We wish to introduce creation and annihilation operators for Fermions, to be denoted $\hat{a}^+_{\bar{k}}$, $\hat{a}_{\bar{k}}$ respectively. They must have the following properties:

$$\left.\begin{array}{lll}
\text{(a)} & \hat{a}^+_{\bar{k}}\,|\ldots\rangle|0_{\bar{k}}\rangle|\ldots\rangle = |\ldots\rangle|1_{\bar{k}}\rangle|\ldots\rangle, \\
\text{(b)} & \hat{a}^+_{\bar{k}}\,|\ldots\rangle|1_{\bar{k}}\rangle|\ldots\rangle = 0, \\
\text{(c)} & \hat{a}_{\bar{k}}|\ldots\rangle|0_{\bar{k}}\rangle|\ldots\rangle = 0, \\
\text{(d)} & \hat{a}_{\bar{k}}|\ldots\rangle|1_{\bar{k}}\rangle|\ldots\rangle = |\ldots\rangle|0_{\bar{k}}\rangle|\ldots\rangle.
\end{array}\right\} \tag{9.54}$$

Equation (9.54b) is the manifestation of the Pauli principle; not more than one particle† can be in any state $\bar{k}$. It follows from (9.54b) and (9.54c) that

$$(\hat{a}_{\bar{k}})^2 = (\hat{a}^+_{\bar{k}})^2 = 0. \tag{9.55}$$

To construct a two-particle state, operate on $|0\rangle$ with two creation operators; if $\bar{k}$ is to the left of $\bar{l}$ in the standard order (9.52), write

$$|0_{\bar{k}_1}\rangle|0_{\bar{k}_2}\rangle \ldots |1_{\bar{k}}\rangle \ldots |1_{\bar{l}}\rangle \ldots = \hat{a}^+_{\bar{k}}\hat{a}^+_{\bar{l}}|0\rangle. \tag{9.56}$$

To achieve antisymmetry, we require

$$\hat{a}^+_{\bar{l}}\hat{a}^+_{\bar{k}}|0\rangle = -\hat{a}^+_{\bar{k}}\hat{a}^+_{\bar{l}}|0\rangle. \tag{9.57}$$

† No normalization factors analogous to (9.31) are required in (9.54), because $n_{\bar{k}}$ cannot exceed unity.

In other words, the interchange of the two creation operators is equivalent to the interchange of $\bar{k}$ and $\bar{l}$ in the Slater determinant (9.51).

Equations (9.54) and (9.57) are satisfied if the $\hat{a}^{+}_{\bar{k}}$, $\hat{a}_{\bar{k}}$ operators satisfy the *anti*commutation rules

$$\left.\begin{aligned} \{\hat{a}_{\bar{k}}, \hat{a}^{+}_{\bar{l}}\}_{+} &\equiv \hat{a}_{\bar{k}}\hat{a}^{+}_{\bar{l}} + \hat{a}^{+}_{\bar{l}}\hat{a}_{\bar{k}} = \delta_{\bar{k},\bar{l}}, \\ \{\hat{a}_{\bar{k}}, \hat{a}_{\bar{l}}\}_{+} &= \quad \{\hat{a}^{+}_{\bar{k}}, \hat{a}^{+}_{\bar{l}}\}_{+} = 0, \end{aligned}\right\} \tag{9.58}$$

instead of the Boson commutation rules (9.44).

Exercise. Derive (9.55) and (9.57) from (9.58).

As in the Boson case (9.46), the number operators

$$\hat{n}_{\bar{k}} \equiv \hat{a}^{+}_{\bar{k}}\hat{a}_{\bar{k}}, \quad \hat{N}_{\mathrm{e}} \equiv \sum_{\bar{k}} \hat{n}_{\bar{k}}, \tag{9.59}$$

may be defined. The appropriateness of the definition follows by observing that, from (9.54),

$$\left.\begin{aligned} \hat{n}_{\bar{k}}|0_{\bar{k}}\rangle &\equiv \hat{a}^{+}_{\bar{k}}\hat{a}_{\bar{k}}|0_{\bar{k}}\rangle = 0, \\ \hat{n}_{\bar{k}}|1_{\bar{k}}\rangle &\equiv \hat{a}^{+}_{\bar{k}}\hat{a}_{\bar{k}}|1_{\bar{k}}\rangle = |1_{\bar{k}}\rangle, \end{aligned}\right\} \tag{9.60}$$

i.e. $\hat{n}_{\bar{k}}$ has eigenvalues 0 and 1, representing the number of particles in the state $\bar{k}$.

In the new formalism the Fermi sea at zero temperature has a state vector

$$\prod_{k<k_F} (\hat{a}^{+}_{\bar{k}})|0\rangle, \tag{9.61}$$

where the product is taken over all one-particle states $\bar{k}$ with momentum $\hbar k$ below the Fermi surface $\hbar k_F$. There is an ambiguity in the sign of (9.61), in so far as an interchange of any two factors in the product leads to a sign change (by (9.58)). This is exactly analogous to the sign ambiguity in the Slater determinant (9.51), and reflects the antisymmetry of the wave function. For definiteness we shall interpret the $\Pi_{\bar{k}}$ to refer to the product with the $\bar{k}$'s in the standard order (9.52) unless otherwise specified.

As in the Boson system, the Hamiltonian for a system of non-interacting Fermions is

$$\hat{H} = \sum_{k} \varepsilon(\bar{k})\hat{n}_{\bar{k}} = \sum_{k} \varepsilon(\bar{k})\, \hat{a}_{\bar{k}}^{+}\, \hat{a}_{\bar{k}}. \tag{9.62}$$

Here $\varepsilon(\bar{k})$ is the kinetic energy $\hbar^2k^2/2m$ of a particle in the state $\bar{k}$. More generally, the Hamiltonian can still be brought to a form like (9.62) when the particles move in an external potential $V(\mathbf{r})$. But now $\hat{a}_{\bar{k}}^{+}$, $\hat{a}_{\bar{k}}$ must be interpreted as creation and annihilation operators for a particle in the $\bar{k}$th eigenstate of the one-particle Schrödinger equation; $\varepsilon(\bar{k})$ is the corresponding one-particle energy. The interpretation of (9.62) is straightforward: the energy of the system is the sum of all the single-particle energies, weighted with the number of particles in each of the one-particle states. In the Fermion case, $\hat{n}_{\bar{k}}$ only has eigenvalues 0 and 1, while in the Boson case the eigenvalues are zero and all the positive integers.

9.5. Interaction between particles: scattering

The previous two sections have described the formalism of creation and annihilation operators for non-interacting particles. For example, for Fermions there is a one-to-one correspondence between the states $\Pi_{\bar{k}}(\hat{a}_{\bar{k}}^{+})|0\rangle$ and the Slater-determinant states in conventional notation. But unless interactions can be included in the new notation the formalism remains empty of physical content.

In conventional notation, a typical spin-independent interaction Hamiltonian for a system of N particles might be of the form

$$\hat{H}_{\text{int}} = \tfrac{1}{2}\sum_{i \neq j} V(|\mathbf{r}_i - \mathbf{r}_j|). \tag{9.63}$$

This particular form, with instantaneous two-body forces only, will serve as an illustration. An interaction such as (9.63) can be characterized by its matrix elements between Slater-determinant states:†

$$(\hat{H}_{\text{int}})_{k,l;\bar{p},\bar{q}} = \frac{1}{2\Omega}\int d^3r_1\, d^3r_2\, V(|\mathbf{r}_1-\mathbf{r}_2|)\exp\{i(\mathbf{k}-\mathbf{q}).\mathbf{r}_1+i(\mathbf{l}-\mathbf{p}).\mathbf{r}_2\}$$

$$= \frac{1}{2\Omega}\sum_{\mathbf{Q}} \mathscr{V}_{\mathbf{Q}}\delta_{\mathbf{k}-\mathbf{q},\mathbf{Q}}\delta_{\mathbf{p}-\mathbf{l},\mathbf{Q}}, \tag{9.64}$$

where

$$\mathscr{V}_{\mathbf{Q}} = \frac{1}{\Omega}\int d^3r\, V(\mathbf{r})\exp(i\mathbf{Q}.\mathbf{r}). \tag{9.65}$$

Thus the effect of H_{int} can be described in terms of two-particle scattering processes, in which a momentum $\mathbf{Q}$ is transferred from one particle to the other. In second-quantized form we may write

$$\hat{H}_{\text{int}} = \frac{1}{2\Omega}\sum_{k,l,\mathbf{Q}} \mathscr{V}_{\mathbf{Q}}\, \hat{a}^{+}_{l-\mathbf{Q}}\hat{a}_{l}\hat{a}^{+}_{k+\mathbf{Q}}\hat{a}_{k}. \tag{9.66}$$

This clearly has the same matrix elements as (9.63) above—these matrix elements are, of course, to be taken between states $\Pi_{\text{occupied }k}(\hat{a}_k)|0\rangle$.

The interaction (9.66), in terms of the Fourier transforms (9.48) and (9.65), is

$$\hat{H}_{\text{int}} = \tfrac{1}{2}\sum_{\sigma,\sigma'}\int d^3r\, d^3r'\, \hat{\Psi}^{+}_{\sigma}(\mathbf{r})\hat{\Psi}_{\sigma}(\mathbf{r})V(|\mathbf{r}-\mathbf{r}'|)\hat{\Psi}^{+}_{\sigma'}(\mathbf{r}')\hat{\Psi}_{\sigma'}(\mathbf{r}'), \tag{9.67}$$

where

$$\left.\begin{aligned}\hat{\Psi}^{+}_{\sigma}(\mathbf{r}) &= \Omega^{-\frac{1}{2}}\sum_{\mathbf{k}} \hat{a}^{+}_{k}\, e^{-i\mathbf{k}.\mathbf{r}}\\ \text{and}\qquad \hat{\Psi}_{\sigma}(\mathbf{r}) &= \Omega^{-\frac{1}{2}}\sum_{\mathbf{k}} \hat{a}_{k}\, e^{i\mathbf{k}.\mathbf{r}}.\end{aligned}\right\} \tag{9.68}$$

† We have already noted that the Slater-determinant wave functions form a complete set. To verify the expression (9.64), expand the determinantal wave functions in terms of their 2 × 2 minors from the $\bar{k}$th and $\bar{l}$th columns (and the conjugate wave function in terms of the minors from the $\bar{p}$th and $\bar{q}$th rows). The term containing $-\delta_{\mathbf{k}-\mathbf{p},\mathbf{Q}}\,\delta_{\mathbf{q}-\mathbf{l},\mathbf{Q}}$ vanishes if the momenta are arranged in the standard order (since if $\bar{k}$ precedes l and $\bar{p}$ precedes $\bar{q}$ then $\bar{k}=\bar{p}$ implies $\bar{q}\neq\bar{l}$).

Interpreting $\hat{\Psi}_\sigma^+(\mathbf{r})\hat{\Psi}_\sigma(\mathbf{r})$ as the density of spin-σ particles at the point $\mathbf{r}$, we have a simple interpretation of (9.67). The integrand is the product of the two-body interaction potential $\frac{1}{2}V(|\mathbf{r}-\mathbf{r}'|)$ and the density of particles at $\mathbf{r}$ and at $\mathbf{r}'$.

Of course, (9.66) admits of generalization to velocity-dependent interactions; $\mathscr{V}_\mathbf{Q}$ is merely replaced by a function of *two* momentum variables. Another straightforward generalization is to the case of an external potential in addition to the

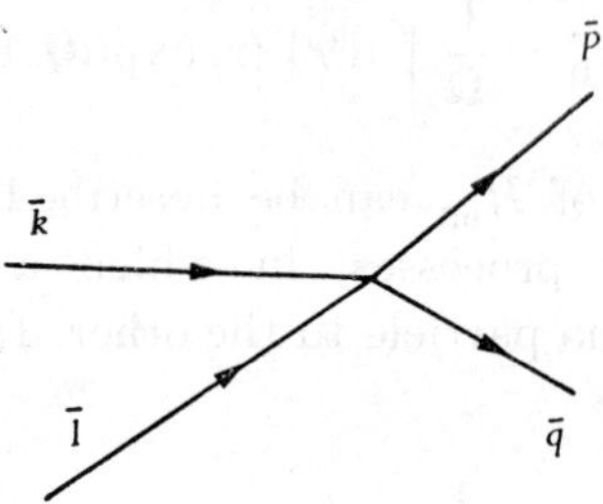

FIG. 9.2. Scattering of two Fermions. The direction of increasing time is to the right.

interaction. In this case we define $\hat{a}_{\bar{k}}^+$, $\hat{a}_{\bar{k}}$ operators which create and annihilate electrons in states which are eigenstates of the one-particle Schrödinger equation. The δ-functions in (9.64) are the manifestation of momentum conservation. In the presence of an external potential, these δ-functions are replaced by more complicated functions, since momentum is conserved only if the Hamiltonian is translation-invariant. For example, in a crystal the suffixes $\bar{k}$, $\bar{l}$, $\bar{p}$, $\bar{q}$ will be 'quasimomenta', and Umklapp collisions exist for which $\mathbf{k}+\mathbf{l} \neq \mathbf{p}+\mathbf{q}$ (see p. 173).

The most general two-body scattering, written in second quantization, is

$$\tfrac{1}{2}\sum_{\bar{k},\bar{l};\bar{p},\bar{q}} V(\bar{k},\, \bar{l};\;\; \bar{p},\, \bar{q})\, \hat{a}_{\bar{p}}^+ \hat{a}_{\bar{l}} \hat{a}_{\bar{q}}^+ \hat{a}_{\bar{k}}, \tag{9.69}$$

where we allow the possibility of spin-flip. Figure 9.2 illustrates a convenient diagrammatic representation of a scattering process. The direction of time is taken from left to right.

Two particles are shown entering the collision in the states $\bar{k}$, $\bar{l}$ and leaving in the states $\bar{p}$, $\bar{q}$.

It is clear from the above discussion that any interaction that can be written in conventional notation can also be written in second quantization. But note that it is also possible to write second-quantized interactions which do *not* have a conventional equivalent. Thus, at least formally, interactions can be constructed *which do not conserve the number* of particles. All that is required to do so is that the number of $\hat{a}^+$ operators should differ from the number of $\hat{a}$ operators.

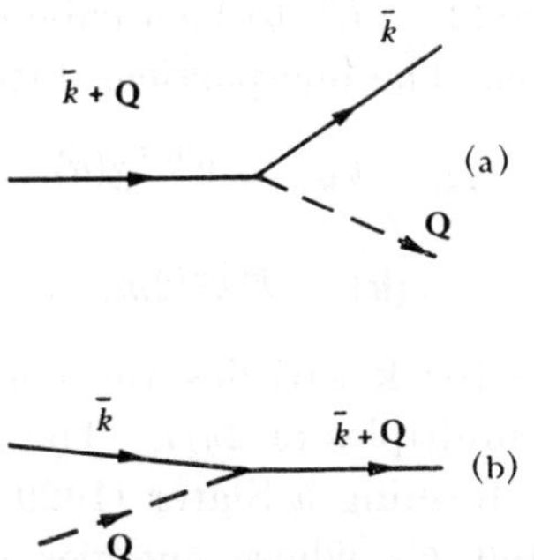

FIG. 9.3. Scattering of a Fermion by (a) emission and (b) absorption of a Boson.

Although the number of *electrons* in a non-relativistic system *is* conserved, it is useful to invent other entities (quasiparticles, see Chapter 12) which are not conserved. Thus for example phonons (the quanta of lattice vibration) may be emitted or absorbed in electron-scattering processes (equation (10.34)). These processes may be represented by such expressions as

$$\sum_{\bar{k},\mathbf{Q}} (\mathscr{C}_{\bar{k},\mathbf{Q}} \hat{a}^+_{\bar{k}} \hat{a}_{\bar{k}+\mathbf{Q}} \hat{b}^+_{\mathbf{Q}} + \mathscr{C}^*_{\bar{k},\mathbf{Q}} \hat{a}^+_{\bar{k}+\mathbf{Q}} \hat{a}_{\bar{k}} \hat{b}_{\mathbf{Q}}). \tag{9.70}$$

These processes are illustrated diagrammatically in Fig. 9.3. Here $\hat{b}^+_{\mathbf{Q}}$, $\hat{b}_{\mathbf{Q}}$ are (Bose–Einstein) creation and annihilation operators for phonons, and $\mathscr{C}_{\bar{k},\mathbf{Q}}$ is the matrix element whose absolute square gives the probability for the process. Note that (9.70) is Hermitian, as it ought to be.

CHAPTER 10

THE BLOCH–FRÖHLICH ELECTRON–PHONON INTERACTION

10.1. The Sommerfeld model

In Sommerfeld's (1928) model of a metal, the conduction electrons are treated as non-interacting particles free to move in a potential well of constant depth. For simplicity, consider the specimen, of volume $\Omega = L^3$, to be a cube with Born–Karman boundary conditions. The one-particle states are plane waves

$$\psi_{\bar{k}} \equiv \psi_{\mathbf{k},\sigma} = e^{i\mathbf{k}.\mathbf{r}}\chi(\sigma), \tag{10.1}$$

with energies

$$\varepsilon(\mathbf{k}) = \hbar^2 k^2/2m, \tag{10.2}$$

where the wave vector $\mathbf{k}$ satisfies the condition that k_x, k_y, and k_z are integral multiples of $2\pi/L$. The ground state of the system is found by forming a Slater (1929) determinant, containing all permitted $\bar{k}$'s whose energies do not exceed the Fermi energy μ_0.

Exercise. Find the one-particle states in a box whose shape is a rectangular parallelepiped, and where the boundary condition is that $\psi = 0$ on the walls.

If N_e is the number of conduction electrons per unit volume, then†

$$N_e\Omega = \sum_{\substack{\varepsilon(k)<\mu_0 \\ \sigma=\pm}} 1 = 2\Omega \int \Xi(\varepsilon)\,d\varepsilon. \qquad (10.3)\dagger$$

† In (10.3), the assumption is made that the levels are so closely spaced that the sum over $\mathbf{k}$ may be replaced by an integral:

$$\Sigma_{\mathbf{k}} \ldots \to (2\pi)^{-3}\Omega \int d^3k \ldots = \Omega \int \Xi(\varepsilon)\,d\varepsilon \ldots.$$

Bloch's model, to be discussed in § 10.3, may be regarded as a generalization of Sommerfeld's model, in which (10.2) does not hold. Instead $\varepsilon(\mathbf{k})$ is a more complicated function of $\mathbf{k}$. Correspondingly, (10.4) is modified. Equations which are still valid under these more general assumptions are indicated by a dagger throughout this section.

Here $2\Xi(\varepsilon)$ is the density of states, i.e. the number of states per unit volume per unit energy range. It is convenient to write the factor 2, for the sum over spin orientations, explicitly. In fact in a Sommerfeld metal,

$$\Xi(\varepsilon) \equiv (2\pi)^{-2}k^2\left\{\left(\frac{\mathrm{d}\varepsilon(k)}{\mathrm{d}k}\right)^{-1}\right\}_{k=k_{\mathrm{F}}} = \tfrac{3}{4}N_{\mathrm{e}}\varepsilon^{\frac{1}{2}}\mu_0^{-\frac{3}{2}}. \tag{10.4}$$

Hence, by (10.3),

$$\mu_0 = \left(\frac{3N_{\mathrm{e}}}{\pi}\right)^{\frac{2}{3}}\frac{\pi^2\hbar^2}{2m}. \tag{10.5}$$

The total energy of the conduction electrons is

$$E_0 = \sum_{\substack{\varepsilon(k)<\mu_0 \\ \sigma=\pm}} \varepsilon(k) = 2\Omega\int \Xi(\varepsilon)\,\varepsilon\,\mathrm{d}\varepsilon \qquad (10.6)\dagger$$

$$= \tfrac{3}{5}N_{\mathrm{e}}\Omega p_{\mathrm{F}}^2/2m, \tag{10.7}$$

where the Fermi momentum $p_F = \hbar k_F$ is defined by

$$\mu_0 = \varepsilon(k_F).$$

At finite temperatures the distribution function is

$$f(\bar{k}) = \langle\hat{n}(\bar{k})\rangle_{\mathrm{av}} = \left\{1+\exp\frac{\varepsilon(k)-\mu}{\mathrm{k}T}\right\}^{-1}, \qquad (10.8)\dagger$$

where the chemical potential μ takes the value μ_0 as $T \rightarrow 0$. To second order in $\mathrm{k}T/\mu_0$, μ is

$$\mu = \mu_0 - \frac{\pi^2}{6}\left\{\frac{(\mathrm{k}T)^2}{\Xi}\frac{\mathrm{d}\Xi}{\mathrm{d}\varepsilon}\right\}_{\varepsilon=\mu_0}. \qquad (10.9)\dagger$$

The Sommerfeld model accounts for the linear term in the specific heat of metals. The internal energy U at a temperature T is

$$U = \sum_{\bar{k}} f(\bar{k})\,\varepsilon(\bar{k}) = 2\Omega\int \Xi(\varepsilon)\,f(\varepsilon)\,\varepsilon\,\mathrm{d}\varepsilon, \qquad (10.10)\dagger$$

† See footnote, p. 164.

so that the specific heat at constant volume (per unit mass) is

$$c_V = \frac{1}{M}\frac{\partial U}{\partial T} = \frac{2}{3}\pi^2 k^2 T \Xi(\mu_0)/\rho. \quad (10.11)\dagger$$

However the success of Sommerfeld's theory is limited. The model does not give any criterion for classifying solids as metals or non-metals. Nor can it account for the finite electrical conductivity of metals. On a microscopic level there are two very obvious omissions: the neglect of (a) the Coulomb interaction between electrons, and (b) the lattice structure.

10.2. Coulomb interaction

The influence of the Coulomb interaction has been the subject of much recent work. The main results are:

(a) There exist collective vibrations of the electron gas ('plasma oscillations'). These are of high frequency,[73]

$$\omega_P^2 = 2\pi e^2 N_e/m,$$

i.e. $\hbar\omega_P \gtrsim 10$ eV, so that normally these degrees of freedom of the electron gas are not excited.

(b) The Coulomb interaction between any pair of electrons is *screened* by the motion of all the other electrons. The effective interaction is one of quite short range instead of the very long range of the unscreened Coulomb forces.

The detailed discussion of Coulomb interactions is beyond the scope of this book.‡ A very rough-and-ready account of the screening length will be sufficient (Mott 1936, Mott and Jones 1936 p. 87).

Consider a charge e placed at the origin of coordinates. In the absence of all other charges, the electrostatic potential at the point $\mathbf{r}$ would be e/r. Let the potential in the presence of all the other electrons be $\phi(\mathbf{r})$ and let the number density of electrons at $\mathbf{r}$ be $n(\mathbf{r})$.

† See footnote, p. 164.

‡ For a fuller account of this topic, see e.g. Pines (1961 pp. 55–65).

In Thomas–Fermi approximation, the number density is given by

$$n(\mathbf{r}) = (3\pi^2\hbar^3)^{-1}(2me\phi+p_{\mathrm{F}}^2)^{\frac{3}{2}}. \qquad (10.12)$$

A second relation between n and ϕ follows from Poisson's equation:

$$\nabla^2\phi = -4\pi e(n-n_0), \qquad (10.13)^{(74)}$$

where n_0 is the unperturbed number density

$$n_0 = k_{\mathrm{F}}^3/3\pi^2 = p_{\mathrm{F}}^3/3\pi^2\hbar^3. \qquad (10.14)$$

We assume

$$\phi \ll p_{\mathrm{F}}^2/2me. \qquad (10.15)$$

Eliminating n between (10.12) and (10.13), retaining only the lowest order in ϕ, we have

$$\nabla^2\phi = 4\pi e\frac{(2me)^{\frac{3}{2}}}{3\pi^2\hbar^3}\{(\phi+p_{\mathrm{F}}^2/2me)^{\frac{3}{2}}-(p_{\mathrm{F}}^2/2me)^{\frac{3}{2}}\}$$

$$= \frac{4me^2p_{\mathrm{F}}}{\pi\hbar^3}\phi\,\{1+O(2me\phi/p_{\mathrm{F}}^2)\}. \qquad (10.16)^{(75)}$$

If the right-hand side of (10.16) were zero, the equation would reduce to Laplace's equation; the solution for a point source of strength e at the origin is then of course[76] $\phi = e/r$. It is readily verified that

$$\phi = \frac{e}{r}\exp(-\kappa r) \qquad (10.17)^{(76)}$$

is the corresponding solution of (10.16), where the screening length κ^{-1} is given by

$$\kappa^2 = 4me^2p_{\mathrm{F}}/\pi\hbar^3. \qquad (10.18)^{(77)}$$

Thus the Coulomb interactions are screened, i.e. they are confined to an effective range κ^{-1}. At electronic densities typical of metals, the numerical value of the inverse screening length is $\kappa \gtrsim k_{\mathrm{F}} = p_{\mathrm{F}}/\hbar$. The Thomas–Fermi screening length is thus rather less than the mean nearest-neighbour distance.

The above argument has neglected the *unperturbed* electrostatic interaction between the electrons. To see this, note that when $n = n_0$ the solution of (10.13) is $\phi = 0$. Of course the electrostatic potential of a system of negative charges can vanish only in the limit of infinite separation; the total Coulomb energy is always positive. In fact the Coulomb energy per unit volume will depend on the number of electrons, and will diverge as the number of electrons becomes infinite. This divergence is unphysical. It arises from our neglect of the positive ions, whose role is to maintain overall electrical neutrality. The role of the lattice will be discussed in the next section. For the present let us try to make the minimum modification of the Coulomb-gas model. The simplest way out of the dilemma is to validate (10.13) *a posteriori*, by postulating a uniform (and rigid) background of positive charge whose density is so chosen as to guarantee the overall neutrality. We are thus led to the so-called 'jellium' model, whose properties are those described above.

10.3. Electrons in a rigid ionic lattice

For a particle moving in a three-dimensional periodic potential†

$$V(\mathbf{r}) = V(\mathbf{r}+n_1\mathbf{a}_1+n_2\mathbf{a}_2+n_3\mathbf{a}_3), \tag{10.19}$$

it can be shown that the eigenstates of the Schrödinger equation exhibit a band structure (Fig. 10.1). The wave functions are of the form (Bloch 1928)

$$\psi_{\bar{k}}(\mathbf{r}) = u_{\mathbf{k}}(\mathbf{r})e^{i\mathbf{k}.\mathbf{r}}\chi(\sigma), \tag{10.20}$$

where $u_{\mathbf{k}}(\mathbf{r})$ is a periodic function with the periodicity of the lattice, and where the **k** can be either real or purely imaginary. The bands of 'allowed' states have real **k** and are separated by 'forbidden' energy regions with imaginary **k**. The forbidden energy states are clearly *surface* states, since from (10.20) their wave functions are evanescent in character. They will be of no further interest here.

† In (10.19), n_1, n_2, n_3 are integers; $\mathbf{a}_1$, $\mathbf{a}_2$, and $\mathbf{a}_3$ are the three independent periods.

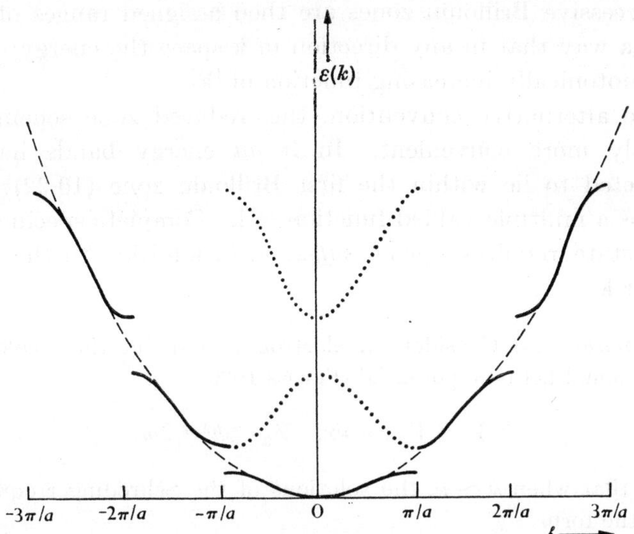

Fig. 10.1. Energy spectrum for an electron in a one-dimensional periodic lattice. The broken curve is the free-electron parabola, the solid curve is the spectrum in the extended zone scheme, and the dotted curve is the spectrum in the reduced zone scheme.

The Bloch factorization (10.20) is of course not unique. If **K** is any vector of the reciprocal lattice† then $e^{i\mathbf{k}.\mathbf{r}}$ has the periodicity of the lattice. Equation (10.20) can therefore be rewritten

$$\psi_{k}(\mathbf{r}) = \{u_{\mathbf{k}}(\mathbf{r})\, e^{i\mathbf{K}.\mathbf{r}}\}\, e^{i(\mathbf{k}-\mathbf{K}).\mathbf{r}}\,\chi(\sigma), \tag{10.21}$$

which is once more of Bloch form. To eliminate the ambiguity a convention must be introduced; two alternatives are in common use. In the first, the 'extended zone scheme', the **k** are so chosen that the *lowest* energy band has

$$0 < \mathscr{K}_i < 1, \tag{10.22}$$

where **k** has been written $\mathscr{K}_1\mathbf{b}_1+\mathscr{K}_2\mathbf{b}_2+\mathscr{K}_3\mathbf{b}_3$.

† We define the reciprocal lattice by the three periods $\mathbf{b}_i (i = 1, 2, 3)$,

$$\mathbf{b}_i.\mathbf{a}_i = 2\pi, \tag{10.23}$$

and write $\mathbf{K} = \sum \nu_i \mathbf{b}_i$, where the ν are integers.

Successive Brillouin zones are then assigned ranges of **k** in such a way that in any direction in **k**-space the energy $\varepsilon(\mathbf{k})$ is a monotonically increasing function of $|\mathbf{k}|$.

The alternative convention, the 'reduced zone scheme', is slightly more convenient. In it *all* energy bands have **k** restricted to lie within the first Brillouin zone (10.22); $\varepsilon(\mathbf{k})$ is thus a multiple-valued function of **k**. Complete specification of a state requires a *band suffix*, n, in addition to the wave vector **k**.

Exercises. (a) Consider an electron moving in the weak one-dimensional periodic potential (Peierls 1930)

$$V = V_0 \cos \kappa x, \quad V_0 \ll \hbar^2\kappa^2/2m.$$

Show that when $k \simeq \kappa$, the solutions of the Schrödinger equation have the form

$$\psi = C_1 \mathrm{e}^{\mathrm{i}kx} + C_2 \mathrm{e}^{\mathrm{i}(k-\kappa)x},$$

and that the energy is

$$E(k) = \tfrac{1}{2}[\epsilon(k)+\epsilon(k-\kappa) \pm \sqrt{\{(\epsilon(k)-\epsilon(k-\kappa))^2+V_0^2\}}],$$

where

$$\epsilon(k) = \hbar^2k^2/2m.$$

(b) Find the Bloch wave functions for an electron of energy $0 < E(k) < V_0$, moving in the one-dimensional periodic potential (Kronig and Penney 1931)

$$\begin{aligned} V &= V_0, \qquad n(a+b) < x < (n+1)a+nb, \\ &= 0, \quad (n-1)a+nb < x < n(a+b). \end{aligned}$$

Show that the energy spectrum is given by the solutions of the transcendental equation

$$\cos k(a+b) = \cosh \beta b \cos \alpha a + \frac{\beta^2-\alpha^2}{2\alpha\beta} \sinh \beta b \sin \alpha a,$$

where

$$\alpha^2 = 2mE/\hbar^2, \; \beta^2 = 2m(V_0-E)/\hbar^2.$$

It can be shown, in the reduced zone notation, that near a zone boundary the energy is a quadratic function of the wave vector:

$$\varepsilon(\mathbf{k}) = C + \sum_{i,j} \hbar^2 (m^{*-1})_{ij} k_i k_j. \tag{10.24}$$

The coefficients $(m^{*-1})_{ij}$ in (10.24) are the components of a tensor whose inverse is called the 'effective mass'. For a cubic lattice, $(m^{*-1})_{ij}$ reduces to $m^{*-1}\delta_{ij}$. The sign of m^* is negative near the top of a band and positive near the bottom.

An improved independent-electron model of a metal may be constructed using the Bloch states (10.20). As in the Sommerfeld model (§ 10.1), the many-electron wave function is a Slater determinant. But the one-electron wave functions out of which the determinant is built are Bloch functions. We can even take the Coulomb interactions into account in a Hartree approximation, since the self-consistent field will itself be periodic with the periodicity of the lattice.

This band model contains a classification of solids into metals and insulators. An insulator has an energy gap Δ between two bands at the Fermi surface, such that in its ground state all the one-electron states in the lower ('valence') band are full and all the states in the next band ('conduction band') are empty. If the energy gap is fairly small ($\Delta \lesssim kT$) then there will be some electrons thermally excited above the gap, and we have a semiconductor. In a metal the Fermi surface is *inside* a band (or several over-lapping bands), so that even at zero temperature there are electrons capable of being accelerated by an applied electric field. Thus we arrive back at something very like the Sommerfeld model, though the electron states are of Bloch form and the energy spectrum is no longer just $\hbar^2k^2/2m$. But we still have an ideal Fermi gas of electrons (or of 'holes', if a band is nearly full). The finiteness of the electrical resistivity is still unexplained.

10.4. Scattering of electrons by lattice vibrations

In a perfect lattice, electrons will not be scattered—the mechanism of resistance has to be sought in deviations from strict periodicity. Randomly distributed lattice defects *will* scatter electrons. This scattering gives rise to the residual resistance (cf. (1.1)). At very low temperatures these defects cannot anneal, so that the residual resistance is independent of temperature as $T \to 0$. The temperature-dependent part of the resistance arises from the vibrations of the crystal lattice. These vibrations constitute time-dependent departures from perfect periodicity. In this section Bloch's calculation of the scattering probabilities will be outlined, but the resistance will not be calculated.† Let $\mathbf{r}$ be the undisturbed coordinates of an element of the medium, and let $\mathbf{x}$ be its instantaneous displacement in the course of a lattice vibration. Then the potential seen by an electron will be modified by an amount

$$\begin{aligned} U(\mathbf{r}) &= V(\mathbf{r}) - V(\mathbf{r} - \mathbf{x}(\mathbf{r})) \\ &= -\mathbf{x}(\mathbf{r}) \,.\, \mathrm{grad}\, V(\mathbf{r}). \end{aligned} \tag{10.25}$$

Equation (10.25) assumes that the ions deform in such a way that the perturbed potential at the displaced position is the same as the original potential at the undisplaced position. This assumption is not essential; Nordheim (1931) proposed instead a rigid-ion model. The results of the rigid-ion model do not differ essentially from those of Bloch's model.‡ Nordheim's model will therefore not be discussed.

To find the scattering of an electron by a single lattice mode $\mathbf{q}$, we note that the generalization of (9.2) to a continuous displacement variable $\mathbf{x}$ is

$$\mathbf{x}(\mathbf{r}) = \xi_{\mathbf{q}} \tilde{\mathbf{q}} \exp(-\mathrm{i}\mathbf{q} \,.\, \mathbf{r}). \tag{10.26}$$

In (10.26), $\tilde{\mathbf{q}} = \mathbf{q}/q$ is the unit vector in the direction of propagation. From (10.25), we note that only longitudinal displacements interact with the electrons. In terms of creation and annihilation operators for phonons (cf. (9.42))

$$\xi_{\mathbf{q}} = \mathrm{i}(2M\omega/\hbar)^{\frac{1}{2}}(\hat{b}_{\mathbf{q}}^{+} - \hat{b}_{-\mathbf{q}}). \tag{10.27}$$

† For a full and lucid account of Bloch's theory, see Sommerfeld and Bethe (1933).

‡ A thorough investigation by Bardeen (1937) suggests that Bloch's model is closer to the truth than Nordheim's. The only effect of Nordheim's assumption here would be to modify the definition of Bloch's constant C in (10.29); see e.g. Wilson (1953 § 9.3).

The matrix element of (10.25) for scattering of a single electron from a state $\bar{k}$ to a state $\bar{k}+\mathbf{Q}$ with a change of the number of phonons from $n_{\mathbf{q}}^{(\mathrm{i})}$ to $n_{\mathbf{q}}^{(\mathrm{f})}$ is

$$\int \mathrm{d}^3r \int \mathrm{d}\xi_{\mathbf{q}_1} \ldots \langle \ldots n_{\mathbf{q}}^{(\mathrm{f})} \ldots |\psi_{\bar{k}+\mathbf{Q}}^{*}(\mathbf{r})\, U(\mathbf{r})\, \psi_{\bar{k}}(\mathbf{r})| \ldots n_{\mathbf{q}}^{(\mathrm{i})} \ldots \rangle. \tag{10.28}$$

By (9.36), (10.28) vanishes unless $n_{\mathbf{q}}^{(\mathrm{i})} = n_{\mathbf{q}}^{(\mathrm{f})}$, $\mathbf{q} \neq \mathbf{Q}$. The $\mathbf{r}$-integration is approximately performed by expressing the $\psi_{\bar{k}}(\mathbf{r})$ in the form (10.20) and remembering that $V(\mathbf{r})$ and the $u_{\mathbf{k}}(\mathbf{r})$ have the periodicity of the lattice. The integral over space may therefore be expressed in terms of an integral over a unit cell. Bloch's approximation consists in considering

$$C \equiv \frac{\hbar^2}{2m} \int_{\text{unit cell}} |\nabla u_{\mathbf{k}}(\mathbf{r})|^2 \, \mathrm{d}^3r \tag{10.29}$$

to be a constant, independent of $\bar{k}$. This assumption reduces the $\mathbf{r}$-integration to a factor

$$N\delta(\mathbf{Q}-\mathbf{q}+\mathbf{K}), \tag{10.30}$$

where $\mathbf{K}$ is any lattice vector in the reciprocal lattice (p. 169). Transitions with $\mathbf{K} \neq 0$ are termed 'Umklapprozesse' (Peierls 1932), and their role in conductivity theory has been carefully studied by Bardeen (1937). For the sake of a very considerable simplification we shall ignore them entirely, and shall consider only normal transitions, with $\mathbf{K} = 0$.

The detailed evaluation of (10.28) finally gives for the matrix elements

$$(\bar{k}, n_{\mathbf{Q}}|\, U(\mathbf{r})\, |\bar{k}\pm\mathbf{Q}, n_{\mathbf{Q}}\pm 1) = \tfrac{2}{3}C\left(\frac{\hbar\omega_{\mathbf{Q}}}{NM\Omega s_0^2}\right)\begin{Bmatrix} \sqrt{n_{\mathbf{Q}}} \\ \sqrt{(n_{\mathbf{Q}}+1)} \end{Bmatrix}. \tag{10.31}$$

Here Ω is the normalization volume, M is the ionic mass, and s_0 is the velocity of sound, unperturbed by the interaction between the electron and the lattice vibrations. We shall denote the observed velocity of sound as s. The upper alternative on the right-hand side of (10.31) goes with the upper sign on the left-hand side.

10.5. The Fröhlich Hamiltonian

Fröhlich (1950, 1952) showed that the Bloch theory of the interaction between electrons and lattice vibrations amounts to a field theory, and that it can be formulated in terms of a field-theoretical Hamiltonian. This application† of field-theoretical concepts revolutionized the conceptual framework of solid-state physics.

Let us assume a model Hamiltonian

$$\hat{H} = \hat{H}_{\text{el}} + \hat{H}_{\text{ph}} + \hat{H}_{\text{int}} + \hat{H}_{\text{Coul}}. \tag{10.32}$$

The electronic term $\hat{H}_{\text{el}}$ is assumed to be of the form (9.62), but the single-electron energies are not necessarily of the free-electron form $\hbar^2k^2/2m$. The phonon Hamiltonian $\hat{H}_{\text{ph}}$ is taken to be

$$\hat{H}_{\text{ph}} = \sum_{\mathbf{q}} \hbar\omega_{\mathbf{q}} \hat{b}_{\mathbf{q}}^{+} \hat{b}_{\mathbf{q}}, \tag{10.33}$$

where we have neglected the (constant) zero-point energy $\sum_{\mathbf{q}} \frac{1}{2}\hbar\omega_{\mathbf{q}}$. The transverse modes have also been omitted, since in Bloch's approximation (§ 10.4) only the longitudinal lattice-vibration modes interact with the electrons.

In second-quantized notation the Bloch interaction is

$$\hat{H}_{\text{int}} = \mathrm{i} \sum_{k,\mathbf{q}} D_{\mathbf{q}} (\hat{a}_{k-\mathbf{q}}^{+} \hat{a}_{k} \hat{b}_{\mathbf{q}}^{+} - \hat{a}_{k}^{+} \hat{a}_{k-\mathbf{q}} \hat{b}_{\mathbf{q}}), \tag{10.34}$$

where

$$D_{\mathbf{q}}^2 = \tfrac{4}{9} C^2 \hbar\omega_{\mathbf{q}} / NM\Omega s_0^2. \tag{10.35}$$

To confirm the correctness of (10.34), it is sufficient to verify that the matrix elements of $\hat{H}_{\text{int}}$ are correctly given by (10.31).

Finally, we have included in (10.32) a Coulomb term whose form and role will be discussed later (p. 190).

† Strictly this was the second application of field theory to solid-state physics. The first was that of Fröhlich *et al.* (1950) on the motion of an electron in an ionic medium (the polaron).

The Fermion operators $\hat{a}_k^+$, $\hat{a}_k$ are operators for Bloch electrons, i.e. they create and annihilate electrons in the states (10.20). But when (as is often approximately true) the single-particle energies $\varepsilon(k)$ are proportional to k^2, we may simplify the model and regard these states as free-particle states (i.e. plane waves) with an effective mass

$$m^* = \hbar^2 k^2/2\varepsilon(k). \tag{10.36}$$

But it should be noted that this 'free-electron' simplification may only be made *after* C has been evaluated. For, if we put $u_{\mathbf{k}}(\mathbf{r}) = 1$ in (10.29), it gives $C \equiv 0$.

The **q**th Fourier coefficient of the electron density is

$$\rho_{\mathbf{q}} = \sum_i \exp(\mathrm{i}\mathbf{q}\,.\,\mathbf{x}_i), \tag{10.37}$$

where the sum is taken over all electrons. In second-quantized notation it takes the form

$$\rho_{\mathbf{q}} = \sum_k \hat{a}_{k+\mathbf{q}}^+ \hat{a}_k. \tag{10.38}$$

Transforming back from **q**-space to configuration space, the local electron density is

$$\rho(\mathbf{r}) = \hat{\Psi}^+(\mathbf{r})\hat{\Psi}(\mathbf{r}). \tag{10.39}$$

In (10.39) the Fermion field operators $\hat{\Psi}^+(\mathbf{r})$, $\hat{\Psi}(\mathbf{r})$ are defined as

$$\left.\begin{aligned} \hat{\Psi}^+(\mathbf{r}) &= \Omega^{-\frac{1}{2}} \sum_k \hat{a}_k^+ \exp(-\mathrm{i}\mathbf{k}\,.\,\mathbf{r}), \\ \hat{\Psi}(\mathbf{r}) &= \Omega^{-\frac{1}{2}} \sum_k \hat{a}_k \exp(\mathrm{i}\mathbf{k}\,.\,\mathbf{r}), \end{aligned}\right\} \tag{10.40}$$

by analogy with the corresponding Boson operators (equation (9.48)). In terms of $\rho_{\mathbf{q}}$, $\hat{H}_{\text{int}}$ takes the form

$$\hat{H}_{\text{int}} = \mathrm{i}\sum_{\mathbf{q}} D_{\mathbf{q}}(\rho_{\mathbf{q}}\hat{b}_{\mathbf{q}}^+ - \rho_{\mathbf{q}}^+\hat{b}_{\mathbf{q}}). \tag{10.41}$$

Exercise. Derive (10.41) from

$$\hat{H}_{\text{int}} = -\int \mathrm{d}^3r\, \rho(\mathbf{r})\, \mathbf{x}(\mathbf{r})\,.\,\mathrm{grad}\, V(\mathbf{r}), \tag{10.42}$$

which follows from (10.25), on multiplying by $\rho(\mathbf{r})$ and integrating over space.

It is convenient to introduce Fröhlich's dimensionless coupling constant $\mathscr{F}$, related to $D_{\mathbf{q}}$ by

$$\mathscr{F} = \Omega \Xi_0 D_{\mathbf{q}}^2/\hbar\omega_{\mathbf{q}}. \tag{10.43}$$

Ξ_0 is the density of energy levels at the Fermi surface, given by

$$\Xi_0 = (2\pi)^{-2}\left\{k^2 \bigg/ \frac{\mathrm{d}\varepsilon(k)}{\mathrm{d}k}\right\}_{k=k_{\mathrm{F}}}, \tag{10.44}$$

i.e. the number of levels of one spin per unit energy range per unit volume. For the quasi-free electron model, Ξ_0 has the value

$$\Xi_0 = \tfrac{3}{4}N_{\mathrm{e}}/\mu_0, \tag{10.45}$$

where N_{e} is the number of free electrons per unit volume and μ_0 is the Fermi energy. $\mathscr{F}$ turns out to be $\lesssim \frac{1}{4}$ in real metals.

10.6. Qualitative indications for superconductivity

Fröhlich (1950) noticed that the electron–phonon interaction (10.34) implies an electron–electron interaction, and also a correction to the electronic energy levels. Although perturbation theory is not valid here, let us apply perturbative methods in the hope of getting some preliminary insight into how such effects can arise. In second order, the perturbation-theoretical correction to the energy is

$$\Delta E = -\sum_{\mathbf{k},\mathbf{q}} \frac{f_{\mathbf{k}}(1-f_{\mathbf{k}-\mathbf{q}})|D_{\mathbf{q}}|^2}{\varepsilon(\mathbf{k})-\varepsilon(\mathbf{k}-\mathbf{q})+\hbar\omega_{\mathbf{q}}}, \tag{10.46}$$

where $f_{\mathbf{k}} = \langle \hat{n}_{\mathbf{k}} \rangle_{\mathrm{av}} = \langle \hat{a}_{\mathbf{k}}^{+}\hat{a}_{\mathbf{k}} \rangle_{\mathrm{av}}$. (By using second-quantized notation we have automatically taken care of the Pauli exclusion principle.) Let us split ΔE into two parts

$$\Delta E = \Delta E_1 + \Delta E_2, \tag{10.47}$$

where

$$\Delta E_1 = -\sum_{\mathbf{k},\mathbf{q}} \frac{f_{\mathbf{k}}|D_{\mathbf{q}}|^2}{\varepsilon(\mathbf{k})-\varepsilon(\mathbf{k}-\mathbf{q})+\hbar\omega} \tag{10.48}$$

and

$$\Delta E_2 = \sum_{\bar{k},\mathbf{q}} \frac{f_{\bar{k}} f_{\bar{k}-\mathbf{q}} |D_{\mathbf{q}}|^2}{\varepsilon(\mathbf{k}) - \varepsilon(\mathbf{k}-\mathbf{q}) + \hbar\omega_{\mathbf{q}}}. \tag{10.49}$$

The motive for this separation is that ΔE_1 represents only a shift ('renormalization') of the one-electron energy levels, but ΔE_2 is an interaction between electrons.

The larger the coupling constant $\mathscr{F}$, the more likely it would appear that a ground state can be found differing from the Bloch ground state. Fröhlich showed that $\mathscr{F} \gtrsim 1$ led to an anomalous ground state, which he tentatively identified with the superconducting state. Thus he suggested a criterion for superconductivity: $\mathscr{F} > 1$.

From the role of $\mathscr{F}$ in the Bloch theory, large values of $\mathscr{F}$ imply that the normal resistance is high. Thus poorly conducting metals (e.g. lead, tin, mercury) should become superconducting, in preference to good conductors (e.g. copper, silver, sodium). On the whole the criterion agrees with experiment; however it is not quantitatively accurate. The BCS criterion (11.32) is not very different from Fröhlich's—no criterion yet exists which can predict reliably whether a given substance will be a superconductor, nor at what temperature T_c (see § 15.1).

The condensation energy can be roughly estimated by noting that most of the terms in the sum are small compared with those for which

$$|\varepsilon(\mathbf{k}) - \varepsilon(\mathbf{k}-\mathbf{q})| \lesssim \hbar\omega_{\mathbf{q}}. \tag{10.50}$$

Replace $\omega_{\mathbf{q}}$ by the Debye cut-off frequency ω_D, and restrict the summation to initial states within $\hbar\omega_D$ of the Fermi surface (since otherwise the final state will have a low probability of being unoccupied). Thus

$$\Delta E_2 \sim \sum_{\substack{|\varepsilon(\mathbf{k}-\mathbf{q})-\mu_0| < \hbar\omega_D \\ |\varepsilon(\mathbf{k})-\mu_0| < \hbar\omega_D}} \frac{D_{\mathbf{q}}^2}{\hbar\omega_D} \sim \frac{N_e \Omega C^2}{\mu_0} \cdot \frac{m}{M}. \tag{10.51}$$

The latter expression follows, remembering that the Debye cut off ω_D is comparable to $s_0 k_F$, and using (10.35).

Equation (10.51) contains an isotope effect. If ΔE_2 is identified with the zero-temperature condensation energy $H_0^2/8\pi$, (10.51) implies that, if M is the mean atomic mass,

$$H_0 \propto M^{-\frac{1}{2}}. \tag{10.52}$$

From the empirical law of corresponding states, H_0 is proportional to T_c, and hence

$$T_c \propto M^{-\frac{1}{2}}. \tag{10.53}$$

The result (10.53) is in excellent agreement with the experimental results of Reynolds *et al.* (1950) for mercury. A direct confirmation of (10.52), not depending on the law of corresponding states, is afforded by the experiments of Allen *et al.* (1950 *a*, *b*) for tin. The dependence of H_0 on M provides very strong evidence for the central role of electron–phonon interaction in superconductivity.

The magnitude of the predicted condensation energy, from (10.51), is smaller than the Fermi energy by a factor $\sim m/M$. But (10.51) is still too large, by a factor between 10 and 100, compared with the observed condensation energies.

Bardeen (1950) attempted to construct a variational theory using the electron–phonon interaction, but subsequently he proved (Bardeen 1951) that this theory was equivalent to Fröhlich's perturbation theory.

The perturbation theory suggests that $\mathscr{F}$ has a critical value $\mathscr{F}_1 \sim 1$, such that superconductivity will not occur if $\mathscr{F} < \mathscr{F}_1$. But if $\mathscr{F}$ were greater than unity, then the perturbation-theoretical correction to the velocity of sound would be negative and so large as to make the velocity of sound itself negative (Wentzel 1951). Huang (1951) showed that the Wentzel instability arises from neglecting the possibility that the lattice can adjust adiabatically to the motion of the electrons. Various authors have studied the problem of renormalization of the velocity of sound (Fröhlich 1952, Nakajima 1953, Bardeen and Pines 1955, Bogoliubov *et al.* 1959 Chapter 3). It emerges that the coupling constant relevant to superconductivity may be as large as we please without lattice instability occurring.

It will be shown (§ 11.2) that the effective coupling constant need not be large; in fact in most superconductors $\mathscr{F}$ is found to be between 0·1 and 0·2. We are therefore justified in using perturbation theory to study the renormalization. The more complicated strong-interaction methods are required only for lead and mercury.

The Fröhlich interaction has as its primitive process the emission or absorption of a phonon by an electron (10.34). Thus it is only in second order that an electron–electron interaction occurs—one electron has to emit a phonon, which a second electron absorbs. The process is illustrated diagramatically in Fig. 10.2. At non-zero temperatures, when thermal phonons

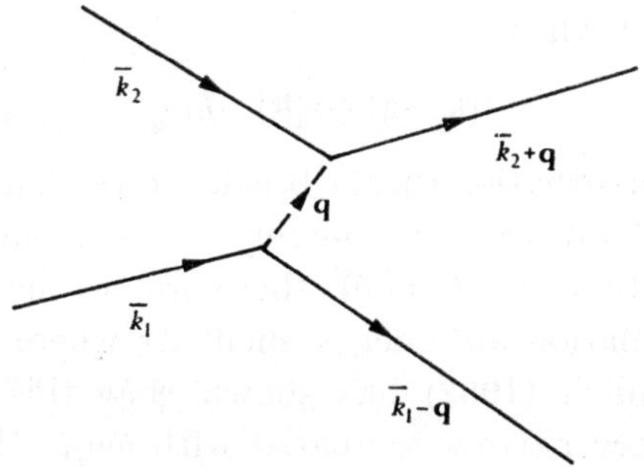

FIG. 10.2. Diagrammatic representation of the Fröhlich electron–electron interaction resulting from the exchange of a virtual phonon.

are present, it is of course possible for one electron to absorb a phonon, and the second electron to emit into the same mode. This process interferes destructively with the emission-followed-by-absorption process, but cannot cancel it completely; the difference is just the spontaneous emission probability. The perturbation-theoretical expression for the interaction is

$$\hat{H}' = \sum_{k,l,\mathbf{q}} \frac{\frac{1}{2}D_{\mathbf{q}}^2}{\varepsilon(\mathbf{k})-\varepsilon(\mathbf{k}-\mathbf{q})+\hbar\omega_{\mathbf{q}}} \hat{a}^+_{k+\mathbf{q}}\hat{a}_k\hat{a}^+_{l-\mathbf{q}}\hat{a}_l(\hat{b}_{\mathbf{q}}\hat{b}^+_{\mathbf{q}}-\hat{b}^+_{\mathbf{q}}\hat{b}_{\mathbf{q}})$$
$$= -\sum_{k,l,\mathbf{q}} \frac{\frac{1}{2}D_{\mathbf{q}}^2}{\varepsilon(\mathbf{k})-\varepsilon(\mathbf{k}-\mathbf{q})+\hbar\omega_{\mathbf{q}}} \hat{a}^+_{k+\mathbf{q}}\hat{a}_k\hat{a}^+_{l-\mathbf{q}}\hat{a}_l, \tag{10.54}$$

since, by (9.44), $\hat{b}_{\mathbf{q}}\hat{b}^{+}_{\mathbf{q}}-\hat{b}^{+}_{\mathbf{q}}\hat{b}_{\mathbf{q}} \equiv -1$. The factor $\frac{1}{2}$ is necessary in order not to count the process twice over, when we sum over all $\bar{k}$ and $\bar{l}$. With a little rearrangement, we may write

$$\hat{H}' = -\sum_{\bar{k},\mathbf{q}} \left(\frac{\mathscr{F}}{\Omega\Xi_0}\right) \frac{\hbar^2\omega_{\mathbf{q}}^2}{\{\varepsilon(\mathbf{k}-\mathbf{q})-\varepsilon(\mathbf{k})\}^2-\hbar^2\omega_{\mathbf{q}}^2} \rho_{\mathbf{q}}\hat{a}^{+}_{\bar{k}}\hat{a}_{\bar{k}-\mathbf{q}}. \tag{10.55}$$

Equation (10.55) differs from the expressions of Fröhlich (1952) and Bardeen and Pines (1955) in that it contains the original $\mathscr{F}$ and $\omega_{\mathbf{q}}$. The more rigorous calculations replace $\mathscr{F}$ and $\omega_{\mathbf{q}}$ by their renormalized values.

Exercise. Derive (10.55) from (10.54).

We note that when

$$\varepsilon(\mathbf{k}-\mathbf{q})-\varepsilon(\mathbf{k})-\hbar\omega_{\mathbf{q}} \tag{10.56}$$

is small, the perturbation theory breaks down. This is the condition for *real*, rather than merely *virtual*, electron–phonon interactions. In using (10.55), therefore, we have to exclude from the calculation an 'energy shell' $\Delta\varepsilon$ where (10.56) holds. However Fröhlich (1952) has shown that this shell can be taken to be very narrow compared with $\hbar\omega_D$. Hence, in such expressions as (10.55), it is permissible to take the limit $\Delta\varepsilon \to 0$. In other words, an integral or sum containing a pole in the integrand, such as appears in (10.55), may be interpreted as meaning the principal value.

CHAPTER 11

THE SUPERCONDUCTING GROUND STATE

11.1. Instability of the normal Bloch ground state: Cooper pairs

THE electron–phonon interaction has been shown in § 10.6 to lead to an effective electron–electron interaction. From (10.55), this interaction is attractive (i.e. the matrix element is negative) when the Bloch energies $\varepsilon(\mathbf{k})$, $\varepsilon(\mathbf{k}-\mathbf{q})$ of the two electrons are nearly equal. It is repulsive (i.e. the matrix element is positive) when the energy difference $|\varepsilon(\mathbf{k})-\varepsilon(\mathbf{k}-\mathbf{q})| > \hbar\omega_{\mathbf{q}}$. The actual form of the matrix element in (10.55), namely

$$\mathscr{W}(\mathbf{k}, \mathbf{k}-\mathbf{q}) = \frac{\hbar\bar{\omega}^2}{\{\varepsilon(\mathbf{k})-\varepsilon(\mathbf{k}-\mathbf{q})\}^2-(\hbar\bar{\omega})^2}, \tag{11.1}$$

is illustrated in Fig. 11.1.

A fundamental step in the development of the theory of superconductivity was the observation by Cooper (1956) that the Fermi sea is unstable under such an attractive interaction. A pair of electrons in the immediate vicinity of the Fermi surface can form a bound quasimolecule. We shall demonstrate the possibility of forming Cooper pairs in a somewhat simplified system. Following Cooper, the phonon spectrum is taken to be of Einstein rather than Debye form:

$$\omega_{\mathbf{q}} = \text{const.} = \bar{\omega}, \tag{11.2}$$

say. As a further simplification, (11.1) will be replaced by

$$\left.\begin{aligned} \mathscr{W}(\mathbf{k}, \mathbf{k}-\mathbf{q}) &= -1 \text{ for } |\varepsilon(\mathbf{k})-\varepsilon(\mathbf{k}-\mathbf{q})| < \hbar\bar{\omega}, \\ &= 0 \quad \text{for } |\varepsilon(\mathbf{k})-\varepsilon(\mathbf{k}-\mathbf{q})| > \hbar\bar{\omega}. \end{aligned}\right\} \tag{11.3}$$

This approximation (broken curve, Fig. 11.1) is good for both large and small values of the energy difference $|\varepsilon(\mathbf{k})-\varepsilon(\mathbf{k}-\mathbf{q})|$.

Near the pole the approximation is of course very poor, but we might hope that the large positive and negative contributions will nearly cancel. This hope is likely to be fulfilled if $\mathscr{W}$ occurs as a factor in an integrand, since the integral across the pole is to be interpreted as a principal value (see p. 180).

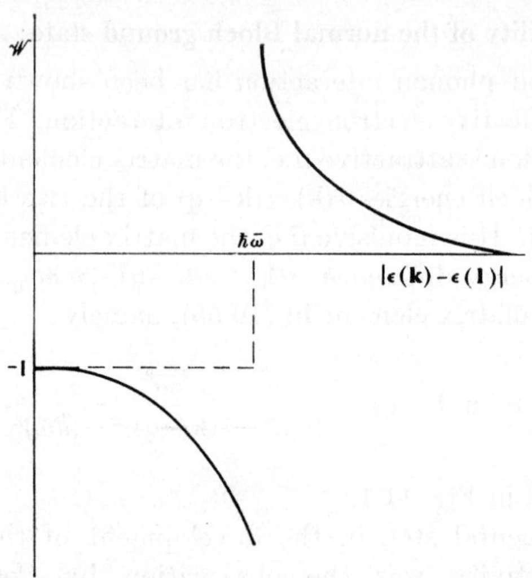

FIG. 11.1. The form of the interaction $\mathscr{W}(\mathbf{k}, \mathbf{l})$: full curve, equation (11.1); broken curve, equation (11.3).

Let us attempt to add two electrons, with energies ε_1, ε_2,($>\mu_0$), to a full Fermi sea. The Schrödinger equation for this pair of electrons with the interaction (11.3) can be solved (Cooper 1956). On account of Pauli's exclusion principle, these electrons can only scatter into states above the Fermi energy, a restriction which greatly influences the solution. For simplicity consider a further approximation: all pair states into which the two electrons can scatter are taken to have the same energy, $\varepsilon_1+\varepsilon_2 = 2\varepsilon$, say. Let there be n such pair states available (where we might expect $n \sim (\Xi_0\hbar\bar{\omega})^2$). Then in an obvious matrix representation the pair Hamiltonian is the $n \times n$ matrix

$$\hat{H}_{\text{pair}} = \begin{pmatrix} 2\varepsilon & -V & -V & . & . & . & . & -V \\ -V & 2\varepsilon & -V & . & . & . & . & -V \\ -V & -V & 2\varepsilon & . & . & . & . & -V \\ . & . & . & . & . & . & . & . \\ . & . & . & . & . & . & . & . \\ . & . & . & . & . & . & . & . \\ -V & -V & -V & . & . & . & . & 2\varepsilon \end{pmatrix}, \tag{11.4}$$

where $V = \mathscr{F}/\Omega\Xi_0$. Since (11.4) is a cyclic matrix, it is diagonalized by the unitary transformation

$$\hat{H}^{\mathrm{T}} = S^{-1}\hat{H}_{\text{pair}}S, \tag{11.5}$$

where

$$S_{jk} = \exp(2\pi \mathrm{i} jk/n). \tag{11.6}$$

The transformation Hamiltonian is

$$\hat{H}^{\mathrm{T}} = \begin{pmatrix} 2\varepsilon - nV & 0 & 0 & . & . & . & . & 0 \\ 0 & 2\varepsilon & 0 & . & . & . & . & 0 \\ 0 & 0 & 2\varepsilon & . & . & . & . & 0 \\ . & . & . & . & . & . & . & . \\ . & . & . & . & . & . & . & . \\ . & . & . & . & . & . & . & . \\ 0 & 0 & 0 & . & . & . & . & 2\varepsilon \end{pmatrix}. \tag{11.7}$$

The n-fold degenerate pair state has been split into an $(n-1)$-fold level and a non-degenerate lower state—the Cooper pair. In the real problem the unperturbed levels are not completely degenerate, but many of them are within the energy range nV. There is thus still enough degeneracy for the pairing phenomenon to survive, though the calculation is more complicated. Instead of the condensation energy of the pair being nV it is smaller by roughly a factor $\exp(-1/\mathscr{F})$.

The above calculation has shown that, under an attractive interaction, *at least one* Cooper pair will be formed. (The two paired electrons were described above as two 'extra' electrons, added to the full Fermi sea. But of course the original full sea could have contained N_e-2 electrons, and the pair could be formed out of the last two electrons to complete the N_e-particle sea.) The present method does not tell us how many pairs will in fact condense. To describe a state with many Cooper pairs is a more complex problem, and forms the subject of the next section.

To the extent that Cooper pairs are Bosons, we might hope to interpret superconductivity in terms of the Bose–Einstein condensation. It has long been believed (and was rigorously proved by Schafroth in 1955) that a condensed Bose gas of charged but non-interacting particles is superconducting. However, the condensed Bose-gas interpretation presents some difficulties. Cooper has shown that the spatial extent of a pair is $\sim 10^{-4}$ cm, of the same order of magnitude as the Pippard length ξ (§ 6.2). Hence unless the density of Cooper pairs is much less than $10^{-12}\,N_e$, the overlap of pairs cannot be neglected, and they cannot be treated in any reasonable approximation as non-interacting Bosons.

11.2. The Bardeen–Cooper–Schrieffer ground state

The microscopic theory (Bardeen *et al.* 1957, hereafter quoted as BCS) makes a variational *Ansatz*, designed to take maximum advantage of the Cooper condensation. In the ground state all Cooper pairs should have zero momentum, by analogy with the ground state of a Bose gas. In BCS the trial ground state is taken to be

$$|\Phi_0\rangle = \prod_{\mathbf{k}} (u_{\mathbf{k}} + v_{\mathbf{k}} \hat{a}^{+}_{\mathbf{k}\uparrow} \hat{a}^{+}_{-\mathbf{k}\downarrow})|0\rangle. \tag{11.8}$$

$|0\rangle$ is the state with no electrons ('electron vacuum'). The normalization condition

$$u_{\mathbf{k}}^2 + v_{\mathbf{k}}^2 = 1 \tag{11.9}$$

is imposed.

In the state (11.8), the total number of electrons is not specified. Expanding the product, we see that there is a non-zero probability amplitude for *any number* of electron pairs. But however many pairs are present, there are no unpaired electrons. The probability of finding a pair in the particular pair state $(\mathbf{k}\uparrow, -\mathbf{k}\downarrow)$ is $v_{\mathbf{k}}^2$, and the probability of this pair state being empty is $u_{\mathbf{k}}^2$. The fact that the state (11.8) does not contain a specified total number of electrons is a weakness of the formalism†; the number of electrons commutes with the Hamiltonian (10.32) and is therefore a constant of the motion. In principle it is possible to project from (11.8) only that part with precisely N_{e} electrons. But in practice the calculation is too complicated. Instead we shall use (11.8) as it stands. However, a constraint will be imposed on the variational parameters $u_{\mathbf{k}}$, $v_{\mathbf{k}}$, to ensure that the *expectation* value of the number of electrons is N_{e}.

The particular choice of variational coefficients

$$\left.\begin{aligned} u_{\mathbf{k}} &= 0,\ v_{\mathbf{k}} = 1 \text{ for } k < k_{\mathrm{F}}, \\ u_{\mathbf{k}} &= 1,\ v_{\mathbf{k}} = 0 \text{ for } k > k_{\mathrm{F}} \end{aligned}\right\} \tag{11.10}$$

reduces $|\Phi_0\rangle$ to the normal Fermi ground state; our *Ansatz* is general enough to contain the non-superconducting ground state. But there is a second solution, qualitatively different from the normal one. In this solution, the condensation energy has an exponential dependence on the interaction strength. The small exponential factor accounts for the fact that superconducting transition temperatures are very much lower than Debye temperatures. The nature of excited states, and the demonstration that the BCS state has superconducting properties, will be deferred to Chapters 12 and 13 respectively.

Let us write the single-particle energy, relative to the Fermi surface,

$$\eta(\mathbf{k}) = \varepsilon(\mathbf{k}) - \mu. \tag{11.11}$$

In terms of the $\eta(\mathbf{k})$, the kinetic energy operator is

$$\begin{aligned} \hat{H}_{\mathrm{el}} &\equiv \sum_{k} \varepsilon(\mathbf{k}) \hat{a}_k^{+} \hat{a}_k \\ &= \tfrac{3}{5} N_{\mathrm{e}} \mu + \sum_{k>k_{\mathrm{F}}} \eta(\mathbf{k}) \hat{a}_k^{+} \hat{a}_k + \sum_{k<k_{\mathrm{F}}} |\eta(\mathbf{k})| \hat{a}_k \hat{a}_k^{+}. \end{aligned} \tag{11.12}$$

† The failure of particle conservation in (11.8) may also be regarded as a symptom of the breaking of gauge symmetry, analogous to the breaking of rotational symmetry in a crystal or a Weiss ferromagnet (e.g. Brout 1965 § 1.2).

Here the first term is just the energy of the normal Fermi sea, and is constant. The second term represents the excess kinetic energy of electrons above the Fermi sea. The third term is the energy of the 'holes' below the Fermi surface, i.e. the energy which has had to be expended to lift electrons out of these states up to the Fermi surface. With the BCS *Ansatz* (11.8), the expectation value of (11.12) is†

$$\begin{aligned}\langle \hat{H}_{\mathrm{el}}\rangle_{\mathrm{BCS}} &\equiv \langle \Phi_0|\hat{H}_{\mathrm{el}}|\Phi_0\rangle \\ &= \tfrac{3}{5}N_{\mathrm{e}}\mu + \sum_{k>k_{\mathrm{F}}} 2v_{\mathbf{k}}^2\eta(\mathbf{k}) + \sum_{k<k_{\mathrm{F}}} 2u_{\mathbf{k}}^2|\eta(\mathbf{k})|. \qquad (11.13)\end{aligned}$$

The interaction energy is the expectation value of (10.55), i.e.

$$\langle \hat{H}'\rangle_{\mathrm{BCS}} = \frac{\mathscr{F}}{\Omega\Xi_0}\sum_{\mathbf{k},\mathbf{l}} u_{\mathbf{k}}v_{\mathbf{k}}u_{\mathbf{l}}v_{\mathbf{l}}\,\mathscr{W}(\eta(\mathbf{k}),\eta(\mathbf{l})), \qquad (11.14)$$

where $\mathscr{F}$ is defined in (10.43), and (cf. (11.1)),

$$\mathscr{W}(\eta(\mathbf{k}),\eta(\mathbf{l})) = \frac{(\hbar\bar{\omega})^2}{\{\eta(\mathbf{k})-\eta(\mathbf{l})\}^2-(\hbar\bar{\omega})^2}.$$

Exercise. Derive (11.14) from (11.8) and (10.55).

Bardeen *et al.* defined a reduced Hamiltonian, by retaining in $\hat{H}'$ only those terms for which

$$\left.\begin{aligned}\mathbf{k}+\mathbf{q} &= \mathbf{l},\\ \sigma' &= -\sigma,\end{aligned}\right\} \qquad (11.15)$$

i.e.

$$\hat{H}_{\mathrm{red}} = \hat{H}_{\mathrm{el}} + \sum_{\mathbf{k},\mathbf{l},\sigma}\frac{\mathscr{F}}{\Omega\Xi_0}\mathscr{W}(\varepsilon(\mathbf{k}),\varepsilon(\mathbf{l}))\hat{a}^{+}_{\mathbf{l},\sigma}\hat{a}_{\mathbf{k},\sigma}\hat{a}^{+}_{-\mathbf{l},-\sigma}\hat{a}_{-\mathbf{k},-\sigma}. \qquad (11.16)$$

The BCS wave function (11.8), with a suitable choice of the parameters $u_{\mathbf{k}}, v_{\mathbf{k}}$, can be shown to be an exact eigenstate of $\hat{H}_{\mathrm{red}}$. Moreover the terms of $\hat{H}'$ which have been dropped in the 'reduction' have vanishing expectation value. To see this, observe that these terms break the pair correlations. They therefore generate, out of $|\Phi_0\rangle$, states which are orthogonal to

† In (11.13), the sum is over all $\mathbf{k}$. But we must not sum over spins, since in (11.8) we have already specified the spins. There is nonetheless a factor 2 in (11.13), since, for every $\mathbf{k}$, *two* electrons (or holes) of energy $|\eta(\mathbf{k})|$ contribute.

$|\Phi_0\rangle$. To some extent the reduction of the Hamiltonian and the *Ansatz* (11.8) justify each other. Admittedly this justification is very weak, since the terms of $\hat{H}'$ omitted outnumber the terms retained by a factor of order N_e. In § 12.1, other reasons for believing (11.8) to be a good *Ansatz* will be given.

With Cooper's simplification (11.3) in the interaction Hamiltonian,† (11.14) becomes

$$\langle \hat{H}' \rangle_{\text{BCS}} = -\frac{\mathscr{F}}{\Omega \Xi_0} \sum_{\substack{\mathbf{k},\mathbf{l} \\ |\varepsilon(\mathbf{k})-\varepsilon(\mathbf{l})|<\hbar\bar{\omega}}} u_\mathbf{k} v_\mathbf{k} u_\mathbf{l} v_\mathbf{l}. \tag{11.17}$$

Varying the total expectation energy with respect to $v_\mathbf{k}$ gives

$$\begin{aligned} 0 &= 4\eta(\mathbf{k})v_\mathbf{k} + \frac{\mathscr{F}}{\Omega\Xi_0} \sum_\mathbf{l} \mathscr{W}(\eta(\mathbf{k}),\eta(\mathbf{l}))\, u_\mathbf{l} v_\mathbf{l} \frac{\partial}{\partial v_\mathbf{k}} \{v_\mathbf{k}(1-v_\mathbf{k}^2)^{\frac{1}{2}}\} \\ &= 4\eta(\mathbf{k})v_\mathbf{k} + \frac{\mathscr{F}}{\Omega\Xi_0} \sum_\mathbf{l} \mathscr{W}(\eta(\mathbf{k}),\eta(\mathbf{l}))\, u_\mathbf{l} v_\mathbf{l} \frac{u_\mathbf{k}^2 - v_\mathbf{k}^2}{u_\mathbf{k}}. \end{aligned} \tag{11.18}$$

The factor 4 in the first term of (11.18) arises because this term is a sum of two contributions. One comes from the first term of (11.13) with $k > k_\mathrm{F}$ and the other comes from the second term of (11.13) with $k < k_\mathrm{F}$. Replacing the sum by an integral in the standard way (see p. 164), (11.18) becomes an integral equation for $v_\mathbf{k}$. Since it is a non-linear equation, its solution is in general difficult. But if the potential is 'factorizable', i.e. if $\mathscr{W}$ can be written in the form

$$\mathscr{W}(\eta(\mathbf{k}), \eta(\mathbf{l})) = \mathscr{U}(\eta(\mathbf{k}))\, \mathscr{U}(\eta(\mathbf{l})), \tag{11.19}$$

then (11.18) simplifies greatly. Bardeen *et al.* accordingly assume

$$\left.\begin{aligned} \mathscr{W} &= -1 \text{ if both} |\eta(\mathbf{k})| < \hbar\bar{\omega},\ |\eta(\mathbf{l})| < \hbar\bar{\omega}, \\ &= 0 \quad \text{otherwise.} \end{aligned}\right\} \tag{11.20}$$

This assumption corresponds to taking $\mathscr{W}$ to be non-zero in the horizontally-shaded square in Fig. 11.2, instead of being

† Note that (11.17) is valid whether the simplified interaction (11.3) is or is not reduced.

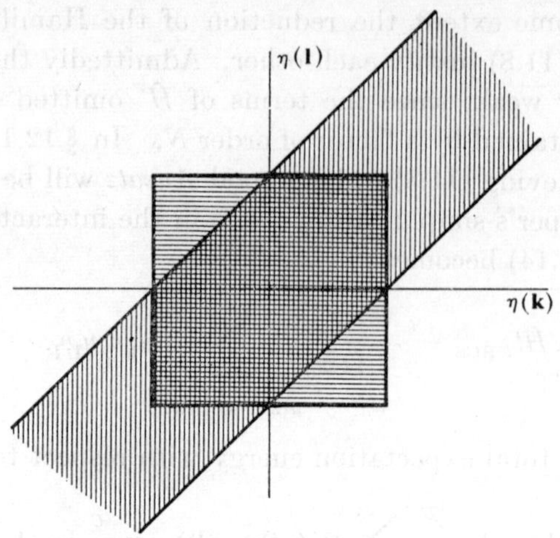

FIG. 11.2. The domain of integration in the $\eta(\mathbf{k})$, $\eta(\mathbf{l})$ plane. The vertically-shaded strip is the domain of integration using the Cooper approximation (11.3); the horizontally-shaded square is the integration region with the BCS factorizable potential (11.20).

non-zero in the vertically-shaded strip according to equation (11.3). In fact (11.20) is not nearly as bad an approximation to (11.3) as it appears at first sight. In the regions of the strip neglected under the approximation (11.20), one or other of the parameters $u_{\mathbf{k}}$, $v_{\mathbf{k}}$ is small.

Let us define

$$\Delta \equiv \tfrac{1}{2}(\mathscr{F}/\Omega\Xi_0)\sum_{\mathbf{l}} u_{\mathbf{l}} v_{\mathbf{l}}. \tag{11.21}$$

This is an important quantity, as it represents the energy gap for quasiparticle excitations (see §§ 12.1, 12.2). Substituting (11.21) in (11.18) gives

$$0 = 4\eta(\mathbf{k})v_{\mathbf{k}} - 2\Delta(u_{\mathbf{k}}^2 - v_{\mathbf{k}}^2)/u_{\mathbf{k}}, \tag{11.22}$$

whence

$$\left.\begin{aligned} u_{\mathbf{k}}^2 &= \tfrac{1}{2}\left[1 \pm \frac{\eta(\mathbf{k})}{\sqrt{\{\eta(\mathbf{k})^2 + \Delta^2\}}}\right], \\ v_{\mathbf{k}}^2 &= \tfrac{1}{2}\left[1 \mp \frac{\eta(\mathbf{k})}{\sqrt{\{\eta(\mathbf{k})^2 + \Delta^2\}}}\right]. \end{aligned}\right\} \tag{11.23}$$

The ambiguity is resolved by choosing the sign such that, in the limit $\Delta \to 0$, (11.23) reduces to the normal Fermi distribution. Thus if $k > k_F$, $u_k \to 1$ and $v_k \to 0$ as $\Delta \to 0$, i.e.

$$\left.\begin{aligned} u_{\mathbf{k}}^2 &= \tfrac{1}{2}\left[1+\frac{\eta(\mathbf{k})}{\surd\{\eta(\mathbf{k})^2+\Delta^2\}}\right], \\ v_{\mathbf{k}}^2 &= \tfrac{1}{2}\left[1-\frac{\eta(\mathbf{k})}{\surd\{\eta(\mathbf{k})^2+\Delta^2\}}\right]. \end{aligned}\right\} \tag{11.24}$$

The opposite signs must be chosen for $k < k_F$.

For (11.24) to be a solution of (11.18), the self-consistency condition must be imposed:

$$\Delta = \tfrac{1}{4}\frac{\mathscr{F}}{\Omega\Xi_0}\sum_k \frac{\Delta}{\surd\{\eta(\mathbf{k})^2+\Delta^2\}}. \tag{11.25}$$

Equation (11.25) follows from substituting (11.24) into the definition (11.21) of Δ. The integral equation (11.25) has as its solution

$$\Delta = \hbar\bar{\omega}\,\mathrm{cosech}(1/\mathscr{F}). \tag{11.26}$$

In the case of a material for which Fröhlich's interaction constant $\mathscr{F}$ is small, (11.26) simplifies to

$$\Delta = 2\hbar\bar{\omega}\exp(-1/\mathscr{F}). \tag{11.27}$$

$\mathscr{F} \lesssim \frac{1}{4}$ for most metals, and the weak-coupling limit (11.27) is adequate. The main exceptions are lead and mercury.

The condensation energy is

$$\begin{aligned} W &= \langle\Phi_0|\hat{H}|\Phi_0\rangle - \langle\Phi_N|\hat{H}|\Phi_N\rangle \\ &= \frac{-\Xi_0\Omega\,(\hbar\bar{\omega})^2}{\exp(2/\mathscr{F})-1}. \end{aligned} \tag{11.28}$$

Here $|\Phi_N\rangle$ is the normal Fermi ground state. In the weak-coupling limit, (11.28) simplifies to

$$W = -\Xi_0\Omega\,(\hbar\bar{\omega})^2\exp(-2/\mathscr{F}). \tag{11.29}$$

The expressions (11.28) and (11.29) contain the isotope effect (10.52), through their dependence on $\bar{\omega}$. Moreover the exponential factor $\exp(-1/\mathscr{F})$ is rather small and is an exceedingly sensitive function of $\mathscr{F}$. Its presence† explains both the smallness of T_c compared to the Debye temperature Θ and the poor correlation of T_c with Θ.

In the BCS theory the Coulomb term of (10.32) must not be neglected. However, it is not entirely clear what expression should be taken for the Coulomb energy. In Hartree approximation, some of the Coulomb interaction has already been absorbed into the Bloch periodic potential (p. 171). Following the discussion of § 10.2, we shall assume a screened form for the Coulomb interaction

$$\hat{H}_{\text{Coul}} = \sum_{\mathbf{q}} \left(\frac{2\pi e^2}{q^2+\kappa^2}\right) \rho_{\mathbf{q}} \rho_{-\mathbf{q}}. \qquad (11.30)^{(73)}$$

The screening length κ^{-1} may be introduced by Mott's (1936) Thomas-Fermi method (10.18), or more rigorously by modern many-body theory (e.g. Nozières and Pines 1958). The introduction of this screened interaction will avoid the mathematical complications arising from the pole in the unscreened Coulomb interaction at $q = 0$.

If it were justifiable to neglect the Coulomb energy *all* metals should become superconductors, since the condensation energy (11.28) is negative for *all* positive $\mathscr{F}$, however small. But the Coulomb potential is purely repulsive, and its operator structure is of the same form as $\hat{H}'$. It therefore acts to reduce $\mathscr{F}$; the requirement that the 'effective' $\mathscr{F}$ is positive leads to a criterion for superconductivity. The screened Coulomb interaction (11.30) will contribute

$$\hat{H}^{\text{red}}_{\text{Coul}} = \frac{\mathscr{F}_{\text{Coul}}}{\Omega \Xi_0} \sum_{\mathbf{k},\mathbf{l},\sigma} \hat{a}^{+}_{\mathbf{l},\sigma} \hat{a}_{\mathbf{k},\sigma} \hat{a}^{+}_{-\mathbf{l},-\sigma} \hat{a}_{-\mathbf{k},-\sigma} \qquad (11.31)$$

† Fröhlich (1954) anticipated the exponential factor, in a one-dimensional Hartree-like theory. However, his model did not treat the lattice dynamically, and the isotope-effect factor $(\hbar\bar{\omega})^2$ was lacking.

to the reduced Hamiltonian, where $\mathscr{F}_{\text{Coul}}/\Omega\Xi_0$ is the value of the matrix element[73] $2\pi e^2/\{(\mathbf{k}-\mathbf{l})^2+\kappa^2\}$, averaged over angles. The calculation may now be taken over intact, with $\mathscr{F}$ replaced by $\mathscr{F}-\mathscr{F}_{\text{Coul}}$. The criterion for superconductivity is that (§ 15.1)

$$\tilde{\mathscr{F}} \equiv \mathscr{F}-\mathscr{F}_{\text{Coul}} > 0. \tag{11.32}$$

To summarize, if the net interaction is attractive, the BCS state (11.8) has a lower energy than the normal Fermi ground state. But clearly there are many unsatisfactory features in the above theory. Since the BCS theory is variational, there is no internal criterion for the accuracy of the approximations. The method of Bogoliubov (see § 12.1 and Appendix D) provides a more systematic approach, as well as simplifying the description of excited states.

CHAPTER 12

QUASIPARTICLES

IN earlier chapters, there have been many references to 'quasiparticles'. The concept will be explained in the present chapter. One particular kind of quasiparticle, which plays a central role in the theory of superconductivity, will be discussed in some detail. Essentially a quasiparticle is simply an excitation of a many-body system, with some of the properties of a free particle. This statement is, of course, much too vague to serve as a definition; for the present a few examples will illustrate the concept. The formal definition of a quasiparticle is most conveniently given in the language of Green functions (see Chapter 14).

Although some specific examples of quasiparticles have been familiar for a long time, the actual word has been in use for only a few years. The main reason for its popularity is that it acts as a unifying concept; it emphasizes the similar aspects of a great variety of diverse problems, and enables us to treat them in a similar way. Historically the concept originated in Landau's (1941) theory of liquid He^4; it was in that work that the terms *phonon* and *roton* were coined. Landau advanced arguments to make it seem plausible that the low-lying energy eigenstates of a gas of *interacting* Bosons could be in one-to-one correspondence with those of a gas of *non*interacting Bosons. More recently Landau (1956) introduced the 'Fermi liquid' model for He^3—here again, despite the strong short-range interactions between the atoms, there is a one-to-one correspondence between the low-lying excitations of the system and those of an ideal (Fermi) gas.

An important landmark in the evolution of the quasiparticle concept was the introduction by Fröhlich *et al.* (1950) of the 'polaron' model of an electron moving through an ionic crystal. An electron injected into an ionic medium is strongly coupled to the optical modes of vibration of the lattice, since these modes are associated with strong electrical polarization of the medium. In comparison, the coupling to the acoustic modes of the medium is very weak and is usually neglected. In its motion through the medium, the electron polarizes the lattice in its vicinity. If the motion is slow enough, the polarization cloud remains symmetric about the electron and accompanies it in its motion. The resulting entity—the electron and its cloud of polarization—behaves in many ways like a free particle. Its charge is of course the electronic charge, but its mass differs from that of the free electron. It is this compound entity that is called the polaron and classified as a 'quasiparticle'.

As the idea of quasiparticles has developed, it has come to permeate the whole of solid-state theory. Thus, for example, the Debye theory of specific heat can be regarded as a quasiparticle theory in which the quasiparticles are phonons, the quanta of excitation of the normal modes of a crystal lattice (§ 9.3). The possibility of such an interpretation follows from the fact (see e.g. Dirac 1958 pp. 227–32) that there is a one-to-one correspondence between the states of a gas of non-interacting Bosons and those of a family of harmonic oscillators. Thus the phonons behave like Bosons and have a Bose–Einstein distribution function. The only novel feature is the energy–momentum relation for a phonon, $E(k) = \hbar ks$.

In principle the concept of the quasiparticle is perhaps redundant. In the field-theoretical approach to the theory of solids, quasiparticles enter the formalism in precisely the same way as 'physical' particles enter the formalism of elementary particle physics. The conduction electrons and holes of semiconductor theory are closely analogous to the electrons and positrons of the Dirac theory. The anomalous magnetic moments of the proton and neutron are ascribed to a cloud of pions around the particle—the analogy with the polaron is evident. But in particle physics the bare particles are inaccessible to observation, while in solid-state physics the bare electron, for example, can easily be studied *outside* the medium. It is only because of our direct observational knowledge of the 'real' particles and their interactions that we are able to degrade phonons, holes, polarons, etc. to the second-class status of quasiparticles.

In the present chapter, the quasiparticle spectrum of superconductors is studied and the 'phenomenological BCS model' (§ 2.3) is derived from the microscopic theory of Bardeen *et al.* (1957). The presentation is not historical. The canonical transformation (12.16) was proposed by Bogoliubov (1958) and Valatin (1958) as an *ad hoc* procedure, and the coefficients $u_{\mathbf{k}}$, $v_{\mathbf{k}}$ of that transformation were optimized according to one of several intuitive criteria. Anderson's (1958) approach, which will be followed here, helps to motivate what might otherwise appear as a rather artificial procedure.

12.1. Fermion excitations in superconductors: the Bogoliubov transformation

The idea of quasiparticles is particularly useful in discussing the excited states of a superconductor. If we assume the existence of an unpaired Bloch electron in the state $\bar{k}$, the pairing interactions between all other Bloch states and the pair of states $\bar{k}$ and $-\bar{k}$ are broken. It is shown in BCS that the energy of such a state with broken pairing exceeds the ground-state energy by at least the energy gap 2Δ. We shall not follow that treatment in detail. Instead the formalism of Bogoliubov (1958) and Valatin (1958) will be employed, since their method provides a much more concise description of the excited states. The new description of the system will be shown to have the following properties.

(a) The ground state is the BCS ground state (11.8).

(b) There are quasiparticle excitations ('Bogoliubons') with Fermion properties.

(c) There is an energy threshold Δ in the excitation spectrum for quasiparticles.

(d) Although the number of Bogoliubons is not a constant of the motion, all terms in the Hamiltonian contain an even number of Bogoliubon field operators. Thus the quasiparticles can only be created or destroyed in pairs, so that the energy gap is 2Δ.

A complication to be faced at the outset is that the BCS ground state $|\Phi_0\rangle$ (11.8) is not an eigenstate of the number of electrons. And after the transformation to Bogoliubons, the number operator for these quasiparticles is not a constant of the motion. But the real physical electrons *are* conserved at these non-relativistic energies. Somehow, the conservation of electrons must be incorporated into the theory. It is most easily done by imposing a *constraint*. We then relax the constraint slightly; it is no longer to be imposed as an identity obeyed by the field operators. Instead it is merely to be a property of the expectation values of these operators, taken with respect to the wave functions of the theory.† The relaxation to this weaker constraint is effected by means of a Lagrangian multiplier μ. More specifically, instead of seeking simultaneous eigenstates of the Hamiltonian $\hat{H}$ and the particle number operator $\hat{N}_e$ (9.59), we minimize

$$\hat{\mathscr{H}} = \hat{H} - \mu\hat{N}_e \tag{12.1}$$

against variations in the trial ground-state wave function $|\Phi_0\rangle$. μ is then determined by requiring

$$\partial\hat{\mathscr{H}}/\partial\mu = \langle\hat{N}_e\rangle. \tag{12.2}$$

From (12.2) it is clear that the Lagrangian multiplier μ plays the role of a chemical potential. On account of the structure of the number operator $\hat{N}_e$, it is convenient to include the term $\mu\hat{N}_e$ together with the kinetic energy term in the Hamiltonian. In fact this procedure is already implicit in (11.11), where the energies of single electrons are expressed relative to the Fermi surface.

Exercise. Calculate the Fermi-Dirac distribution for a grand ensemble, and show that the chemical potential plays the role of the Fermi energy at $T = 0$.

† In the language of statistical mechanics we consider a grand ensemble. See, e.g., Kadanoff and Baym (1962 § 1.1).

For a system of noninteracting particles with a Hamiltonian (cf. (9.62), (12.1)),

$$\hat{\mathscr{H}}^{\text{Aug}}_{\text{el}} = \sum_{k} \{\varepsilon(\mathbf{k}) - \mu\} \hat{a}^{+}_{k} \hat{a}_{k} = \sum_{k} \eta(\mathbf{k}) \hat{a}^{+}_{k} \hat{a}_{k}. \tag{12.3}$$

The Heisenberg equations of motion† for the creation and annihilation operators have the simple form

$$\left.\begin{aligned} \mathrm{i}\hbar\dot{\hat{a}}^{+}_{k}(t) &= -[\hat{\mathscr{H}}_{\text{el}}, \hat{a}^{+}_{k}(t)] = \eta(\mathbf{k})\hat{a}^{+}_{k}(t), \\ \mathrm{i}\hbar\dot{\hat{a}}_{k}(t) &= -[\hat{\mathscr{H}}_{\text{el}}, \hat{a}_{k}(t)] = -\eta(\mathbf{k})\hat{a}_{k}(t). \end{aligned}\right\} \tag{12.4}$$

Equations (12.4) are immediately integrable, to give

$$\left.\begin{aligned} \hat{a}^{+}_{k}(t) &= \hat{a}^{+}_{k}(0)\exp\{-\mathrm{i}\eta(\mathbf{k})t/\hbar\}, \\ \hat{a}_{k}(t) &= \hat{a}_{k}(0)\exp\{\mathrm{i}\eta(\mathbf{k})t/\hbar\}. \end{aligned}\right\} \tag{12.5}$$

Thus the creation and annihilation operators $\hat{a}^{+}_{k}(t)$, $\hat{a}_{k}(t)$ are oscillatory functions of time, with a frequency $\eta(\mathbf{k})/\hbar$.

In the presence of interactions, (12.4) are replaced by a more complicated set of equations—$\dot{\hat{a}}_{k}(t)$ does not depend linearly on $\hat{a}_{k}(t)$, but rather on a product of three quantized field operators. Thus if we add to $\hat{\mathscr{H}}_{\text{el}}$ of (12.3) an interaction term of the form (9.66)‡,

$$\hat{\mathscr{H}} = \sum_{k} \eta(\mathbf{k}) \hat{a}^{+}_{k} \hat{a}_{k} + \tfrac{1}{2} \sum_{k,l,\mathbf{q}} V_{\mathbf{q}} \hat{a}^{+}_{k+\mathbf{q}} \hat{a}_{k} \hat{a}^{+}_{l-\mathbf{q}} \hat{a}_{l}, \tag{12.6}$$

then the equations of motion for $\hat{a}^{+}_{k}(t)$, $\hat{a}_{k}(t)$ read:

$$\left.\begin{aligned} \mathrm{i}\hbar\dot{\hat{a}}^{+}_{k}(t) &= \eta(\mathbf{k})\hat{a}^{+}_{k}(t) + \sum_{l,\mathbf{q}} V_{\mathbf{q}}, \hat{a}^{+}_{k+\mathbf{q}} \hat{a}^{+}_{l-\mathbf{q}} \hat{a}_{l}, \\ \mathrm{i}\hbar\dot{\hat{a}}_{k}(t) &= -\eta(\mathbf{k})\hat{a}_{k}(t) - \sum_{l,\mathbf{q}} V_{\mathbf{q}}, \hat{a}^{+}_{l} \hat{a}_{l-\mathbf{q}} \hat{a}_{k+\mathbf{q}}. \end{aligned}\right\} \tag{12.7}$$

Exercise. Calculate the commutator of $\hat{a}_{k}(t)$ with $\hat{\mathscr{H}}$, and hence verify (12.7).

† In these equations, we use the Heisenberg picture in which stationary states are represented by time-independent state vectors. The relation between the Heisenberg and Schrödinger pictures is described briefly in Appendix C; for a fuller account, see Dirac (1958 pp. 108–18).

‡ The interaction (10.55) is somewhat more complicated than (9.66), insofar as the matrix element does not *only* depend on $\mathbf{q}$. The simpler interaction (9.66) is, however, adequate to illustrate the principles involved.

By formal integration of (12.7) and repeated iteration, we can generate an infinite regression of increasingly complicated expressions, involving multiple products of field operators at different times. However, this is not a profitable procedure; a more fruitful approach is to use some approximation to terminate the regression after at most a few steps. The simplest of these schemes is the random-phase approximation (Bohm and Pines 1953, Anderson 1958), in which it is assumed that

$$\langle \hat{a}_{\bar{l}}^{+} \hat{a}_{\bar{p}} \rangle = 0, \; \bar{l} \neq \bar{p}. \tag{12.8}$$

Expressions of the form $\hat{a}_{\bar{l}}^{+} \hat{a}_{\bar{l}}$ on the right-hand side of (12.7) are replaced by their expectation values

$$\langle \hat{a}_{\bar{l}}^{+} \hat{a}_{\bar{l}} \rangle. \tag{12.9}$$

Under the random-phase approximation, the equations of motion (12.7) for the Fermion creation and annihilation operators become

$$\left.\begin{aligned} i\hbar\dot{\hat{a}}_{\bar{k}}^{+} &= \eta(\mathbf{k})\hat{a}_{\bar{k}}^{+} + \hat{a}_{\bar{k}}^{+}\{V_0 \sum_{\bar{l}} \langle \hat{a}_{\bar{l}}^{+}\hat{a}_{\bar{l}}\rangle - \sum_{\mathbf{q}} V_{\mathbf{q}} \langle \hat{a}_{\bar{k}+\mathbf{q}}^{+}\hat{a}_{\bar{k}+\mathbf{q}}\rangle\} \\ &\qquad\qquad = \tilde{\eta}(\mathbf{k})\hat{a}_{\bar{k}}^{+}, \\ i\hbar\dot{\hat{a}}_{\bar{k}} &= -\eta(\mathbf{k})\hat{a}_{\bar{k}} - \hat{a}_{\bar{k}}\{V_0 \sum_{\bar{l}} \langle \hat{a}_{\bar{l}}^{+}\hat{a}_{\bar{l}}\rangle - \sum_{\mathbf{q}} V_{\mathbf{q}} \langle \hat{a}_{\bar{k}+\mathbf{q}}^{+}\hat{a}_{\bar{k}+\mathbf{q}}\rangle\} \\ &\qquad\qquad = -\tilde{\eta}(\mathbf{k})\hat{a}_{\bar{k}}. \end{aligned}\right\} \tag{12.10}$$

These (cf. (12.4)) are immediately integrable, to give

$$\left.\begin{aligned} \hat{a}_{\bar{k}}^{+}(t) &= \hat{a}_{\bar{k}}^{+}(0)\exp\{-i\tilde{\eta}(\mathbf{k})t/\hbar\}, \\ \hat{a}_{\bar{k}}(t) &= \hat{a}_{\bar{k}}(0)\exp\{i\tilde{\eta}(\mathbf{k})t/\hbar\}, \end{aligned}\right\} \tag{12.11}$$

where

$$\tilde{\eta}(\mathbf{k}) = \eta(\mathbf{k}) + \{\sum_{\mathbf{q}} - V_{\mathbf{q}} \langle \hat{a}_{\bar{k}+\mathbf{q}}^{+}\hat{a}_{\bar{k}+\mathbf{q}}\rangle + V_0 \sum_{\bar{l}} \langle \hat{a}_{\bar{l}}^{+}\hat{a}_{\bar{l}}\rangle\}. \tag{12.12}$$

Thus, for the single-particle energies, the random-phase approximation is equivalent to the familiar Hartree–Fock approximation.† Its self-consistency is readily seen: on

† The Bohm–Pines formalism is not entirely equivalent to the Hartree–Fock approximation, but is an improvement on it. The improvement arises because the method contains collective (plasma) degrees of freedom as well as single-particle modes.

substituting for the time-dependent field operators (12.11) in (12.12), the phase factors cancel, and the expression multiplying $\hat{a}_{\bar{k}}^{+}(t)$ or $\hat{a}_{\bar{k}}(t)$ on the right-hand side of (12.10) is indeed independent of time.

Comparison of (12.11) with (12.5) shows that the Hartree–Fock approximation leads to quasiparticle excitations, since the equations of motion for the $\hat{a}_{\bar{k}}^{+}$, $\hat{a}_{\bar{k}}$ operators have the same form as in the non-interacting system. Thus $\hat{a}_{\bar{k}}^{+}(t)$ and $\hat{a}_{\bar{k}}(t)$ behave as creation and annihilation operators for quasiparticles of energy $\tilde{\eta}(\mathbf{k})$. From the structure of (12.12), there is no reason to expect an energy gap; detailed calculation confirms that there is indeed *no* gap.

In the above method of linearizing the equations of motion (12.7), the crucial assumption is that *all* quadratic expressions in the field operators, except those of the form (12.9), have zero expectation value. This postulate is indeed true for a normal Fermi gas, but it is *not* true of a superconductor. In particular it is easy to see that it is untrue for the BCS ground state (11.8). The *Ansatz* (11.8) does indeed satisfy (12.8), and has a non-zero expectation value (12.9). But the additional expressions†

$$\left.\begin{array}{c}\langle\Phi_0|\hat{a}_{\bar{l}}\hat{a}_{-\bar{l}}|\Phi_0\rangle,\\ \langle\Phi_0|\hat{a}_{\bar{l}}^{+}\hat{a}_{-\bar{l}}^{+}|\Phi_0\rangle\end{array}\right\} \tag{12.13}$$

are non-zero. Apart from (12.9) and (12.13), all *other* quadratic field expressions have vanishing expectation values:

$$\left.\begin{array}{r}\langle\Phi_0|\hat{a}_{\bar{k}}^{+}\hat{a}_{\bar{l}}|\Phi_0\rangle = 0,\\ \langle\Phi_0|\hat{a}_{\bar{k}}^{+}\hat{a}_{-\bar{l}}^{+}|\Phi_0\rangle = 0,\\ \langle\Phi_0|\hat{a}_{-\bar{k}}\hat{a}_{\bar{l}}|\Phi_0\rangle = 0,\end{array}\right\}\quad \bar{k}\neq\bar{l}. \tag{12.14}$$

† These expressions can be nonzero because $|\Phi_0\rangle$ is not an eigenstate of the particle number operator. If we wish to use states which *do* conserve particles, and if we denote the projection of $|\Phi_0\rangle$ onto the space of N_e particles by $|\Phi_0(N_e)\rangle$, then these expressions must be written

$$\langle\Phi_0(N_e)|\hat{a}_{\bar{l}}\hat{a}_{-\bar{l}}|\Phi_0(N_e+2)\rangle,$$

etc., cf. equation (14.29′).

Thus for superconductors we are led to a natural generalization of the random-phase approximation: *replace all possible quadratic products of field operators by their expectation values, and make use of* (12.14) *to retain only expressions containing the forms* (12.9) *and* (12.13). The resulting linearized equations of motion are

$$\left.\begin{aligned} i\hbar\dot{\hat{a}}_{\bar{k}}^{+} &= \eta(\mathbf{k})\hat{a}_{\bar{k}}^{+} - \{\sum_{\mathbf{q}} V_{\mathbf{q}}\langle \hat{a}_{\bar{k}-\mathbf{q}}^{+}\hat{a}_{\bar{k}-\mathbf{q}}\rangle - V_0 \sum_{l}\langle \hat{a}_l^{+}\hat{a}_l\rangle\}\hat{a}_{\bar{k}}^{+} \\ &\qquad - \sum_{\mathbf{q}} V_{\mathbf{q}}\langle \hat{a}_{\bar{k}-\mathbf{q}}^{+}\hat{a}_{-\bar{k}+\mathbf{q}}^{+}\rangle \hat{a}_{-\bar{k}}, \\ i\hbar\dot{\hat{a}}_{\bar{k}} &= -\eta(\mathbf{k})\hat{a}_{\bar{k}} + \{\sum_{\mathbf{q}} V_{\mathbf{q}}\langle \hat{a}_{\bar{k}-\mathbf{q}}^{+}\hat{a}_{\bar{k}-\mathbf{q}}\rangle - V_0 \sum_{l}\langle \hat{a}_l^{+}\hat{a}_l\rangle\}\hat{a}_{\bar{k}} \\ &\qquad + \sum_{\mathbf{q}} V_{\mathbf{q}}\langle \hat{a}_{\bar{k}-\mathbf{q}}\hat{a}_{-\bar{k}+\mathbf{q}}\rangle \hat{a}_{-\bar{k}}^{+}. \end{aligned}\right\} \tag{12.15}$$

We see that the equations of motion for the one-particle states $\bar{k}$, $-\bar{k}$ are coupled. To decouple them, introduce as new variables Bogoliubov's (1958) mixtures of the field operators for $\bar{k}$ and $-\bar{k}$:

$$\left.\begin{aligned} \hat{\alpha}_{\mathbf{k},\sigma}^{+} &= u_{\mathbf{k}}\hat{a}_{\mathbf{k},\sigma}^{+} - \sigma v_{\mathbf{k}}\hat{a}_{-\mathbf{k},-\sigma}, \\ \hat{\alpha}_{\mathbf{k},\sigma} &= u_{\mathbf{k}}\hat{a}_{\mathbf{k},\sigma} - \sigma v_{\mathbf{k}}\hat{a}_{-\mathbf{k},-\sigma}^{+}. \end{aligned}\right\} \tag{12.16}$$

The coefficients $u_{\mathbf{k}}$, $v_{\mathbf{k}}$ are real and satisfy

$$\left.\begin{aligned} u_{\mathbf{k}}^2 + v_{\mathbf{k}}^2 &= 1; \\ u_{\mathbf{k}} = u_{-\mathbf{k}};\ v_{\mathbf{k}} &= v_{-\mathbf{k}}. \end{aligned}\right\} \tag{12.17}$$

Exercises. (a) In the BCS ground state (11.8), verify that (12.14) holds, and that (12.13) and (12.9) are indeed non-zero.

(b) Verify that the transformation (12.16) is canonical, i.e. that it preserves all commutation relations.

(c) Solve (12.16) for $\hat{a}_{\bar{k}}^{+}$, $\hat{a}_{\bar{k}}$ in terms of $\hat{\alpha}_{\bar{k}}^{+}$ and $\hat{\alpha}_{\bar{k}}$. The solution is

$$\left.\begin{aligned} \hat{a}_{\mathbf{k},\sigma}^{+} &= u_{\mathbf{k}}\hat{\alpha}_{\mathbf{k},\sigma}^{+} + \sigma v_{\mathbf{k}}\hat{\alpha}_{-\mathbf{k},-\sigma}, \\ \hat{a}_{\mathbf{k},\sigma} &= u_{\mathbf{k}}\hat{\alpha}_{\mathbf{k},\sigma} + \sigma v_{\mathbf{k}}\hat{\alpha}_{-\mathbf{k},-\sigma}^{+}. \end{aligned}\right\} \tag{12.18}$$

Let us define

$$\Delta(\mathbf{k}) = \sum_{\mathbf{q}} V_{\mathbf{q}} v_{\mathbf{k}-\mathbf{q}} u_{\mathbf{k}-\mathbf{q}}, \tag{12.19}$$

in order to simplify the form of (12.15):

$$\left.\begin{aligned} i\hbar\dot{a}_{\bar{k}}^{+} &= \quad \tilde{\eta}(\mathbf{k})\,a_{\bar{k}}^{+} - \Delta(\mathbf{k})\,a_{-\bar{k}}, \\ i\hbar\dot{a}_{\bar{k}} &= -\tilde{\eta}(\mathbf{k})\,a_{\bar{k}} + \Delta(\mathbf{k})\,a_{-\bar{k}}^{+}. \end{aligned}\right\} \tag{12.20}$$

The particular choice of the coefficients $u_{\mathbf{k}}$, $v_{\mathbf{k}}$ required to reduce (12.20) to oscillator form is (cf. (11.24))

$$\left.\begin{aligned} u_{\mathbf{k}}^{2} &= \tfrac{1}{2}\{1+\tilde{\eta}(\mathbf{k})/E(\mathbf{k})\}, \\ v_{\mathbf{k}}^{2} &= \tfrac{1}{2}\{1-\tilde{\eta}(\mathbf{k})/E(\mathbf{k})\}, \end{aligned}\right\} \tag{12.21}$$

where

$$E(\mathbf{k}) = \sqrt{\{\tilde{\eta}(\mathbf{k})^{2}+\Delta(\mathbf{k})^{2}\}}. \tag{12.22}$$

We must however demand that our procedure be self-consistent. From (12.19), (12.21) and (12.22), the condition for self-consistency is

$$\begin{aligned} \Delta(\mathbf{k}) &= \sum_{\mathbf{q}} V_{\mathbf{q}} v_{\mathbf{k}-\mathbf{q}} u_{\mathbf{k}-\mathbf{q}} \\ &= \sum_{\mathbf{q}} V_{\mathbf{q}} \Delta(\mathbf{k}-\mathbf{q})/E(\mathbf{k}-\mathbf{q}). \end{aligned} \tag{12.23}$$

This integral equation determines the energy-gap function $\Delta(\mathbf{k})$. If (cf. (10.55), (11.1), (11.19)) the matrix element V is a function of the energies $\tilde{\eta}(\mathbf{k})$, $\tilde{\eta}(\mathbf{k}-\mathbf{q})$, then (12.23) takes the simpler form

$$\Delta(\tilde{\eta}) = \Xi_{0}\int \mathrm{d}\tilde{\eta}'\; V(\tilde{\eta},\tilde{\eta}')\,\Delta(\tilde{\eta}')/\sqrt{\{\tilde{\eta}'^{2}+\Delta^{2}(\tilde{\eta}')\}}. \tag{12.24}$$

In particular, with the simple interaction (11.20), Δ is independent of $\bar{k}$. The integral equation reduces to (11.25), and Δ takes the value $\hbar\bar{\omega}$ cosech $(1/\mathscr{F})$, as found in (11.26).

Under the transformation (12.16), with the above choice (12.21) of the coefficients, (12.20) reduces to the simple form

$$\left.\begin{aligned} i\hbar\dot{\alpha}_{\bar{k}}^{+} &= E(\mathbf{k})\alpha_{\bar{k}}^{+}, \\ i\hbar\dot{\alpha}_{\bar{k}} &= -E(\mathbf{k})\alpha_{\bar{k}}, \end{aligned}\right\} \tag{12.25}$$

analogous to (12.4). The operators $\alpha_{\bar{k}}^{+}$, $\alpha_{\bar{k}}$ are creation and annihilation operators for quasiparticles with the energy spectrum (12.22)—and in particular with an energy gap $2\Delta(k_F)$.

An alternative criterion for choosing the coefficients $u_{\mathbf{k}}$, $v_{\mathbf{k}}$ is that of Valatin (1958).† The transformation (12.16) is regarded as merely a convenient way of applying the BCS variational *Ansatz* (11.8). For it is easily verified that the state (11.8) is the quasiparticle 'vacuum', i.e. that

$$\hat{\alpha}_{\bar{k}}|\Phi_0\rangle = 0. \tag{12.26}$$

Thus the BCS variational calculation of the energy may be performed by first calculating the vacuum expectation value of the transformed Hamiltonian (substituting (12.16) into (12.6)), and minimizing the resulting expression with respect to $u_{\mathbf{k}}$ (or $v_{\mathbf{k}}$).

Note that the Bogoliubov scheme contains the normal Bloch states as a special case ($u_{\mathbf{k}} = 1$, $v_{\mathbf{k}} = 0$). Another special choice with a direct interpretation is

$$u_{\mathbf{k}} = 1,\ v_{\mathbf{k}} = 0 \text{ for } k > k_F,$$
$$u_{\mathbf{k}} = 0,\ v_{\mathbf{k}} = 1 \text{ for } k < k_F.$$

This is the transformation which converts the normal Fermi sea into the quasiparticle vacuum. For Bloch states above the Fermi surface, the operator $\hat{\alpha}^{+}_{\bar{k}} = \hat{a}^{+}_{\bar{k}}$ is the creation operator for an electron. On the other hand for states below the Fermi surface, $\hat{\alpha}^{+}_{\bar{k}} = \pm\hat{a}_{-\bar{k}}$ is interpreted as a *creation* operator for a 'hole' $-\bar{k}$, instead of an annihilation operator for an electron $\bar{k}$. However a more general choice of $u_{\mathbf{k}}$, $v_{\mathbf{k}}$, such as (12.21), destroys the correspondence between quasiparticle states and Bloch states; some caution is needed in interpreting the transformation. In particular the *charge* associated with a quasiparticle is not a good quantum number, since the creation

† Both the Valatin and Anderson criteria for choosing the coefficients $u_{\mathbf{k}}$, $v_{\mathbf{k}}$ are applicable only to a Hamiltonian like (12.6), with an explicit electron–electron interaction. Bogoliubov (1958) had previously introduced another criterion, that of 'compensation of dangerous diagrams'. This method has the advantage that it can be applied directly to Fröhlich's electron–phonon Hamiltonian (10.32), (10.34)—an important feature in discussing the validity of some of the approximations. Bogoliubov's method is described briefly in Appendix D; for a fuller account see Bogliubov *et al.* (1959).

operator $\hat{\alpha}_k^+$ is a superposition of creation and annihilation operators for Bloch electrons. Depending on the values of $u_{\mathbf{k}}$, $v_{\mathbf{k}}$, the quasiparticle can have an expectation value of charge anywhere in the range from $-e$ to e. For this reason the semiconductor model can be misleading (see p. 32). It is adequate in problems where the charge of a quasiparticle is irrelevant, and it also works for the tunnelling of single quasiparticles. However, it does *not* work (cf. pp. 139–40) when we study the correlated tunnelling of two or more quasiparticles, and a model more closely related to the microscopic theory is required (Adkins 1964).

12.2. Finite temperatures

At very low temperatures the thermally excited quasiparticles are so few and so well separated that they may be regarded as non-interacting particles. In other words the excitations may be thought of as being analogous to those of an intrinsic semiconductor—though some caution is required (see footnote, p. 32). Following BCS, the theory is extended to higher temperatures by assuming that the excitations are statistically independent. The energy gap, or some other equivalent order parameter, is the analogue of the internal field of a Weiss ferromagnet. We shall assume that this 'internal field' enters the quasiparticle distribution only parametrically. That is, the probability of occupation of a given single-particle state will depend parametrically on the *whole* distribution of quasiparticles, but not on the detailed question of whether some other particular state is occupied.

We introduce the quasiparticles through a Bogoliubov transformation, as in § 12.1. But now the coefficients $u_{\mathbf{k}}$, $v_{\mathbf{k}}$ are allowed to depend on temperature. This means that the actual Bogoliubov transformation depends on temperature. Hence the complete set of states

$$|\Phi_0\rangle, \quad \hat{\alpha}_k^+|\Phi_0\rangle, \quad \hat{\alpha}_k^+\hat{\alpha}_l^+|\Phi_0\rangle, \quad \ldots, \tag{12.27}$$

which forms our basis, is also temperature-dependent.

The procedure is as follows. The Bogoliubov transformation (12.16) is used to calculate the Helmholtz free energy

$$F = \langle \hat{\mathscr{H}} - TS \rangle_{\mathrm{av}}. \tag{12.28}$$

(Here and henceforth the symbol $\langle \ldots \rangle_{\mathrm{av}}$ will denote the statistical average of the quantity within the bracket, taken over a grand canonical ensemble.)† Minimizing F with respect to both the quasiparticle distribution and the Bogoliubov coefficients determines the optimum values of these functions. Finally, the argument is completed by imposing a condition of self-consistency, analogous to (12.23). Hence we find an expression for the temperature-dependent energy gap. The resulting quasiparticle energy spectrum is of a suitable form to construct an energy-gap model; indeed it is precisely the spectrum which was postulated in § 2.3, in constructing the phenomenological 'BCS model'.

Let $f_{\bar{k}}$ denote the expectation value of the number $\hat{\nu}_{\bar{k}}$ of quasiparticles in the state $\bar{k}$:

$$f_{\bar{k}} \equiv \langle \hat{\nu}_{\bar{k}} \rangle_{\mathrm{av}} \equiv \langle \hat{\alpha}_{\bar{k}}^{+} \hat{\alpha}_{\bar{k}} \rangle_{\mathrm{av}}. \tag{12.29}$$

By the assumption of the statistical independence of excitations, the entropy of the system is given by the usual expression for Fermions,

$$S = \mathrm{k} \sum_{\bar{k}} \{ f_{\bar{k}} \ln f_{\bar{k}} + (1 - f_{\bar{k}}) \ln (1 - f_{\bar{k}}) \}, \tag{12.30}$$

where k is Boltzmann's constant (see, e.g. Landau and Lifshitz, 1959 p. 155).

At a temperature T the internal energy of the system, i.e. the expectation value of the Hamiltonian, may be dissected into two parts: (a) the contribution from the kinetic energy operator $\langle \hat{H}_{\mathrm{el}} \rangle_{\mathrm{av}}$:

$$\hat{H}_{\mathrm{el}} = \sum_{k} \eta(\mathbf{k}) \hat{a}_{k}^{+} \hat{a}_{k}, \tag{12.31}$$

† We must remember that the states of the quasiparticle gas are taken to be (12.27). None of these states are eigenstates of the true particle number operator $\hat{N}_{\mathrm{e}} = \sum_{k} \hat{a}_{k}^{+} \hat{a}_{k}$. In taking statistical averages of operators, we are therefore obliged to use a grand canonical ensemble. That is to say we must study the states of the 'grand' Hamiltonian (12.1), rather than the actual Hamiltonian.

and (b) the interaction energy contribution $\langle\hat{H}'\rangle_{\mathrm{av}}$:

$$\hat{H}' = \sum_{\bar{k},\bar{l},\mathbf{q}} V_{\mathbf{q},\bar{k},\bar{l}}\hat{a}^{+}_{\bar{k}+\mathbf{q}}\hat{a}_{\bar{k}}\hat{a}^{+}_{\bar{l}-\mathbf{q}}\hat{a}_{\bar{l}}. \tag{12.32}$$

From (12.18) and (12.31), the expectation kinetic energy is

$$\langle\hat{H}_{\mathrm{el}}\rangle_{\mathrm{av}} = \sum_{\bar{k}}\langle\eta(\mathbf{k})(u_{\mathbf{k}}\hat{\alpha}^{+}_{\bar{k}}+\sigma v_{\mathbf{k}}\hat{\alpha}_{-\bar{k}})(u_{\mathbf{k}}\hat{\alpha}_{\bar{k}}+\sigma v_{\mathbf{k}}\hat{\alpha}^{+}_{-\bar{k}})\rangle_{\mathrm{av}}$$

$$= \sum_{\bar{k}}\eta(\mathbf{k})\{v^{2}_{\mathbf{k}}(1-f_{\mathbf{k}})+u^{2}_{\mathbf{k}}f_{\mathbf{k}}\}, \tag{12.33}$$

since, by the assumption of the statistical independence of quasiparticles, the expectation value of the quasiparticle correlation $\hat{\alpha}^{+}_{\bar{k}}\hat{\alpha}^{+}_{-\bar{k}}$ is zero. We may interpret the first term of (12.33) as the contribution from complexions which do not have a quasiparticle in the state $\bar{k}$—and for which the probability of electronic occupation is $v^{2}_{\mathbf{k}}$. The second term is the contribution from complexions *with* a quasiparticle in the state $\bar{k}$. The two terms carry the appropriate weight factors $1-f_{\mathbf{k}}$, $f_{\mathbf{k}}$ respectively.

The expectation value of the interaction energy (12.32) is similarly calculated:

$$\langle\hat{H}'\rangle_{\mathrm{av}} = \sum_{\bar{k},\bar{l}} V_{\mathbf{l}-\mathbf{k}\,\bar{k},\bar{l}}(1-2f_{\mathbf{k}})(1-2f_{\mathbf{l}})u_{\mathbf{k}}u_{\mathbf{l}}v_{\mathbf{k}}v_{\mathbf{l}}. \tag{12.34}$$

With the BCS assumption (11.19) that the potential is factorizable, $V_{\mathbf{l}-\mathbf{k}\,\bar{k},\bar{l}} = \mathscr{U}_{\bar{k}}\mathscr{U}_{\bar{l}}$, (12.34) factorizes:

$$\langle H'\rangle_{\mathrm{av}} = \bar{\mathscr{U}}^{2}, \tag{12.35}$$

$$\bar{\mathscr{U}} = \sum_{\bar{k}}\mathscr{U}_{\bar{k}}u_{\mathbf{k}}v_{\mathbf{k}}(1-2f_{\mathbf{k}}). \tag{12.36}$$

Equations (12.35) and (12.36) have a direct interpretation. The factor $\bar{\mathscr{U}}$ is a sum over single-particle states $\bar{k}$, each term in that sum being capable of dissection into two terms:

$$\bar{\mathscr{U}}^{(1)}_{\bar{k}} = \mathscr{U}_{\bar{k}}u_{\mathbf{k}}v_{\mathbf{k}}(1-f_{\mathbf{k}}), \tag{12.37}$$

$$\bar{\mathscr{U}}^{(2)}_{\bar{k}} = -\mathscr{U}_{\bar{k}}u_{\mathbf{k}}v_{\mathbf{k}}f_{\mathbf{k}}. \tag{12.38}$$

Here $\bar{\mathscr{U}}^{(1)}_{\bar{k}}$ is the contribution to $\bar{\mathscr{U}}$ from complexions *not* containing a quasiparticle in the state $\bar{k}$. It carries the correct weight factor $1-f_{\mathbf{k}}$. Similarly $\bar{\mathscr{U}}^{(2)}_{\bar{k}}$ is the contribution arising from complexions in which there *is* a quasiparticle in the state $\bar{k}$; it has the weight $f_{\mathbf{k}}$.

Substituting from (12.30), (12.33), and (12.35) into (12.28), the Helmholtz free energy is a functional of $v_{\mathbf{k}}$ (or equivalently of $u_{\mathbf{k}}$) and of $f_{\mathbf{k}}$, to be minimized with respect to variations of both these functions:

$$\begin{aligned} F &= \langle \hat{H}_{\text{el}} \rangle_{\text{av}} + \langle \hat{H}' \rangle_{\text{av}} - TS \\ &= \sum_{k} \eta(\mathbf{k})\{u_{\mathbf{k}}^2 f_{\mathbf{k}} + v_{\mathbf{k}}^2(1-f_{\mathbf{k}})\} - \sum_{k,l} \mathscr{U}_{k}\mathscr{U}_{l} u_{\mathbf{k}} u_{\mathbf{l}} v_{\mathbf{k}} v_{\mathbf{l}} (1-2f_{\mathbf{k}})(1-2f_{\mathbf{l}}) \\ &\qquad - \mathrm{k}T \sum_{k} \{f_{\mathbf{k}} \ln f_{\mathbf{k}} + (1-f_{\mathbf{k}}) \ln(1-f_{\mathbf{k}})\}. \end{aligned} \tag{12.39}$$

First minimize (12.39) with respect to $v_{\mathbf{k}}$. The calculation proceeds exactly as in the zero-temperature case (pp. 187–9), and leads to (cf. (11.24))

$$\left. \begin{aligned} u_{\mathbf{k}}^2 &= \tfrac{1}{2}[1 + \eta(\mathbf{k})/\surd\{\eta(\mathbf{k})^2 + \Delta(T)^2\}], \\ v_{\mathbf{k}}^2 &= \tfrac{1}{2}[1 - \eta(\mathbf{k})/\surd\{\eta(\mathbf{k})^2 + \Delta(T)^2\}]. \end{aligned} \right\} \tag{12.40}$$

Here

$$\Delta(T) = \sum_{k} \frac{\mathscr{F}}{\Omega \Xi_0} u_{\mathbf{k}} v_{\mathbf{k}} (1-2f_{\mathbf{k}}) \tag{12.41}$$

is the half energy gap, now allowed to depend on temperature. In (12.41), $\mathscr{F}$ is Fröhlich's constant defined in (10.43), and Ξ_0 is the (normal) density of states at the Fermi surface.

Minimizing now with respect to $f_{\mathbf{k}}$ gives

$$0 = \mathrm{k}T\{\ln f_{\mathbf{k}} + \ln(1-f_{\mathbf{k}})\} + \surd\{\eta(\mathbf{k})^2 + \Delta(T)^2\}, \tag{12.42}$$

i.e.

$$f_{\mathbf{k}} = \frac{1}{1 + \exp\{E(\mathbf{k})/\mathrm{k}T\}}. \tag{12.43}$$

Thus we recover the usual Fermi–Dirac distribution function. However, the quasiparticle energy

$$E(\mathbf{k}) = \surd\{\eta(\mathbf{k})^2 + \Delta(T)^2\} \tag{12.44}$$

is now temperature dependent, through the temperature dependence of Δ.

By substituting (12.40) and (12.43) into (12.41), the self-consistency condition is found:

$$\mathscr{F}^{-1} = \tfrac{1}{2}\int\limits_0^{\hbar\bar{\omega}} \mathrm{d}\eta\, \frac{\tanh(E/2\mathrm{k}T)}{E}. \tag{12.45}$$

From (12.45) the dependence of the energy gap on temperature can be computed. $\Delta(T)$ cannot be found in closed form, but extensive tables have been published (Mühlschlegel 1959). The behaviour of $\Delta(T)$ is shown in Fig. 12.1. We see from the

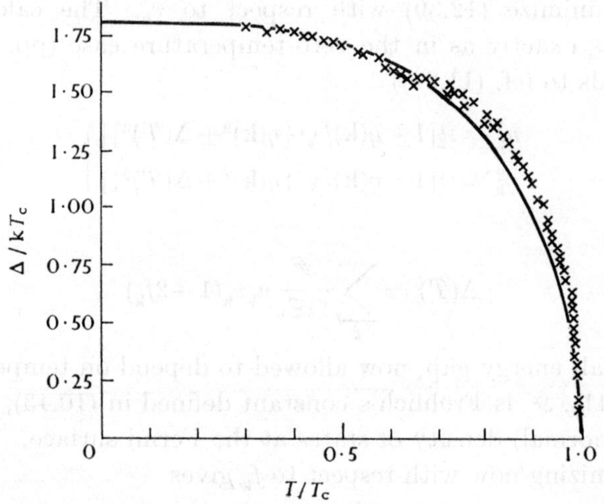

FIG. 12.1. The temperature-dependence of the BCS energy gap (Bardeen *et al.* 1957). The experimental points are those of Morse and Bohm (1957) for tin.

figure that at low temperatures the gap has very little temperature dependence, but at somewhat higher temperatures it is a decreasing function of temperature. At the transition temperature T_c, the curve meets the temperature axis with infinite slope ($\Delta \propto (T_c - T)^{\frac{1}{2}}$). The transition temperature is defined by the condition

$$\Delta(T_c) = 0, \tag{12.46}$$

which, with (12.45), gives

$$kT_c = 1{\cdot}14\hbar\bar{\omega}\exp(-1/\mathscr{F}). \tag{12.47}$$

When (12.46) is satisfied, of course, $\eta(\mathbf{k})$ and $E(\mathbf{k})$ become identical. The Bogoliubov transformation (12.16) reduces to the trivial case previously discussed; the quasiparticle and particle states coincide. From the behaviour of Δ near T_c, it can be shown that F has a discontinuity in its second derivative; thus there is an Ehrenfest second-order transition. The specific heat has a discontinuity of magnitude

$$\left[\frac{c_s - c_n}{c_n}\right]_{T_c} = 1{\cdot}43, \tag{12.48}$$

which is fairly close to the observed values for many superconductors (cf. Table 2.1, p. 36). The transition temperature shows the correct isotopic-mass dependence (10.53), since $\bar{\omega}$, the average phonon frequency, is proportional to the acoustic velocity s, and hence to $M^{-\frac{1}{2}}$. Moreover the appearance of the factor $\exp(-1/\mathscr{F})$ accounts for the smallness of the transition temperature T_c compared with the Debye temperature Θ.

Thus the microscopic theory leads to a temperature-dependent energy gap and quasiparticle spectrum. These have precisely the properties required for the construction of the phenomenological energy-gap model of § 2.3.

Exercises. (a) By substituting (12.40) and (12.43) into (12.39), derive (2.33).

(b) Use the one-to-one correspondence between Bloch states of energy $\eta(\mathbf{k})$ and quasiparticle states of energy $E(\mathbf{k}) = \sqrt{(\eta(\mathbf{k})^2 + \Delta^2)}$ to derive the expression (2.28) for the density of quasiparticle states in a superconductor.

12.3. Coherence effects: the BCS matrix elements for a perturbation

One of the successes of the microscopic theory has been the prediction that the attenuation coefficient of ultrasound falls off very sharply with temperature below the transition temperature T_c. This behaviour is *not* shared by, for example, the high-frequency electromagnetic loss; one cannot detect the onset of superconductivity optically.† This marked difference between the ultrasonic and electromagnetic attenuation was not foreshadowed by any of the phenomenological models. Thus its experimental confirmation by Morse and Bohm (1957) gave strong support to the BCS microscopic theory. In Fig. 12.2 the results of Morse and Bohm are compared with theory.

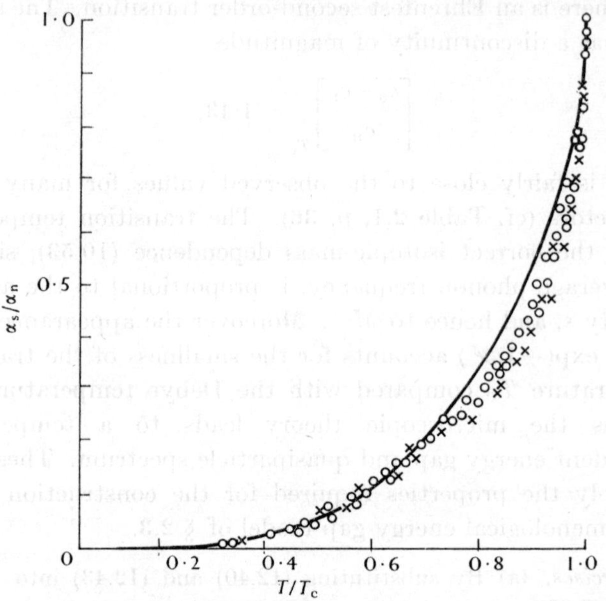

FIG. 12.2. The ultrasonic attenuation of superconductors. The ordinate is the ratio of the attenuation in the superconductor to the attenuation in the normal material. The full curve is theoretical (Bardeen *et al.* 1957) and the experimental points are those of Morse and Bohm (1957). o Tin; × Indium.

In order to discuss problems of this kind theoretically, the transition matrix elements between Bogoliubon states will be calculated for various kinds of external perturbation. This

† At microwave frequencies there *is* a marked change in the surface resistance between n- and s-phases. However, this is due not to changes in the matrix elements but to screening (see e.g. Waldram 1964).

calculation reveals the existence of coherence properties. For some perturbations there is a *constructive* interference between the Cooper pairs, e.g. for externally applied electromagnetic fields. But for a second class of perturbations, including ultrasonic propagation, the interference is *destructive* and the transition probability is greatly reduced.

In the absence of an external perturbation, the Bogoliubov–Valatin transformation (12.16) is assumed to diagonalize the Hamiltonian. This assumption is not rigorous, but provides a good enough approximation. An external field, acting on a normal metal, scatters single electrons from one Bloch state to another. Whether it interacts with the charge density or the current density,† the interaction can be written

$$\hat{H}_{\text{ext}} = \sum_{\mathbf{k},\mathbf{l},\sigma} \Gamma_{\mathbf{k},\mathbf{l},\sigma} \hat{a}^{+}_{\mathbf{k},\sigma} \hat{a}_{\mathbf{l},\sigma}. \tag{12.49}$$

However, the symmetry properties of the matrix elements $\Gamma_{\mathbf{k},\mathbf{l},\sigma}$ will depend on the nature of the perturbing field. The $\Gamma_{\mathbf{k},\mathbf{l},\sigma}$ will in general contain quantized field operators of the (Boson) perturbing field, but they do not contain any Fermion $\hat{a}^{+}$ or $\hat{a}$ operators. Here we may treat them as if they were pure c-numbers.

In particular for the problem of ultrasonic attenuation the interaction is the usual Bloch–Fröhlich interaction‡ (10.41), for which

$$\Gamma_{\mathbf{k},\mathbf{l},\sigma} = \Gamma_{-\mathbf{l},-\mathbf{k},-\sigma}. \tag{12.50}$$

In contrast with (12.50), the matrix elements for an electromagnetic perturbation have the property (see pp. 221–2)

$$\Gamma_{\mathbf{k},\mathbf{l},\sigma} = -\Gamma_{-\mathbf{l},-\mathbf{k},-\sigma}. \tag{12.51}$$

† We neglect the possibility of 'spin flip'. For the generalization to include this possibility, see e.g. Schrieffer (1964 pp. 69–72). This generalization is necessary for the explanation of the nuclear relaxation properties (Hebel and Slichter 1959).

‡ Since we are dealing with an acoustic wave of macroscopic amplitude, the number of phonons is so large that the phonon field can be treated in a classical approximation, with $\hat{b}_{\mathbf{q}}\exp(i\mathbf{q}.\mathbf{r}) = \hat{b}^{+}_{\mathbf{q}}\exp(-i\mathbf{q}.\mathbf{r}) = \sqrt{n_{\mathbf{q}}}$. This amounts to neglect of the spontaneous emission probability.

We shall refer to the symmetries (12.50) and (12.51) as Case (A) and Case (B) respectively.

Before Bogoliubov-transforming (12.49), we rearrange it in the form

$$\hat{H}_{\text{ext}} = \sum_{\mathbf{k},\mathbf{l}} \Gamma_{\mathbf{k},\mathbf{l},\uparrow} (\hat{a}^{+}_{\mathbf{k}\uparrow} \hat{a}_{\mathbf{l}\uparrow} \pm \hat{a}^{+}_{-\mathbf{l}\downarrow} \hat{a}_{-\mathbf{k}\downarrow}). \tag{12.52}$$

In (12.52) and subsequent expressions, the upper sign is to be taken when the perturbation has (A) symmetry and the lower sign when the perturbation has (B) symmetry. Applying the transformation (12.16), we have

$$\begin{aligned}\hat{H}_{\text{ext}} = \sum_{\mathbf{k},\mathbf{l}} \Gamma_{\mathbf{k},\mathbf{l},\uparrow} \{ &(u_{\mathbf{k}} v_{\mathbf{l}} \pm v_{\mathbf{k}} u_{\mathbf{l}})(\hat{\alpha}^{+}_{\mathbf{k}\uparrow} \hat{\alpha}^{+}_{-\mathbf{l}\downarrow} \pm \hat{\alpha}_{-\mathbf{k}\uparrow} \hat{\alpha}_{\mathbf{l}\downarrow}) \\ &+ (u_{\mathbf{k}} u_{\mathbf{l}} \mp v_{\mathbf{k}} v_{\mathbf{l}})(\hat{\alpha}^{+}_{\mathbf{k}\uparrow} \hat{\alpha}_{\mathbf{l}\uparrow} \pm \hat{\alpha}_{-\mathbf{k}\downarrow} \hat{\alpha}^{+}_{-\mathbf{l}\downarrow}) \\ &+ (v_{\mathbf{k}}^2 \pm v_{\mathbf{k}}^2)\delta_{\mathbf{k},\mathbf{l}} \}. \end{aligned} \tag{12.53}$$

On substitution from (12.21), this gives

$$\begin{aligned}\hat{H}_{\text{ext}} = \tfrac{1}{2} \sum_{\mathbf{k},\mathbf{l}} \Gamma_{\mathbf{k},\mathbf{l},\uparrow} \Bigg[&\left\{ 1 - \frac{\eta(\mathbf{k})\eta(\mathbf{l}) \mp \Delta^2}{E(\mathbf{k})E(\mathbf{l})} \right\} (\hat{\alpha}^{+}_{\mathbf{k}\uparrow} \hat{\alpha}^{+}_{-\mathbf{l}\downarrow} \pm \hat{\alpha}_{-\mathbf{k}\uparrow} \hat{\alpha}_{\mathbf{l}\downarrow}) \\ &+ \left\{ 1 + \frac{\eta(\mathbf{k})\eta(\mathbf{l}) \mp \Delta^2}{E(\mathbf{k})E(\mathbf{l})} \right\} (\hat{\alpha}^{+}_{\mathbf{k}\uparrow} \hat{\alpha}_{\mathbf{l}\uparrow} \pm \hat{\alpha}_{-\mathbf{k}\downarrow} \hat{\alpha}^{+}_{-\mathbf{l}\downarrow}) \\ &+ \{1 - \eta(\mathbf{k})/E(\mathbf{k})\}(1 \pm \delta_{\mathbf{k},\mathbf{l}}) \Bigg]. \end{aligned} \tag{12.54}$$

The transition probability P is required for a transition from an initial state containing a Bogoliubon of energy $E(\mathbf{k})$ to a final state in which it is replaced by a Bogoliubon of energy $E(\mathbf{l})$ (and all other quasiparticles are unaltered). P is proportional to the product of the density $\Xi(E(\mathbf{k}))$ of initial states, the density $\Xi(E(\mathbf{l}))$ of final states, the statistical factors $f(E(\mathbf{k}))$ and $1 - f(E(\mathbf{l}))$, and the square of the transition matrix element. Writing the arguments $\mathbf{k}$ and $\mathbf{l}$ for convenience as subscripts, we have

$$\begin{aligned} P = \int \frac{2\pi}{\hbar} \delta(E_{\mathbf{k}} - E_{\mathbf{l}} \pm \hbar\omega)\, \mathrm{d}E_{\mathbf{k}} \mathrm{d}E_{\mathbf{l}} \\ |(\hat{H}_{\text{ext}})_{\mathbf{k},\mathbf{l}}|^2 \Xi(E_{\mathbf{k}}) \Xi(E_{\mathbf{l}})\, f(E_{\mathbf{k}}) \{1 - f(E_{\mathbf{l}})\}. \end{aligned} \tag{12.55}$$

With the simplifying assumption that the matrix elements $\Gamma_{\mathbf{k},\mathbf{l},\sigma}$ are symmetrical against reflection in the Fermi surface, P may be reduced to

$$P = \int \frac{2\pi}{\hbar} \mathrm{d}E_{\mathbf{k}} \mathrm{d}E_{\mathbf{l}}\, \Xi(E_{\mathbf{k}})\Xi(E_{\mathbf{l}})\, 4|\Gamma_{\mathbf{k},\mathbf{l},\sigma}|^2 \{1 \mp \Delta^2/E_{\mathbf{k}}E_{\mathbf{l}}\} f(E_{\mathbf{k}})\{1 - f(E_{\mathbf{l}})\}. \quad (12.56)$$

The distinction between Cases (A) and (B) is now clear. For Case (A), e.g. for ultrasonic absorption, the upper sign is appropriate. The density-of-states factors $\Xi(E_{\mathbf{k}})$ and $\Xi(E_{\mathbf{l}})$ cancel the factor $1 - \Delta^2/E_{\mathbf{k}}E_{\mathbf{l}}$. The rate of absorption of energy can be shown to be of the form

$$\alpha_{\mathrm{s}}/\alpha_{\mathrm{n}} = 2f(\Delta),$$

where α_{n} is the absorption in the normal state. On the other hand, in the electromagnetic case the perturbing Hamiltonian has type (B) symmetry, corresponding to the lower sign in (12.56). The factor $1 + \Delta^2/E_{\mathbf{k}}E_{\mathbf{l}}$ does *not* cancel with the density-of-states factors, and is always greater than unity.

Thus, in addition to justifying the assumptions of the phenomenological energy-gap theory, the BCS theory has successfully predicted the appearance of some striking and unexpected coherence properties.

CHAPTER 13

ELECTROMAGNETIC PROPERTIES OF SUPERCONDUCTORS

In Chapter 12 the BCS–Bogoliubov theory has been shown to give a good account of the energy gap, and to justify the postulates of the 'BCS energy-gap model' (§ 2.3). Moreover, through the coherence factors described in § 12.3, the microscopic theory was able to break new ground. The theory predicted the hitherto unsuspected difference between the electromagnetic and ultrasonic attenuation at high frequencies and low temperatures. But we still have to exhibit the low-frequency or static electromagnetic properties; in other words it remains to be shown *that the BCS-condensed phase is superconducting.*

The two fundamental electromagnetic properties of a superconductor are (a) perfect conductivity and (b) perfect diamagnetism. In § 13.1, the phenomenon of persistent currents is discussed. Wave functions are introduced to describe states carrying a persistent current. These states are shown to lie within the energy gap. However, the persistence of the current is not easily explained without a fuller theoretical framework. The discussion of this topic is taken up in § 14.3.1. The derivation of the Meissner effect by Bardeen *et al.* (1957) is described in § 13.2. This treatment leads to Pippard's non-local equation (6.11). However, the BCS theory is not gauge-invariant. The lack of gauge invariance leads to some fundamental difficulties, which are discussed in § 13.3. This discussion indicates why the BCS theory, despite its lack of gauge invariance, leads to the correct penetration depth and Pippard length.

The chapter closes with a brief account (§ 13.4) of the Knight shift (Knight 1949, Reif 1957). This is one effect for which the theoretical predictions are *not* in accord with the results of experiment. Some possible reasons for the discrepancy are given. The subject is still somewhat controversial, and the discussion must therefore be regarded as provisional.

13.1. Wave functions for persistent currents

The first property of superconductors to be discovered was the vanishing of the electric resistance (§ 1.1). The present section is an introduction to the microscopic theory of this phenomenon. But first it should be noted that perfect conductivity manifests itself in two somewhat different experimental situations. (a) A simply-connected superconductor is supplied with current through non-superconducting external leads. The potential difference between the points of contact with the superconductor is zero even when a current is flowing. (b) A multiply-connected superconductor without external leads has magnetic flux trapped in the non-superconducting inclusions. There is a persistent current which maintains the trapped flux, and suffers no attenuation. In case (b) the situation is complicated by the condition (3.45) of flux quantization. On the other hand in case (a) the flux is not quantized, and the magnetic field does not even appear to be relevant. This does not mean that case (a) is the simpler theoretically, for it has its own complicating features. There are necessarily junctions between the superconductor and the normal leads, and at these points there may be contact resistances, potential discontinuities, and tunnelling effects.

It is difficult to give a satisfactory account of either of these two geometries. We evade both sets of difficulties by studying a simpler problem. Consider an infinitely long superconductor or, equivalently, a superconducting ring which is so large that the quantization of flux is unimportant. In a finite ring, the quantization of flux appears to provide an independent mechanism for suppressing the resistive scattering processes, by imposing an additional barrier against small changes in the current. Therefore, if currents can persist in an infinite ring, *a fortiori* they can persist in a finite ring.

The previous chapter was largely concerned with the properties of a class of electron-like quasiparticles, the Bogoliubons.

These excitations are separated from the ground state by an energy gap. But besides these there are many other possible kinds of excitation. In particular a class of current-carrying states will be constructed for which there is *no* energy gap. These excitations of the system are 'collective', in the sense that they involve the correlated motion of many of the conduction electrons. Because they are collective, and because there is an energy gap separating the Bogoliubon states from them, the usual electron–phonon or electron–impurity scattering processes are incapable of destroying the current they carry.

Consider the state

$$|\Phi_{\mathbf{Q}}\rangle = \prod_{\mathbf{k}} (u_{\mathbf{k}} + v_{\mathbf{k}} \hat{a}^{+}_{\mathbf{k}+\mathbf{Q},\uparrow} \hat{a}^{+}_{-\mathbf{k}\downarrow})|0\rangle, \tag{13.1}$$

where $\mathbf{Q}$ is a constant vector, the same for *all* Cooper pairs. The interaction energy $\langle\Phi_{\mathbf{Q}}|\hat{H}'_{\mathrm{BCS}}|\Phi_{\mathbf{Q}}\rangle$ is substantially the same for $|\Phi_{\mathbf{Q}}\rangle$ as for $|\Phi_0\rangle$. Consequently the two states $|\Phi_{\mathbf{Q}}\rangle$ and $|\Phi_0\rangle$ differ in energy only by the kinetic energy of the electron current of the state (13.1).

The total 'momentum' operator†

$$\hat{\mathbf{P}} = \sum_{k} \hbar \mathbf{k} \hat{a}^{+}_{k} \hat{a}_{k} \tag{13.2}$$

commutes with the BCS Hamiltonian (12.6), and is therefore a constant of the motion. Thus the energy and the momentum of the system can be simultaneously quantized. We wish to find the lowest-energy state belonging to a specified momentum value $\mathbf{P}$; this state will be shown to be of the form (13.1).

To take account of the constraint (13.2) on the momentum, let us introduce a Lagrangian multiplier $\mathbf{v}$, and look for the lowest-energy eigenstate of the augmented Hamiltonian

$$\hat{\mathscr{H}}_{\mathrm{aug}} = \hat{H}_{\mathrm{el}} - \mu \hat{N}_{\mathrm{e}} - \mathbf{v}.\hat{\mathbf{P}} + \hat{H}'_{\mathrm{BCS}} \tag{13.3}$$

† This conservation law is strictly a law of conservation of the wave vector rather than the momentum. For a Bloch electron, the two are not identical. Moreover, the law is not a rigorous one. The wave vector in a crystal lattice is only conserved *modulo* $\mathbf{K}$, where $\mathbf{K}$ is a vector of the reciprocal lattice (cf. § 10.4). Provided that the Brillouin zone boundary does not touch the Fermi surface, (13.2) will hold at low enough temperatures. For, in this case we can neglect Umklapp processes.

With the substitutions

$$\left.\begin{aligned} \mathbf{k}' &= \mathbf{k}-m\mathbf{v}, \\ \mu' &= \mu+\tfrac{1}{2}mv^2, \end{aligned}\right\} \tag{13.4}$$

the kinetic and Lagrange-multiplier terms of (13.3) may be written

$$\hat{H}_{\text{el}}-\mu\hat{N}_{\text{e}}-\mathbf{v}.\hat{\mathbf{P}} = \sum_{k'}\{\varepsilon(\mathbf{k}')-\mu'\}\hat{a}^{+}_{k'}\hat{a}_{k'}. \tag{13.5}$$

The only effect of transformation (13.4) on the BCS interaction Hamiltonian is to relabel the momentum states, without affecting the matrix elements. The expectation value of $\hat{H}'_{\text{BCS}}$ in the state (13.1) will differ from that for the BCS ground state (11.8) only in the selection of the *reduced* terms from the Hamiltonian. The differences will only arise for pairs $(\mathbf{k},\uparrow;-\mathbf{k},\downarrow)$ with the following property: *one* of the Bloch states $-\mathbf{k},\downarrow$; $\mathbf{k}+\mathbf{Q},\uparrow$ must have energy $|\eta|<\hbar\bar{\omega}$, and the other member $|\eta|>\hbar\bar{\omega}$. If the vector $\mathbf{Q}$ is small enough, the states affected are numerically unimportant compared with the other states. Thus we may write

$$\langle\Phi_{\mathbf{Q}}|\hat{H}'_{\text{BCS}}|\Phi_{\mathbf{Q}}\rangle \simeq \langle\Phi_0|\hat{H}'_{\text{BCS}}|\Phi_0\rangle. \tag{13.6}$$

In this approximation, the energy of the state (13.1) differs from that of the ground state by an amount

$$(\mu'-\mu)N_{\text{e}} = \tfrac{1}{2}N_{\text{e}}mv^2. \tag{13.7}$$

Equation (13.7) has a natural interpretation: the kinetic energy of the uniform motion of the electron gas *as a whole*. The current appears in the role of a single degree of freedom, with a macroscopic mass.

The above argument is quite general, and has not used any of the properties of the condensate. The only special property of the BCS Hamiltonian which we have used has been the approximate translation invariance of the potential energy. But the mere existence of these collective current-carrying states is not sufficient to explain their *persistence*. Precisely the same transformation can be used to describe states of uniform motion of a rigid body—indeed for such a system the translation invariance of the Hamiltonian is exact. A rifle bullet travelling through the air may be described by this transformation (the ordinary Galilean transformation in a slightly unfamiliar notation). Moreover, the rifle bullet will hardly be affected in its motion by single collisions with air molecules. But besides having macroscopic mass, the bullet also has a macroscopic *cross-section* for collision with air molecules, and of course its kinetic energy of motion will be dissipated rapidly.

In a superconductor the important new feature is the existence of the energy gap. Although the current-carrying collective states (13.1) lie in the gap, they are separated from other states (12.27) (containing at most a few quasiparticles) by an energy $\geqslant \Delta$. In order to scatter an electron out of the condensate into a quasiparticle state, at least this energy Δ is needed. This large threshold energy will strongly inhibit such processes, since the scattering matrix element is not much affected by the condensation.

Thus we might expect the states (13.1) to be unusually stable. But this mechanism cannot be the whole story. It clearly fails to account for perfect conductivity at temperatures near T_c.† The true explanation is related to the phase of the Ginzburg–Landau wave function; see § 14.3.1.

13.2. The Meissner effect

To test whether any microscopic theory of superconductivity indeed shows a Meissner effect, Schafroth's criterion (§ 4.2) should be used. That is, the response function K_{ij} (defined in (4.25)) is evaluated, and the limit

$$\lim_{q\to 0} q^2 \mathscr{K}_1(q^2) = a_{-1} \tag{13.8}$$

is calculated. If the limit exists and is non-zero, the system *does* expel the magnetic field. Although the above prescription is straightforward in principle, in practice the difficulties are

† I am indebted to Dr B. D. Josephson for stressing this limitation of energy-gap 'proofs' of zero resistivity.

severe. It is much simpler to work in a suitably selected gauge rather than in an arbitrary one. But then (cf. p. 64) approximations which are not gauge-invariant are dangerous.

In the presence of an electromagnetic field, the Hamiltonian is modified: wherever in the absence of the field the momentum $\hat{\mathbf{p}}_i$ of an electron appears, it must be replaced[26] by $\hat{\mathbf{p}}_i-(e/c)\mathbf{A}(\mathbf{r}_i)$, where $\mathbf{r}_i$ is the position variable conjugate to $\hat{\mathbf{p}}_i$ and $\mathbf{A}(\mathbf{r})$ is the vector potential at the point $\mathbf{r}$. In addition there is a new term $-e\phi$ in the Hamiltonian, where ϕ is the scalar potential. Since we are interested in the effect of magnetic rather than electric fields, it is natural to choose the gauge in such a way that $\phi = 0$ in the absence of an electric field. In second-quantized language the current density is

$$\hat{\mathbf{J}}(\mathbf{r}) = ne\mathbf{v} = \sum_\sigma \hat{\Psi}^+_\sigma(\mathbf{r})\frac{e}{m}\left(\mathbf{p}-\frac{e}{c}\mathbf{A}(\mathbf{r})\right)\hat{\Psi}_\sigma(\mathbf{r})$$

$$= \sum_\sigma \left[\frac{ie\hbar}{m}\{\hat{\Psi}^+_\sigma\nabla\hat{\Psi}_\sigma-(\nabla\hat{\Psi}^+_\sigma)\hat{\Psi}_\sigma\}-\frac{e^2}{mc}\hat{\Psi}^+_\sigma\mathbf{A}(\mathbf{r})\hat{\Psi}_\sigma\right], \quad (13.9)^{(26)}$$

and the electronic kinetic energy term is replaced by

$$\sum_\sigma \int d^3r\, \hat{\Psi}^+_\sigma(\mathbf{r})\left\{-i\hbar\nabla-\frac{e}{c}\mathbf{A}(\mathbf{r})\right\}^2\hat{\Psi}_\sigma(\mathbf{r}). \quad (13.10)^{(26)}$$

The field operators $\hat{\Psi}^+_\sigma(\mathbf{r}), \hat{\Psi}_\sigma(\mathbf{r})$ (cf. (9.68)) are

$$\left.\begin{aligned}\hat{\Psi}^+_\sigma(\mathbf{r}) &= \Omega^{-\frac{1}{2}}\sum_{\mathbf{k}}\hat{a}^+_{\mathbf{k},\sigma}\exp(-i\mathbf{k}.\mathbf{r}),\\ \hat{\Psi}_\sigma(\mathbf{r}) &= \Omega^{-\frac{1}{2}}\sum_{\mathbf{k}}\hat{a}_{\mathbf{k},\sigma}\exp(i\mathbf{k}.\mathbf{r}).\end{aligned}\right\} \quad (13.11)$$

We write

$$\mathbf{A}_\mathbf{q} = (2\pi)^{-3}\int d^3r\,\mathbf{A}(\mathbf{r})\exp(-i\mathbf{q}.\mathbf{r}). \quad (13.12)$$

On taking the Fourier transform of (13.10), the electronic kinetic energy (9.62) is replaced by the expression

$$\hat{H}_{\text{el}} = \sum_k \frac{\hbar^2k^2}{2m}\hat{a}^+_k\hat{a}_k+\frac{e\hbar}{\Omega mc}\sum_{k,\mathbf{q}}(\mathbf{A}_\mathbf{q}.\mathbf{k})\hat{a}^+_{k+\mathbf{q}}\hat{a}_k+O(A^2).$$

$$(13.13)^{(26)}$$

Hence in the presence of the field the total Hamiltonian may be written

$$\hat{H} = \hat{H}_{\text{Fröhlich}} + \hat{H}_{\text{e.m.}} + O(A^2)$$

where

$$\hat{H}_{\text{e.m.}} = \frac{e\hbar}{\Omega mc} \sum_{k,\mathbf{q}} (\mathbf{A}_{\mathbf{q}} \,.\, \mathbf{k}) \hat{a}^{+}_{k+\mathbf{q}} \hat{a}_{k}. \qquad (13.14)^{(26)}$$

Fourier-transforming (13.9), the current density is

$$\hat{\mathbf{J}}(\mathbf{r}) = \sum_{k,\,\mathbf{q}} \hat{a}^{+}_{k+\mathbf{q}} \hat{a}_{k} \exp(-i\mathbf{q}\,.\,\mathbf{r}) \left\{ \frac{e\hbar}{2\Omega mc}(2\mathbf{k}+\mathbf{q}) - \frac{1}{2\Omega m}\mathbf{A}_{\mathbf{q}} \right\},$$
$$= \hat{\mathbf{J}}_{\mathrm{P}} + \hat{\mathbf{J}}_{\mathrm{D}}, \qquad (13.15)^{(26)}$$

where the terms $\hat{\mathbf{J}}_{\mathrm{P}}$ and $\hat{\mathbf{J}}_{\mathrm{D}}$ may be interpreted respectively as paramagnetic and diamagnetic contributions to the total current.

The above expressions (13.14), (13.15) are gauge-invariant provided we restrict our attention to time-independent gauge transformations only. This will be general enough.

Starting from (13.14), we may now eliminate the electron–phonon interaction in favour of an electron–electron interaction. This is done by the canonical transformation of Fröhlich (1952) and Bardeen and Pines (1955). The electromagnetic contribution (13.14) must be included in the Hamiltonian before transformation. But there is an important proviso: *simultaneously the same canonical transformation must be applied to the current density operator* (13.15).

However, even in the absence of the electromagnetic term (13.14), the Fröhlich transformation leads to the rather complicated interaction Hamiltonian (10.55). It would be preferable to work with the simplified BCS form (11.20) for the electron–electron interaction. For the present we shall assume that it is legitimate to do so, while retaining the expression (13.15) for the current density. But the simplified Hamiltonian (11.20) of the BCS theory is not gauge-invariant. Hence even if (13.15) and (11.20) are mutually consistent in some gauge, they *cannot* be consistent in every gauge. Our assumption postulates the existence of a London gauge (§ 3.2), in which div $\mathbf{A} = 0$ and in which $\mathbf{H} = 0$ corresponds uniquely to $\mathbf{A} = 0$. In this gauge we shall derive Pippard's form (6.11) for the response function.

In a normal metal, the wave function Ψ adjusts to the applied field in such a way that in the 'd.c. limit' $\mathbf{q} \to 0$, $\mathbf{J}_\mathrm{P}$ just cancels $\mathbf{J}_\mathrm{D}$. For a Meissner effect this cancellation must be incomplete; in other words the response, for a uniform static magnetic field, is to be diamagnetic. For this to happen, i.e. for $\mathbf{J}_\mathrm{D}$ to be larger than $\mathbf{J}_\mathrm{P}$, the wave function must in some sense be 'stiff' against the action of the magnetic field (London 1950 §§ 25, 26), and its adjustment must be incomplete. The BCS calculation, to be outlined here, finds such a stiffening of the wave function to be a consequence of the energy gap.

The calculation is a perturbative one, in which the magnetic vector potential $\mathbf{A}$ is treated as small, and powers of $\mathbf{A}$ higher than the first are neglected. To first order the ground-state wave function, in the presence of the field, is

$$|\Phi\rangle = |\Phi_0\rangle + \sum_i \frac{\langle\Phi_i|\hat{H}_{\mathrm{e.m.}}|\Phi_0\rangle}{E_0 - E_i}|\Phi_i\rangle, \tag{13.16}$$

where $|\Phi_0\rangle$ is the ground state and $|\Phi_i\rangle$ are excited states (of energies E_0 and E_i respectively) in the absence of the field.

The paramagnetic current density $\hat{\mathbf{J}}_\mathrm{P}$ has the expectation value

$$\langle\hat{\mathbf{J}}_\mathrm{P}(\mathbf{r})\rangle = \langle\Phi|\hat{\mathbf{J}}_\mathrm{P}|\Phi\rangle \tag{13.17}$$

$$= \frac{e^2\hbar}{2m^2c\Omega}\sum_{k,l,\mathbf{q}}\{(2\mathbf{k}+\mathbf{q})\,.\,\mathbf{l}\}\mathbf{A}_\mathbf{q}\,\mathrm{e}^{-i\mathbf{q}.\mathbf{r}}$$

$$\left\{\frac{\langle\Phi_0|\hat{a}^+_{l-\mathbf{q}}\hat{a}_l|\Phi_i\rangle\langle\Phi_i|\hat{a}^+_{k+\mathbf{q}}\hat{a}_k|\Phi_0\rangle}{E_0-E_i}+\text{h.c.}\right\}, \tag{13.18}$$ [26]

on substituting the form of the wave function (13.16). The only excited states which enter the sum $\sum_i$ are those formed by operating on $|\Phi_0\rangle$ with expressions such as $\hat{a}^+_{k+\mathbf{q}}\hat{a}_k$; they are all states containing two quasiparticles. At finite temperatures, the quantum-mechanical expectation values must be replaced by ensemble averages. The finite-temperature sum is then over intermediate virtual states containing two particles less, as well as two particles more, than the original state. After Fourier transformation and some lengthy but straightforward reduction, the final expression for the paramagnetic current at finite temperature is

$$\langle \hat{\mathbf{J}}_{\mathrm{P}} \rangle_{\mathbf{q}} = \frac{e^2\hbar}{(2\pi)^3 m^2 c} \sum_{k} \{(2\mathbf{k}+\mathbf{q}).\mathbf{k}\}\mathbf{A}_{\mathbf{q}}\, L(\eta(\mathbf{k}), \eta(\mathbf{k}+\mathbf{q}), \Delta), \tag{13.19}$$ [26]

where

$$L(\eta_1, \eta_2, \Delta) = \tfrac{1}{2}\left\{\left(\frac{1-f_1-f_2}{E_1+E_2}\right)\left(1-\frac{\eta_1\eta_2+\Delta^2}{E_1E_2}\right) + \left(\frac{f_1-f_2}{E_1-E_2}\right)\left(1+\frac{\eta_1\eta_2+\Delta^2}{E_1E_2}\right)\right\}. \tag{13.20}$$

If there were no energy gap, $\Delta = 0$, then the quasiparticle energy E would be identical with the Bloch energy η, and the first term in $L(\eta_1, \eta_2, \Delta)$ would vanish. The second term in $L(\eta_1, \eta_2, \Delta)$ would survive, and compensate the diamagnetic current $\mathbf{J}_{\mathrm{D}}$ exactly. But when there *is* an energy gap, $E \neq \eta$. Then the first term in (13.20) no longer vanishes, and $\mathbf{J}_{\mathrm{D}}$ is not completely compensated. The presence of the energy gap inhibits the appearance of virtual quasiparticles in (13.16), and this is the stiffening of the wave function which we require for a Meissner effect.

Transforming (13.19) back to configuration space, it is found that†

$$\mathbf{J}(\mathbf{r}) = -\frac{3}{4\pi c\Lambda}\left(\frac{\pi\Delta(0)}{\hbar v_{\mathrm{F}}}\right)\int \mathrm{d}^3r' \frac{K(|\mathbf{r}-\mathbf{r}'|, T)\,(\mathbf{r}-\mathbf{r}')\,(\mathbf{r}-\mathbf{r}').\mathbf{A}(\mathbf{r}')}{|\mathbf{r}-\mathbf{r}'|^4}, \tag{13.21}$$ [26]

† For a derivation of (13.21), see Bardeen *et al.* (1957).

with a kernel

$$K(R, T) = \frac{\Lambda(T)\Delta(T)}{\Lambda(0)\Delta(0)}\left\{\tanh\frac{\Delta(T)}{2kT} - \frac{2\Delta(T)}{\pi}\int\limits_0^\infty d\eta\left(\frac{1-2f(\eta)}{\eta E}\right)\sin\left(\frac{2R\eta}{\hbar v_F}\right)\right\}. \tag{13.22}$$

While (13.21) is not quite identical with Pippard's expression (6.11), it is very similar. In Fig. 13.1, the BCS kernel $K(R, 0)$

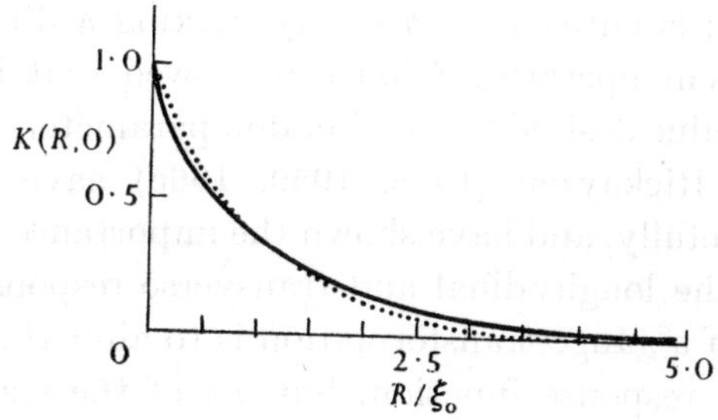

FIG. 13.1. Comparison of the BCS kernel (13.22) (continuous curve) with the Pippard kernel (dotted curve). From Bardeen, Cooper, and Schrieffer, *Phys. Rev.* **108**, 1175 (1957).

is compared with the Pippard kernel $\exp(-R/\xi_0)$; it will be seen that they agree closely throughout the range of R.

In particular, the London limit arises when **A** is nearly constant over the coherence distance

$$\xi_0 = \hbar v_F/\pi\Delta(0). \tag{13.23}$$

Equation (13.21) then reduces to London's second equation (3.15), with[19]

$$\lambda_L^2(0) = mc^2/4\pi ne^2.$$

Superficially, we seem to have proved too much. The existence of an energy gap appears to be a sufficient condition for a Meissner effect. If this were so, all insulators would be superconductors! But in fact the above derivation depends on the detailed structure of the function $L(\eta_1, \eta_2, \Delta)$. This function in turn depends on the coherence factors for Case (B) symmetry (§ 12.3). In an insulator these coherence factors are different, and no Meissner effect results. The real weakness in the above derivation lies elsewhere, and will be discussed of the next section.

13.3. Gauge transformations and the Meissner effect

As was stressed at the beginning of the previous section, the BCS simplified Hamiltonian (11.20) is not gauge-invariant. Schafroth (1958) showed that in this situation there is no unique prescription for defining the current operator. The BCS choice (13.9) is entirely *ad hoc*. By making a different choice for the current operator, Schafroth showed that it is possible to get any value desired for the London parameter Λ. Anderson (1958) and Rickayzen (1958, 1959, 1965) have studied this problem carefully, and have shown the importance of discussing separately the longitudinal and transverse response functions. The effect of a gauge transformation is to alter the form of the longitudinal response function, but *not* of the transverse one. In order to keep the longitudinal response invariant, it is necessary to take into account the collective excitations of the electron gas—in particular the plasma modes. When these modes are included, gauge invariance is approximately restored.

However, the mathematical development is complicated. Sewell (1959) has shown that *in all orders* of perturbation theory, the Meissner effect is absent. Sewell's result, coupled with Rickayzen's, shows that even if the perturbation series converges the convergence cannot be uniform. Consequently Rickayzen's method does not give much insight into the mechanism by which the superconducting material excludes a magnetic field.

An alternative approach, due to Blatt (1960), gives a clearer picture of the physics of the effect. Define the pair creation operator

$$\hat{c}_{\mathbf{k}}^{+} = \hat{a}_{\mathbf{k}\uparrow}^{+}\hat{a}_{-\mathbf{k}\downarrow}^{+}. \tag{13.24}$$

In terms of (13.24), the BCS ground state (11.8) can be written

$$|\Phi_0\rangle = \prod_{\mathbf{k}} (u_{\mathbf{k}}+v_{\mathbf{k}}\hat{c}_{\mathbf{k}}^{+})|0\rangle,$$

or, apart from a normalization factor,

$$|\Phi_0\rangle = \prod_{\mathbf{k}} (1+\phi_{\mathbf{k}}\hat{c}_{\mathbf{k}}^{+})|0\rangle. \tag{13.25}$$

Expanding the product, we have

$$|\Phi_0\rangle = \left(1+\sum_{\mathbf{k}}\phi_{\mathbf{k}}\hat{c}_{\mathbf{k}}^{+}+\frac{1}{2!}\sum_{\mathbf{k},\mathbf{l}}{}'\phi_{\mathbf{k}}\phi_{\mathbf{l}}\hat{c}_{\mathbf{k}}^{+}\hat{c}_{\mathbf{l}}^{+}+\ldots\right)|0\rangle, \tag{13.26}$$

where the primes on the summation signs mean that terms with $\mathbf{k} = \mathbf{l}$, etc., are to be omitted. Note, however, that

$$(\hat{c}_{\mathbf{k}}^{+})^2 \equiv 0, \tag{13.27}$$

and hence that these restrictions on the summation are ineffective. The primes may therefore be omitted, and we may write

$$|\Phi_0\rangle = \left(1+\hat{W}^{+}+\frac{1}{2!}\hat{W}^{+2}+\ldots\right)|0\rangle, \tag{13.28}$$

where

$$\hat{W}^{+} = \sum_{\mathbf{k}}\phi_{\mathbf{k}}\hat{c}_{\mathbf{k}}^{+}. \tag{13.29}$$

Already in the BCS theory, $u_{\mathbf{k}}$ and $v_{\mathbf{k}}$ had to be cut off for $\eta<-\hbar\bar{\omega}$ and $\eta>\hbar\bar{\omega}$ respectively. With this in mind, let us reinterpret the vacuum state $|0\rangle$ in (13.28) to mean a Fermi sea, *full up to the energy* $\mu-\hbar\bar{\omega}$, and empty above it. Let the number of electrons contained in this state $|0\rangle$ be $N_{\mathrm{e}}-2\nu$. Then $\hat{W}^{+}$ may be regarded as the operator which creates a single Cooper pair outside this full Fermi sea. The wave function of this Cooper pair in momentum space is just $\phi_{\mathbf{k}}$. $(\hat{W}^{+})^2$ represents two Cooper pairs, etc. The projection of the BCS ground state (11.8) onto the space of exactly N_{e} electrons will be proportional to

$$(\hat{W}^{+})^{\nu}|0\rangle. \tag{13.30}$$

The operators $\hat{W}^{+}$ are, to lowest approximation, the creation operators for *Bosons*, and it is well known that a gas of charged (but noninteracting) Bosons shows a Meissner effect.† The

† For a rigorous demonstration, see Schafroth (1955). However, in a real superconductor, the 'bosons' overlap strongly, and the present argument is inadequate. The crucial step is to show that Gor'kov's anomalous propagator F (equation (14.29)) can be non-zero.

difficulty with the BCS derivation is that *almost any erroneous calculation will give the effect*. In this situation it is dangerous to make any approximations. In Blatt's picture the problem is less acute, since the Meissner effect appears in zero order. Moreover, it is not *necessary* to have an energy gap in order to have a Meissner effect; all that is needed is a BCS condensation. The distinction becomes important for a class of 'gapless' superconductors which have been discovered (Reif and Woolf 1962). These have a BCS ground-state wave function, but the gap function $\Delta(\mathbf{k})$ vanishes for $\eta(\mathbf{k}) \ll \hbar\bar{\omega}$.

In the Blatt wave function (13.30), the form of the Cooper pair function $\phi_{\mathbf{k}}$ depends on the gauge, but the *number* ν of Cooper pairs present is not affected by gauge transformations. The change of $\phi_{\mathbf{k}}$ under gauge transformations is necessary to preserve the invariance of physically observable quantities.

The way in which London's stiffness of the state manifests itself is through the gauge-invariant quantity ν. Under an applied magnetic field, the number ν of Cooper pairs will in general change. In the presence of an energy gap, however, any change in the number of pairs requires a considerable expenditure of energy. For small enough fields we thus expect the number ν of pairs to be *independent* of the field. For a gapless superconductor the situation is more complicated, particularly in the BCS formalism. The Blatt method is capable of describing a gapless superconductor more easily; ν is merely required to be a sufficiently slow function of the field strength.

13.4. Pauli spin paramagnetism

The Pauli spin paramagnetism of a superconductor is not directly measurable by d.c. techniques, since it is masked by the Meissner diamagnetism. However, the Knight shift provides an indirect method of measuring it. By the Knight shift is meant the shift of the nuclear magnetic resonance frequency in a medium, compared with the free atom (Knight 1949).

At $T = 0$, an energy $>2\Delta$ is required to 'flip' the spin of even one electron. It therefore follows that the Pauli paramagnetism should be zero, since if the magnetic field H is less than $2\Delta/\mu_B$ (where μ_B is the Bohr magneton), there will be insufficient energy available to flip any spins. At finite temperatures $T > 0$, there will be some Bogoliubons present, whose spins *can* be flipped. The theoretical paramagnetism will therefore rise continuously from zero at $T = 0$ to the Pauli value at $T = T_c$. The solid curve in Fig. 13.2 shows Yosida's (1958) prediction, based on the BCS theory.

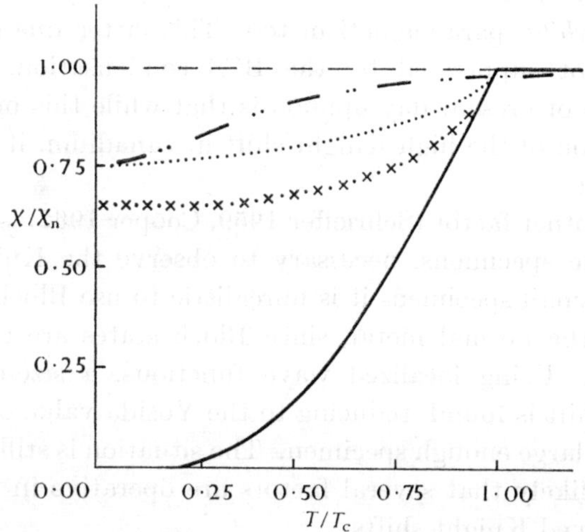

Fig. 13.2. The Knight shift. The full curve shows Yosida's (1958) theoretical prediction, and the other curves represent smoothed and extrapolated experimental data:

— — — — Vanadium films (Noer and Knight 1964);
··——··——·· Aluminium films (Hammond and Kelly 1964);
·········· Tin films (Androes and Knight 1961);
·×·×·×·×·×· Mercury colloid (Reif 1957).

Experiments have been performed on a number of materials and geometries. It is of course necessary to use specimens with at least one dimension small compared with the penetration depth, so that the magnetic field will penetrate the specimen. Some of these experimental results are compared with theory in Fig. 13.2, and it is seen that the agreement could hardly be worse!

Various suggestions have been made to explain the disagreement. Ferrell (1959) and Anderson (1959) believe that the explanation lies in spin-orbit scattering. But, if this were the only factor, we should expect to find a much lower Knight shift in the light metals aluminium and vanadium than in the heavier tin and mercury. In fact, vanadium shows an exceptionally high superconducting Knight shift.

Clogston *et al.* (1962*a*) believe that the Knight shift measurements are not detecting the Pauli paramagnetism alone, but the *orbital* paramagnetism too. This latter quantity is, of course, not suppressed by the BCS condensation. But the consensus of present-day opinion is that while this may be the explanation of the high Knight shift in vanadium, it is not the whole story.

Yet another factor (Schrieffer 1959, Cooper 1962) is the small size of the specimens, necessary to observe the Knight shift. In these small specimens it is unrealistic to use Bloch states to describe the normal metal, since Bloch states are completely non-local. Using localized wave functions, a size-dependent Knight shift is found, reducing to the Yosida value only in the limit of a large enough specimen. The situation is still not clear, but it is likely that several factors are operative in producing the observed Knight shifts.

CHAPTER 14

THE GOR'KOV METHOD: PROPAGATORS IN SUPERCONDUCTORS

THE method of 'propagators' or 'Green functions' has recently become popular in many branches of solid-state physics. The concept of the propagator was originally developed in the study of quantum electrodynamics, but has since been applied to many other problems. It provides a useful set of mathematical techniques for deriving physical results with minimum use of auxiliary and unobservable functions. In principle the wave function or state vector of a quantum state contains all the physical information about the state, as well as some unphysical information (the absolute phase of the wave function). The propagator is a closely related auxiliary function, which represents the correlation between the states of the system at different times.

The Green function formalism is, in effect, a new language for describing physical processes. It is a more economical language, in the sense that physically meaningful results are derived more directly using Green functions than using more traditional methods. But, as with any other language, it is necessary to learn its 'syntax' before using it. The present volume is not intended to be a text on propagator techniques, and the background material has therefore been kept to a minimum. To the reader who is already familiar with Green-function methods, § 14.1 will serve merely to introduce the notation and sign conventions. But many readers may be unfamiliar with these techniques. It is hoped that the introductory section § 14.1 will enable the uninitiated reader to capture something of the spirit of this approach, and to follow the subsequent application to superconductors though the individual steps in the calculation may appear to lack motivation.†

In § 14.1 the one-particle propagator is defined and it is shown that most information of physical interest is contained in it. We shall calculate this propagator for an ideal Fermi gas, and briefly discuss its properties. An auxiliary function, the spectral density $\tilde{\rho}(\mathbf{k}, \omega)$, is defined. It has the useful property that its poles on the real-ω axis correspond to particle-like excitations of the system. A pole *not* on the real axis corresponds to an excitation which decays exponentially in time. Thus we can extend the quasiparticle concept to include 'particles' with a limited lifetime.

In § 14.2, the Green function is calculated for a model Hamiltonian with pairing interactions. Perturbation-theoretical methods are not applicable to this system, but Gor'kov (1958) has shown how the Green function may be calculated—the method is analogous to Anderson's (pp. 196–200). In the calculation, a second auxiliary function $F(\mathbf{k}, \omega)$ is introduced, which is found to be closely related to the energy-gap function. In § 14.3, the Ginzburg–Landau equations (6.23), (6.24) are derived and the microscopic interpretation of the Ginzburg–Landau parameters is found (Gor'kov 1959).

Formally, the Ginzburg–Landau equations imply the Meissner effect and many other phenomena, including flux quantization (§ 3.4.1) and Josephson tunnelling (§ 8.2). Gor'kov's method thus constitutes an essentially complete and self-contained account of the microscopic theory. But the evaluation of the one-particle propagator requires an approximation equivalent to Anderson's generalized random-phase approximation (§ 12.1), and it is difficult to establish a criterion for the validity of the approximation. This is one of the reasons why we preferred to discuss the electrodynamic properties separately in Chapter 13, rather than to rely on the Gor'kov method. A second reason is

† To supply the necessary motivation, we should need a considerable scaffolding of 'grammatical rules', etc. The reader who wishes to follow such a programme through should consult, for example, Kadanoff and Baym (1962). For an account of Green functions with special emphasis on their application to superconductivity, see Schrieffer (1964).

that propagator methods are still relatively unfamiliar, so that the methods of Chapter 13 may help to give more insight into the mechanism of field exclusion. Thirdly, the Pippard non-local equation has not yet been derived by the Gor'kov method, although it has been obtained by more conventional techniques.

14.1. Green functions for normal Fermi systems

The one-particle propagator $\mathscr{G}_\sigma(\mathbf{r}', t'; \mathbf{r}, t)$ is defined as

$$\mathscr{G}_\sigma(\mathbf{r}', t'; \mathbf{r}, t) = -\mathrm{i}\langle \mathrm{T}\{\hat{\Psi}_\sigma(\mathbf{r}', t'), \hat{\Psi}^+_\sigma(\mathbf{r}, t)\}\rangle_{\mathrm{av}}, \qquad (14.1)$$

where the average is to be taken over a grand canonical ensemble, and where T is Wick's time-ordering operator. T is defined to arrange the quantized-field operators following it in their chronological order, with the earliest time on the right. If the chronologically-ordered product is an odd permutation of the product of Fermion field operators as written, a change of sign is required. Thus in (14.1)

$$\left.\begin{aligned}\mathrm{T}\{\hat{\Psi}_\sigma(\mathbf{r}', t'), \hat{\Psi}^+_\sigma(\mathbf{r}, t)\} &= \hat{\Psi}_\sigma(\mathbf{r}', t')\hat{\Psi}^+_\sigma(\mathbf{r}, t), \quad t < t', \\ &= -\hat{\Psi}^+_\sigma(\mathbf{r}', t)\hat{\Psi}_\sigma(\mathbf{r}', t'), \; t > t'.\end{aligned}\right\} \qquad (14.2)$$

In the subsequent development, it will be useful to introduce an imaginary time variable, in order to exploit the formal analogy between the Boltzmann factor $\exp-\beta\hat{H}$ (appearing in the statistical averages) and the time-evolution factor $\exp-\mathrm{i}\hat{H}t/\hbar$. A definition is therefore required of the Wick chronological operator for imaginary times; the convention adopted is that smaller values of the real variable $\tau = it$ are 'earlier'.

In (14.1), one of the two time arguments may be taken to be zero. Provided the Hamiltonian is independent of time, the function $\mathscr{G}_\sigma(\mathbf{r}', t'; \mathbf{r}, t)$ can depend only on the time *difference* $t'-t$, so that taking $t = 0$ involves no loss of generality. Similarly, if the Hamiltonian is invariant against translations in space, the Green function $\mathscr{G}_\sigma(\mathbf{r}', t'; \mathbf{r}, 0)$ can depend only on the *relative* coordinates $\mathbf{r}'-\mathbf{r}$, and not on $\mathbf{r}$, $\mathbf{r}'$ separately.

Because the definition (14.1) of the propagator is discontinuous at $t = t'$, it is convenient to define the two continuous auxiliary functions

$$\left.\begin{aligned} \mathscr{G}^{>}_{\sigma}(\mathbf{r}', t'; \mathbf{r}, t) &= +\mathrm{i}\langle\hat{\Psi}^{+}_{\sigma}(\mathbf{r}, t)\hat{\Psi}_{\sigma}(\mathbf{r}', t')\rangle_{\mathrm{av}}, \\ \mathscr{G}^{<}_{\sigma}(\mathbf{r}', t'; \mathbf{r}, t) &= -\mathrm{i}\langle\hat{\Psi}_{\sigma}(\mathbf{r}', t')\hat{\Psi}^{+}_{\sigma}(\mathbf{r}, t)\rangle_{\mathrm{av}}, \end{aligned}\right\} \tag{14.3}$$

where, of course, from (14.1)

$$\left.\begin{aligned} \mathscr{G} &= \mathscr{G}^{>} \text{ for } t > t', \\ &= \mathscr{G}^{<} \text{ for } t < t'. \end{aligned}\right\} \tag{14.4}$$

Writing the ensemble averages explicitly, we have

$$\mathscr{G}^{>}_{\sigma}(\mathbf{r}', t'; \mathbf{r}, t) = \frac{-\mathrm{i}\,\mathrm{Tr}\,\{\mathrm{e}^{-\beta(\hat{H}-\mu\hat{N}_{\mathrm{e}})}\,\mathrm{e}^{\mathrm{i}t\hat{H}/\hbar}\,\hat{\Psi}^{+}_{\sigma}(\mathbf{r}, 0)\,\mathrm{e}^{-\mathrm{i}t\hat{H}/\hbar}\,\hat{\Psi}_{\sigma}(\mathbf{r}', t')\}}{\mathrm{Tr}\,\mathrm{e}^{-\beta(\hat{H}-\mu\hat{N}_{\mathrm{e}})}}, \tag{14.5}$$

with a similar expression for $\mathscr{G}^{<}$. After a little manipulation, it is easily shown that

$$\mathscr{G}^{<}_{\sigma}(\mathbf{r}', t'; \mathbf{r}, 0) = -\mathrm{e}^{\beta\mu}\,\mathscr{G}^{>}_{\sigma}(\mathbf{r}', t'; \mathbf{r}, -\mathrm{i}\beta\hbar). \tag{14.6}$$

Hence the boundary condition

$$\mathscr{G}_{\sigma}(\mathbf{r}', t'; \mathbf{r}, 0) = -\mathrm{e}^{\beta\mu}\,\mathscr{G}_{\sigma}(\mathbf{r}', t'; \mathbf{r}, -\mathrm{i}\beta\hbar), \quad 0 \leqslant \mathrm{i}t' \leqslant \beta\hbar, \tag{14.7}$$

follows, linking the values of $\mathscr{G}$ at $t = 0$ and $t = -\mathrm{i}\beta\hbar$.

For subsequent use, define the Fourier transforms

$$\mathscr{G}^{\gtrless}_{\sigma}(\mathbf{k}, t-t') = \mathrm{i}\int \mathrm{d}^3r\,\mathrm{e}^{-\mathrm{i}\mathbf{k}.(\mathbf{r}-\mathbf{r}')}\mathscr{G}^{\gtrless}_{\sigma}(\mathbf{r}', t';\ \mathbf{r}, t), \tag{14.8}$$

and

$$\mathscr{G}^{\gtrless}_{\sigma}(\mathbf{k}, \omega_n) = \int_0^{-\mathrm{i}\beta\hbar} \mathrm{d}t\,\mathscr{G}^{\gtrless}_{\sigma}(\mathbf{k}, t)\,\mathrm{e}^{\mathrm{i}\omega_n t}, \tag{14.9}$$

noting that (14.9) automatically satisfies the boundary conditions (14.7) if

$$\hbar\omega_n = \mathrm{i}\pi n/\beta + \mu, \quad n = \pm 1, \pm 3, \pm 5, \ldots \tag{14.10}$$

We shall also introduce the spectral density function $\tilde{\rho}_\sigma(\mathbf{k}, \omega)$, defined on the real frequency axis by

$$\tilde{\rho}_\sigma(\mathbf{k}, \omega) = \mathrm{i} \lim_{\theta \to +0} \{\mathscr{G}_\sigma(\mathbf{k}, \omega+\mathrm{i}\theta) - \mathscr{G}_\sigma(\mathbf{k}, \omega-\mathrm{i}\theta)\}. \quad (14.11)$$

In (14.11) $\mathscr{G}(\mathbf{k}, \omega)$ means the analytic continuation of the $\mathscr{G}(\mathbf{k}, \omega_n)$ defined by (14.9).

It is easily shown (e.g. Kadanoff and Baym 1962 § 2.1) that in thermal equilibrium

$$\mathscr{G}^{>}(\mathbf{k}, \omega) = \tilde{\rho}(\mathbf{k}, \omega)\{1-f(\omega)\},$$

and

$$\mathscr{G}^{<}(\mathbf{k}, \omega) = \tilde{\rho}(\mathbf{k}, \omega) f(\omega),$$

where $f(\omega)$ is the Fermi–Dirac factor

$$f(\omega) = \{1+\exp \beta(\hbar\omega-\mu)\}^{-1}.$$

Thus $\mathscr{G}^{>}(\mathbf{k}, \omega)$ measures the density of states (with momentum $\hbar\mathbf{k}$ and energy $\hbar\omega$) available for an additional particle; $\mathscr{G}^{<}(\mathbf{k}, \omega)$ similarly measures the density of particles. The boundary condition (14.7) is equivalent to the equation

$$\frac{\mathscr{G}^{<}(\mathbf{k}, \omega)}{\mathscr{G}^{>}(\mathbf{k}, \omega)} = \frac{f(\omega)}{1-f(\omega)} = \exp -\beta(\hbar\omega-\mu),$$

i.e. the density of particles is related to the density of states through the Boltzmann factor $\exp -\beta(\hbar\omega-\mu)$.

The equation of motion for the propagator follows from the equations of motion (cf. (12.7)) for the field operators $\hat{\Psi}^{+}$, $\hat{\Psi}$:

$$\left.\begin{aligned}
\mathrm{i}\hbar\dot{\hat{\Psi}}^{+}_{\sigma}(\mathbf{r}, t) &= -\frac{\hbar^2}{2m}\nabla^2\hat{\Psi}^{+}_{\sigma}(\mathbf{r}, t) \\
&\quad -\sum_{\sigma'}\int \mathrm{d}^3r'\, V(\mathbf{r}-\mathbf{r}')\hat{\Psi}^{+}_{\sigma}(\mathbf{r}, t)\hat{\Psi}^{+}_{\sigma'}(\mathbf{r}', t)\hat{\Psi}_{\sigma'}(\mathbf{r}', t), \\
-\mathrm{i}\hbar\dot{\hat{\Psi}}_{\sigma}(\mathbf{r}, t) &= -\frac{\hbar^2}{2m}\nabla^2\hat{\Psi}_{\sigma}(\mathbf{r}, t) \\
&\quad -\sum_{\sigma'}\int \mathrm{d}^3r'\, V(\mathbf{r}-\mathbf{r}')\hat{\Psi}_{\sigma}(\mathbf{r}, t)\hat{\Psi}^{+}_{\sigma'}(\mathbf{r}', t)\hat{\Psi}_{\sigma'}(\mathbf{r}', t).
\end{aligned}\right\} \quad (14.12)$$

The complication is that the operator $\partial/\partial t$ does not commute with the time-ordering operator T; when t passes through zero, $\mathscr{G}_\sigma(\mathbf{r}', 0; \mathbf{r}, t)$ has a discontinuity†

$$\mathscr{G}_\sigma(\mathbf{r}', 0; \mathbf{r}, +0) - \mathscr{G}_\sigma(\mathbf{r}', 0; \mathbf{r}, -0) = \delta(\mathbf{r}-\mathbf{r}'). \quad (14.13)$$

Hence, from (14.12), (14.13), the equation of motion for $\mathscr{G}_\sigma(\mathbf{r}', t'; \mathbf{r}, t)$ is

$$\left(-\mathrm{i}\hbar\frac{\partial}{\partial t}+\frac{\hbar^2}{2m}\nabla_{\mathbf{r}}^2\right)\mathscr{G}_\sigma(\mathbf{r}', t'; \mathbf{r}, t) = \delta(\mathbf{r}-\mathbf{r}')\,\delta(t-t')$$
$$-\mathrm{i}\lim_{\tau\to t'+0}\int \mathrm{d}^3R\; \mathscr{G}_2(\mathbf{r}', t', \mathbf{R}, \tau-0; \mathbf{r}, t, \mathbf{R}, \tau)V(\mathbf{R}-\mathbf{r}). \quad (14.14)$$

The two-particle propagator $\mathscr{G}_2(\mathbf{r}_1', t_1', \mathbf{r}_2', t_2'; \mathbf{r}_1, t_1, \mathbf{r}_2, t_2)$ in (14.14) is defined in a manner analogous to (14.1):

$$\mathscr{G}_2(\mathbf{r}_1', t_1', \mathbf{r}_2', t_2'; \mathbf{r}_1, t_1, \mathbf{r}_2, t_2) =$$
$$(-\mathrm{i})^2\langle \mathrm{T}\{\hat{\Psi}(\mathbf{r}_1', t_1'), \hat{\Psi}(\mathbf{r}_2', t_2'), \hat{\Psi}^+(\mathbf{r}_2, t_2), \hat{\Psi}^+(\mathbf{r}_1, t_1)\}\rangle_{\mathrm{av}}. \quad (14.15)$$

The spin suffixes have been omitted, but can be restored by inspection.

For free particles, the potential $V(\mathbf{R}-\mathbf{r})$ in the equation of motion (14.14) vanishes, and (14.14) simplifies to

$$\left(-\mathrm{i}\hbar\frac{\partial}{\partial t}+\frac{\hbar^2}{2m}\nabla^2\right)\mathscr{G}_0(0, 0; \mathbf{r}, t) = \delta(\mathbf{r})\,\delta(t). \quad (14.16)‡$$

On Fourier transformation of (14.16),

$$\mathscr{G}_0(\mathbf{k}, \omega_n) = \frac{1}{\hbar\omega_n - \hbar^2k^2/2m}. \quad (14.17)$$

† Equation (14.13) follows from the definition (14.1) of the Green function, together with the equal-time commutation relations (9.58).

‡ The form of (14.16) explains why propagators are also called 'Green functions'; the solution of (14.16) is the Green function for the inhomogeneous differential equation

$$\left(-\mathrm{i}\hbar\frac{\partial}{\partial t}+\frac{\hbar^2}{2m}\nabla^2\right)\phi(\mathbf{r}, t) = \chi(\mathbf{r}, t)$$

(see, e.g. Courant and Hilbert 1953).

Hence the spectral function is

$$\tilde{\rho}_0(\mathbf{k}, \omega) = 2\pi\hbar\, \delta(\hbar\omega - \hbar^2k^2/2m), \tag{14.18}$$

using the relation (see Appendix E)

$$\lim_{\theta\to+0} \frac{1}{x+\mathrm{i}\theta} = \mathscr{P}\left(\frac{1}{x}\right) - \mathrm{i}\pi\delta(x). \tag{14.19}$$

The interpretation of the spectral function (14.18) is that free particles have an excitation spectrum $\varepsilon = p^2/2m$. There is no line width, i.e. for a specified momentum the energy is uniquely defined.

For more complicated systems, with $V \neq 0$, an approximate method is needed to evaluate $\mathscr{G}_2(\mathbf{r}_1', t_1', \mathbf{r}_2', t_2'; \mathbf{r}_1, t_1, \mathbf{r}_2, t_2)$ in order to determine $\mathscr{G}(\mathbf{r}', t'; \mathbf{r}, t)$ from (14.14). The simplest† such scheme is the Hartree–Fock approximation in which the interactions between the particles (contained in $\mathscr{G}_2(\mathbf{r}_1', t_1', \mathbf{r}_2', t_2'; \mathbf{r}_1, t_1, \mathbf{r}_2, t_2)$) is neglected. Thus

$$\mathscr{G}_2(\mathbf{r}_1', t_1', \mathbf{r}_2', t_2'; \mathbf{r}_1, t_1, \mathbf{r}_2, t_2) = \\ \mathscr{G}(\mathbf{r}_1', t_1'; \mathbf{r}_1, t_1)\mathscr{G}(\mathbf{r}_2', t_2'; \mathbf{r}_2, t_2) - \mathscr{G}(\mathbf{r}_2', t_2'; \mathbf{r}_1, t_1)\mathscr{G}(\mathbf{r}_1', t_1'; \mathbf{r}_2, t_2). \tag{14.20}$$

Substituting (14.20) into (14.14) leads to an integrodifferential equation for $\mathscr{G}(0, 0; \mathbf{r}, t)$:

$$\left(-\mathrm{i}\hbar\frac{\partial}{\partial t} + \frac{\hbar^2}{2m}\nabla_1^2\right)\mathscr{G}(0, 0; \mathbf{r}_1, t) + \int \mathrm{d}^3r_2\, \mathscr{G}(0, 0; \mathbf{r}_2, t)\tilde{V}(\mathbf{r}_1, \mathbf{r}_2) \\ = \delta(\mathbf{r}_1)\,\delta(t), \tag{14.21}$$

where

$$\tilde{V}(\mathbf{r}_1, \mathbf{r}_2) = \delta(\mathbf{r}_1 - \mathbf{r}_2) \int \mathrm{d}^3r_3\, V(\mathbf{r}_1 - \mathbf{r}_3)\langle n(\mathbf{r}_3)\rangle \\ + \mathrm{i}\,V(\mathbf{r}_1 - \mathbf{r}_2)\mathscr{G}^{<}(0, 0; \mathbf{r}_1, +0) \tag{14.22}$$

† The Hartree approximation is even simpler, but does not take proper account of the Fermion properties of the particles. The Hartree–Fock approximation, which is only slightly more complicated, treats the Pauli exclusion principle consistently.

is the average selfconsistent potential in which the particles move. For a spatially homogenous system, the Fourier transform exists:

$$(\omega_n - \tilde{\varepsilon}(\mathbf{k})/\hbar)\,\mathscr{G}(\mathbf{k}, \omega_n) = 1, \tag{14.23}$$

where the quasiparticle energy $\tilde{\varepsilon}(\mathbf{k})$ is

$$\tilde{\varepsilon}(\mathbf{k}) = \frac{\hbar^2 k^2}{2m} + \frac{1}{(2\pi)^3}\int \mathrm{d}^3k'\{V(\mathbf{k}-\mathbf{k}') + \tfrac{1}{2}V(0)\}\langle n(\mathbf{k}')\rangle. \tag{14.24}$$

The spectral function is once more a δ-function,

$$\tilde{\rho}(\mathbf{k}, \omega) = 2\pi\hbar\,\delta(\hbar\omega - \tilde{\varepsilon}(\mathbf{k})), \tag{14.25}$$

i.e. the momentum again suffices to define the energy precisely.

It is only when we try to improve on the Hartree–Fock approximation that the spectral function takes a more complicated form. Many elaborate perturbation-theoretical methods have been described in the extensive literature on the many-body problem. In particular there are several methods which sum certain classes of Feynman diagrams to all orders of perturbation. A description of these methods would be outside the scope of this book. We merely remark that they can lead to Green functions of form similar to (14.23), *but with the pole not on the real axis*. Let us examine the implications of such a complex pole.

Suppose, for example, that the result of some calculation were

$$\left.\begin{aligned}\mathscr{G}^{>}(\mathbf{k}, \omega) &\propto \{\hbar\omega - E(\mathbf{k}) + \mathrm{i}\gamma(\mathbf{k})\}^{-1},\\ \mathscr{G}^{<}(\mathbf{k}, \omega) &\propto -\{\hbar\omega - E(\mathbf{k}) - \mathrm{i}\gamma(\mathbf{k})\}^{-1}.\end{aligned}\right\} \tag{14.26}$$

From (14.11), the spectral function is found to be of the form

$$\tilde{\rho}(\mathbf{k}, \omega) \propto \frac{\gamma(\mathbf{k})}{\{\hbar\omega - E(\mathbf{k})\}^2 + \gamma(\mathbf{k})^2}\,. \tag{14.27}$$

That is, the relation between energy and momentum is no longer sharp; the spectrum contains a line with Lorentzian shape. Moreover, inverting the Fourier transformations, the time-dependent propagator (in *real* time) is

$$\mathscr{G}^{>}(\mathbf{k}, t) \propto e^{iE(\mathbf{k})t/\hbar}\, e^{-\gamma t/\hbar}. \qquad (14.28)$$

The excitation described by such a propagator is a radioactive state, decaying exponentially with time. If the lifetime is long enough, this will nevertheless be a useful quasiparticle state. Indeed, we may formally *define* a quasiparticle by the appearance of a simple pole (e.g. (14.26)) in the Green functions $\mathscr{G}^{\gtrless}(\mathbf{k}, \omega)$.

14.2. Propagators for superconductors

Superconducting behaviour can never emerge in a perturbation-theoretical calculation. Perturbation theory amounts to an expansion in powers of Fröhlich's dimensionless interaction parameter $\mathscr{F}$. But the ground-state energy (11.28) has been shown to be proportional to $\exp(-2/\mathscr{F})$. This function has an essential singularity at $\mathscr{F} = 0$; if we attempt to expand it in a Taylor series about $\mathscr{F} = 0$, all the coefficients are zero! It is therefore clear that no perturbative calculation can ever reproduce the correct form (11.28) for the condensation energy.

The method of Anderson, outlined in § 12.1, provides a clue. The essential point is that the ground-state expectation values (12.13) do not vanish. The temperature-dependent generalizations of the expressions (12.13) are Gor'kov's (1958) 'anomalous Green functions'

$$F(\mathbf{r}', t'; \mathbf{r}, t) = \langle \mathrm{T}\{\hat{\Psi}_{\sigma}(\mathbf{r}', t'), \hat{\Psi}_{-\sigma}(\mathbf{r}, t)\}\rangle_{\mathrm{av}} \qquad (14.29)\dagger$$

† There is a hidden complication here: with more exact wave functions which are eigenstates of the number of electrons (e.g. (13.30)), (14.29) would vanish identically. A state with $N_e \pm 2$ particles is orthogonal to all states with N_e particles. What Gor'kov in fact does is to define F by

$$\langle N_e|\mathrm{T}\{\hat{\Psi}_{\sigma}(\mathbf{r}', t'), \hat{\Psi}_{-\sigma}(\mathbf{r}, t)\}|N_e+2\rangle = e^{-2i\mu(t'-t)/\hbar} F(\mathbf{r}', t'; \mathbf{r}, t), \quad (14.29')$$

where the thermal average is to be taken over a *canonical* ensemble only. The chemical potential μ makes its appearance through the energy difference $U(N_e+2) - U(N_e)$, between analogous states in the N_e-particle system and the (N_e+2)-particle system. Use of the grand ensemble, with BCS states, leads to many of the correct results, but the phase factor in (14.29′) plays an essential role in the time dependence of the Ginzburg–Landau function; see § 14.3.1.

These anomalous Green functions enable us to generalize the Hartree–Fock approximation. The procedure is equivalent to that used in writing the equations of motion (12.15) for the field operators. Thus the Gor'kov approximation for $\mathscr{G}_2$ is

$$\begin{aligned}\mathscr{G}_2(\mathbf{r}_1', t_1', \mathbf{r}_2', t_2'; \mathbf{r}_1, t_1, \mathbf{r}_2, t_2) = {}& \mathscr{G}(\mathbf{r}_1', t_1'; \mathbf{r}_1, t_1)\mathscr{G}(\mathbf{r}_2', t_2'; \mathbf{r}_2, t_2) \\ & - \mathscr{G}(\mathbf{r}_1', t_1'; \mathbf{r}_2, t_2)\mathscr{G}(\mathbf{r}_2', t_2'; \mathbf{r}_1, t_1) \\ & + F(\mathbf{r}_1, t_1; \mathbf{r}_2, t_2)F^+(\mathbf{r}_1', t_1'; \mathbf{r}_2', t_2'), \quad (14.30)\dagger\end{aligned}$$

and the equation of motion (14.14) takes the form

$$\begin{aligned}&\left(-i\hbar\frac{\partial}{\partial t}+\frac{\hbar^2}{2m}\nabla_1^2\right)\mathscr{G}(0, 0; \mathbf{r}_1, t) \\ &\quad + \int d^3r_2\, \mathscr{G}(0, 0; \mathbf{r}_2, t)\tilde{V}(\mathbf{r}_1, \mathbf{r}_2)\, \mathscr{G}(0, 0; \mathbf{r}_1, t) \\ &\quad - i\frac{\mathscr{F}}{\Xi_0}F(+0)F^+(0, 0; \mathbf{r}_1, t) = \delta(\mathbf{r}_1)\delta(t), \qquad (14.31)\end{aligned}$$

where

$$F(+0) = \lim_{t\to+0} F(\mathbf{r}, t; \mathbf{r}, 0).$$

The equation of motion for the anomalous propagator $F(0, 0; \mathbf{r}, t)$ is similarly derived:

$$\begin{aligned}&\left(-i\hbar\frac{\partial}{\partial t}+\frac{\hbar^2}{2m}\nabla_1^2\right)F^+(0, 0; \mathbf{r}_1, t,) \\ &\quad + \int d^3r_2\, \mathscr{G}(0, 0; \mathbf{r}_2, t)\tilde{V}(\mathbf{r}_1, \mathbf{r}_2)F^+(0, 0; \mathbf{r}_1, t) \\ &\quad + i\frac{\mathscr{F}}{\Xi_0}F^+(+0)\mathscr{G}(0, 0; \mathbf{r}_1, t) = 0. \qquad (14.32)\end{aligned}$$

Equation (14.32) is homogeneous, with no δ-function on the right-hand side, because the anticommutator between $\hat{\Psi}_\sigma(\mathbf{r}, t)$ and $\hat{\Psi}_{-\sigma}(\mathbf{r}', 0)$ has no δ-function at $t = 0$.

† *See footnote* on p. 235.

On Fourier transformation, Gor'kov's equations (14.31), (14.32) become algebraic equations:

$$\left.\begin{aligned}\{\hbar\omega-\tilde{\eta}(\mathbf{k})\}\mathscr{G}(\mathbf{k},\omega)-\mathrm{i}\frac{\mathscr{F}}{\Xi_0\Omega}F(+0)\,F^{+}(\mathbf{k},\omega)&=1,\\ \{\hbar\omega+\tilde{\eta}(\mathbf{k})\}F^{+}(\mathbf{k},\omega)+\mathrm{i}\frac{\mathscr{F}}{\Xi_0\Omega}F(+0)\,\mathscr{G}(\mathbf{k},\omega)&=0.\end{aligned}\right\} \tag{14.33}$$

Their solutions are

$$\left.\begin{aligned}F^{+}(\mathbf{k},\omega)&=\frac{-\mathrm{i}\mathscr{F}J\hbar/\Xi_0\Omega}{\{\hbar\omega-E(\mathbf{k})+\mathrm{i}\theta\}\{\hbar\omega-E(\mathbf{k})-\mathrm{i}\theta\}},\\ \mathscr{G}(\mathbf{k},\omega)&=\frac{\hbar u_{\mathbf{k}}^2}{\hbar\omega-E(\mathbf{k})+\mathrm{i}\theta}+\frac{\hbar v_{\mathbf{k}}^2}{\hbar\omega-E(\mathbf{k})-\mathrm{i}\theta},\end{aligned}\right\} \tag{14.34}$$

where θ is a positive infinitesimal, $E(\mathbf{k})$ is given by (12.22), and

$$J=(2\pi)^{-4}\int \mathrm{d}^3k\,\mathrm{d}\omega\,F^{+}(\mathbf{k},\omega). \tag{14.35}$$

In (14.34), $u_{\mathbf{k}}$ and $v_{\mathbf{k}}$ are the Bogoliubov coefficients (12.16). The infinitesimal imaginary $\mathrm{i}\theta$ is introduced to provide a conventional contour for avoiding the poles at $\hbar\omega = E(\mathbf{k})$; its sign follows from the analytic continuation of $\mathscr{G}(\mathbf{k},\omega_n)$. Thus all the results of Chapter 12 may be rederived. In particular the energy spectrum is given by (12.22), and the equation for the energy gap is once again the integral equation (12.24). Moreover, the thermal properties are also contained in the formalism, since the Green functions are defined as ensemble averages.

Nambu (1960) has cast Gor'kov's theory into an elegant form, by considering the pair of fields $\hat{\Psi}_\sigma(\mathbf{r},t)$, $\hat{\Psi}^{+}_{-\sigma}(\mathbf{r},t)$ to form a two-component spinor. The two Gor'kov functions $\mathscr{G}_\sigma(\mathbf{k},t)$ and $F(\mathbf{k},t)$ now appear as components of the 2×2 matrix

$$\mathscr{G}^{\mathrm{Nambu}}(\mathbf{k},t)=\begin{pmatrix}\mathscr{G}_{\uparrow}(\mathbf{k},t) & F(\mathbf{k},t)\\ F^{+}(\mathbf{k},t) & \mathscr{G}_{\downarrow}(\mathbf{k},t)\end{pmatrix}, \tag{14.36}$$

and the pair of equations (14.33) reduce to a single matrix equation. For a fuller account, see Schrieffer (1964 pp. 169–80).

14.3. The Ginzburg–Landau equations

To study the effect of an applied electromagnetic field, Gor'kov replaces the momentum operator $\hat{\mathbf{p}}$ $(=-i\hbar\nabla)$ by[26] $\hat{\mathbf{p}}-e\mathbf{A}/c$, where $\mathbf{A}$ is the vector potential (cf. p. 217). In general we should also consider the effect of the scalar potential ϕ, but for a purely magnetic field the gauge can always be so chosen that $\phi = 0$. This substitution is satisfactory and unique provided the electron–electron potential is local. However, the Fröhlich interaction is non-local; it depends on the momenta of the interacting electrons as well as on their spatial separation. A calculation such as Rickayzen's (1959) is needed to establish the gauge-invariance.

In terms of the imaginary-time Fourier transforms, Gor'kov's equations (14.31), (14.32) are generalized to†

$$\left.\begin{aligned}\left\{\hbar\omega_n'+\frac{\hbar^2}{2m}\left(\nabla_1-i\frac{e}{c}\mathbf{A}(\mathbf{r}_1)\right)^2+\mu\right\}\mathscr{G}(\mathbf{r}_1,\mathbf{r}_2;\omega_n)&\\ +\Delta(\mathbf{r}_1)F^+(\mathbf{r}_1,\mathbf{r}_2;\omega_n)&=\hbar\delta(\mathbf{r}_1-\mathbf{r}_2),\\ \left\{\hbar\omega_n'+\frac{\hbar^2}{2m}\left(\nabla_1+i\frac{e}{c}\mathbf{A}(\mathbf{r}_1)\right)^2+\mu\right\}F^+(\mathbf{r}_1,\mathbf{r}_2;\omega_n)&\\ -\Delta^*(\mathbf{r}_1)\mathscr{G}(\mathbf{r}_1,\mathbf{r}_2;\omega_n)&=0,\end{aligned}\right\} \qquad (14.37)^{(26}$$

where the energy-gap function $\Delta(\mathbf{r})$ is

$$\Delta(\mathbf{r})=\frac{\mathscr{F}}{\Omega\Xi_0}\frac{1}{\hbar\beta}\sum_n F(\mathbf{r},\mathbf{r};\omega_n). \qquad (14.38)$$

† It is convenient to put $\hbar\omega_n = \hbar\omega_n'+\mu$. Allowed values of n are given by (14.10).

These equations have the formal solution

$$\left.\begin{aligned}\mathscr{G}(\mathbf{r}_1,\mathbf{r}_2;\omega_n) &= \mathscr{G}^0(\mathbf{r}_1,\mathbf{r}_2;\omega_n)\\ &\quad -\int \mathrm{d}^3r_3\,\mathscr{G}^0(\mathbf{r}_1,\mathbf{r}_3;\omega_n)\Delta(\mathbf{r}_3)\mathrm{F}^+(\mathbf{r}_3,\mathbf{r}_2;\omega_n),\\ F^+(\mathbf{r}_1,\mathbf{r}_2;\omega_n) &= \int \mathrm{d}^3r_3\,\mathscr{G}^0(\mathbf{r}_1,\mathbf{r}_3;\omega_n)\Delta^*(\mathbf{r}_3)\mathscr{G}(\mathbf{r}_3,\mathbf{r}_2;\omega_n),\end{aligned}\right\} \tag{14.39}$$

where $\mathscr{G}^0$ is the Green function of a normal metal in the presence of the applied field. $\mathscr{G}^0$ satisfies the equation of motion†

$$\left\{\hbar\omega_n'+\frac{\hbar^2}{2m}\left(\nabla_1-\mathrm{i}\frac{e}{c}\mathbf{A}(\mathbf{r}_1)\right)^2+\mu\right\}\mathscr{G}^0(\mathbf{r}_1,\mathbf{r}_2;\omega_n) = \hbar\delta(\mathbf{r}_1-\mathbf{r}_2). \tag{14.40}$$

(26)

In the absence of a magnetic field, the solution of (14.40) to first order in $\hbar\omega_n'/\mu$ is

$$\mathscr{G}^0(\mathbf{r}_1,\mathbf{r}_2;\omega_n) = -\frac{m}{2\pi\hbar|\mathbf{r}_1-\mathbf{r}_2|}\exp\left(\left\{\mathrm{i}k_\mathrm{F}\frac{\omega_n'}{|\omega_n'|}-\frac{|\omega_n'|}{v_\mathrm{F}}\right\}|\mathbf{r}_1-\mathbf{r}_2|\right). \tag{14.41}$$

The structure of (14.39) is too complicated to hope that in general this pair of equations will have a simple solution. But Gor'kov (1959) has shown how an approximate solution may be found, valid near the transition temperature. Note that the magnetic field will be small for T near T_c. This is evident since, for superconductivity, H must be less than $\max(H_\mathrm{c}, H_{\mathrm{c}2})$. But H_c and $H_{\mathrm{c}2}$ both vanish at the transition temperature T_c. Provided the gauge is so chosen that $\mathbf{A} = 0$ when $\mathbf{H} = 0$, then near the transition the smallness of H_c ensures that $\mathbf{A}$ will also be small. Moreover the anomalous propagator $F(\mathbf{r}, t; \mathbf{r}', t')$ (and hence, by (14.38), the gap function Δ) is also small for temperatures near T_c.

† *See footnote* on opposite page.

By expanding the equations (14.37) in powers of $\mathbf{A}$ and Δ, and solving them to fourth order in Δ and to first order in the product $\mathbf{A}|\Delta^2|$, Gor'kov shows that

$$\left\{\frac{\hbar^2}{2m}\left(\nabla+2\mathrm{i}\frac{e}{c}\mathbf{A}(\mathbf{r})\right)^2+\left(\frac{12\pi T_\mathrm{c}^2}{7\mu\zeta(3)}\right)\frac{T_\mathrm{c}-T}{T_\mathrm{c}}-\frac{3}{2\mu}|\Delta(\mathbf{r})|^2\right\}\Delta^*(\mathbf{r})=0, \tag{14.42}$$

(14.42)[26]

and

$$\mathbf{J}(\mathbf{r})=\frac{\mathrm{i}e}{m}\left\{\Delta(\nabla\Delta^*)-\Delta^*(\nabla\Delta)-\frac{4e^2}{mc}|\Delta(\mathbf{r})|^2\left(\frac{7\zeta(3)N_\mathrm{e}}{16\pi^2T_\mathrm{c}^2}\right)\mathbf{A}\right\}. \tag{14.43}$$

(14.43)[26]

Here ζ denotes the Riemann ζ-function: $\zeta(3)=1{\cdot}202$. Provided we make the identifications

$$\varepsilon=2e, \tag{14.44}$$

and

$$\Psi_{\mathrm{G-L}}(\mathbf{r})=\left(\frac{7\zeta(3)N_\mathrm{e}}{4\pi^2T_\mathrm{c}}\right)\Delta(\mathbf{r}), \tag{14.45}$$

equations (14.42) and (14.43) are just the Ginzburg–Landau equations (6.23) and (6.24).

The Cooper-pair aspects of the theory are clearly reflected in equation (14.44)—the current carriers have twice the electronic charge. Apart from a normalizing constant, (14.45) identifies the Ginzburg–Landau 'wave function' $\Psi_{\mathrm{G-L}}$ with the energy gap function Δ. Since Δ is itself proportional to the anomalous propagator F (by (14.38)), this exhibits the microscopic meaning of the Ginzburg–Landau function: it is a measure of the amplitude of the condensed-pair propagator F.

14.3.1. Anderson's 'extra Ginzburg–Landau equation'

The BCS wave function $|\Phi_0\rangle$, (11.8), is not an eigenstate of the particle number operator; moreover the Bogoliubov transformation (12.16) does not conserve particles. In the Gor'kov formalism it is however possible to take rigorous account of the conservation of particles (cf. footnote, p. 235).

The effect of doing so is to introduce the phase factor $\exp-2i\mu(t-t')/\hbar$ of (14.29′) into the anomalous propagator F.

The energy gap function for the *actual* Hamiltonian $\hat{\mathscr{H}}$ of the system may thus be found from the gap function of the 'grand' Hamiltonian $\hat{\mathscr{H}}-\mu\hat{N}_{\mathrm{e}}$:

$$\Delta_{\hat{\mathscr{H}}} = \Delta_{\hat{\mathscr{H}}-\mu\hat{N}_{\mathrm{e}}}\exp(-2i\mu t/\hbar). \tag{14.46}$$

Anderson *et al.* (1965)† have shown that this phase factor has important consequences. Using (14.45), (14.46) may be expressed in the form

$$\mu = \frac{i\hbar}{4\Psi^*\Psi}\left\{\Psi^*\frac{\partial\Psi}{\partial t}-\frac{\partial\Psi^*}{\partial t}\Psi\right\}. \tag{14.47}$$

Let us write the total chemical potential μ as a sum of a 'purely chemical' contribution $\mu_{\mathrm{c}}(N_{\mathrm{e}}, T) = \partial F/\partial N_{\mathrm{e}}$, and an electrostatic part $e\phi$. Under isothermal conditions, a variation $\delta\mu$ may be written

$$\begin{aligned}\delta\mu &= (\partial\mu_{\mathrm{c}}/\partial N_{\mathrm{e}})_T\delta N_{\mathrm{e}}+e\delta\phi\\ &= -\frac{e}{4\pi}\left(\frac{\partial\mu_{\mathrm{c}}}{\partial N_{\mathrm{e}}}\right)_T\nabla^2(\delta\phi)+e\delta\phi,\end{aligned} \tag{14.48}$$

using Poisson's equation. Hence, from (14.47), the time dependence of the Ginzburg–Landau order parameter Ψ is governed by the 'extra Ginzburg–Landau equation'

$$\lambda_{\mathrm{D}}^2\nabla^2(\delta\phi) = \frac{1}{4e}\left\{\Psi^*\left(i\hbar\frac{\partial}{\partial t}-2e\delta\phi\right)\Psi+\text{c.c.}\right\}+\frac{1-x_{\mathrm{s}}}{e}(\delta\mu-e\delta\phi). \tag{14.49}$$

Here λ_{D} is the Debye screening length ($\lambda_{\mathrm{D}}^2 = (4\pi e^2)^{-1}(\partial\mu_{\mathrm{c}}/\partial\rho)$), and x_{s} is the superconducting fraction $|\Psi/\Psi_0|^2$.

The extreme smallness (~ 1Å) of the Debye length enables us to neglect the left-hand side of (14.49). Under these conditions, macroscopic spatial variations of the scalar potential can occur if and only if Ψ is time dependent. This observation

† See also Anderson and Dayem (1964), Stephen and Suhl (1964), and Josephson (1965).

enables the discussion in § 13.1 to be completed: a time-independent current-carrying state such as $|\Phi_{\mathbf{Q}}\rangle$, (13.1), cannot be associated with a potential gradient, and therefore experiences no resistance.

14.4. The transition temperature

In this section, some results of § 12.2 are rederived, as an illustration of the conciseness of the Gor'kov method.

The self-consistency condition for the gap function Δ follows from (14.38). On substituting for F from (14.39), (14.38) is reduced to

$$\Delta = \frac{\mathscr{F}}{\beta} \sum_n \frac{\pi}{|\hbar\omega_n'|} \Delta. \tag{14.50}$$

The sum in this equation is formally divergent. To avoid the infinity, the BCS cut-off prescription is applied:

$$|\omega_n'| < \bar{\omega} \sim \mathrm{k}\Theta/\hbar. \tag{14.51}$$

At high enough temperatures, the only solution of (14.50) is the trivial one $\Delta = 0$. The threshold condition that a non-trivial solution should exist is

$$\frac{\mathscr{F}}{\beta} \sum_n \frac{\pi}{|\hbar\omega_n'|} = 1. \tag{14.52}$$

But this condition has to be equivalent to the BCS equation for the transition temperature (equation (12.45)). It clearly *is* equivalent, if we make the replacement

$$\sum_n \frac{\pi}{|\hbar\omega_n'|} \to \beta \ln\left(\frac{1{\cdot}14\hbar\bar{\omega}}{\mathrm{k}T}\right). \tag{14.53}$$

This standard replacement is readily justified (e.g. de Gennes 1964).

Exercise. Apart from the magnitude of the numerical constant 1·14, verify that (14.53) follows from (14.10), when the sum is replaced by an integral over n.

14.5. The proximity effect

In the neighbourhood of a boundary between a superconducting and a non-superconducting metal, the properties of the electrons on both sides of the boundary are modified (Cooper 1961, Parmenter 1963, de Gennes and Guyon 1963, de Gennes 1964, Silvert and Cooper 1966). In the s-material the energy gap is reduced, and in extreme cases the system may even lose its superconducting properties. On the other side of the boundary, Cooper pairs (see § 11.1) can exist in the n-material, i.e. the n-material may have superconducting behaviour induced in it. As this is a problem in which the spatial inhomogeneity of the system is essential, the Gor'kov propagator formalism is particularly appropriate.

The theory of the proximity effect has been extensively developed by de Gennes (1964), for various geometrical configurations. For simplicity the present account will be restricted to the case of a two-layer system. The layers are assumed to be thin in comparison with the coherence lengths ξ_{01}, ξ_{02}. In addition, the materials will be assumed to be 'dirty', i.e. the mean free paths l_1, l_2 are small compared with ξ_{01}, ξ_{02}, respectively.

When (14.39) is substituted into (14.38), the resulting integral equation

$$\Delta(\mathbf{r}) = \frac{\mathscr{F}}{\Omega\Xi_0}\frac{1}{\hbar\beta}\sum_n \int \mathrm{d}^3r'\,\Delta(\mathbf{r}')\mathscr{G}(\mathbf{r},\mathbf{r}';\omega_n)\mathscr{G}^0(\mathbf{r}',\mathbf{r};\omega_n) \tag{14.54}$$

is equivalent to the BCS self-consistency equation (12.45). de Gennes shows that the kernel $K_{\omega_n}(\mathbf{r},\mathbf{r}')$ of this equation,

$$K_{\omega_n}(\mathbf{r},\mathbf{r}') = \mathscr{G}(\mathbf{r},\mathbf{r}';\omega_n)\mathscr{G}^0(\mathbf{r}',\mathbf{r};\omega_n), \tag{14.55}$$

obeys the differential equation†

$$2|\omega_n'|K_{\omega_n}(\mathbf{r},\mathbf{r}') - D(\mathbf{r}')\nabla^2_{\mathbf{r}'}K_{\omega_n}(\mathbf{r},\mathbf{r}') = 2\pi\Xi_0(\mathbf{r})\,\delta(\mathbf{r}-\mathbf{r}'). \tag{14.56}$$

† In a dirty material, the Green function $\mathscr{G}(\mathbf{r}',0;\mathbf{r},t)$ (and hence also the Fourier-transformed kernel $K_t(\mathbf{r},\mathbf{r}')$ obeys a diffusion equation

$$\frac{\partial}{\partial t}\mathscr{G}(\mathbf{r}',0;\mathbf{r},t) - D\nabla^2\mathscr{G}(\mathbf{r}',0;\mathbf{r},t) = 2\pi\Xi_0\,\delta(\mathbf{r}-\mathbf{r}')\,\delta(t).$$

Hence, on taking the Fourier transform, (14.56) follows.

In this equation, the diffusion coefficient D has the value

$$D = \tfrac{1}{3}lv_{\mathrm{F}}, \tag{14.57}$$

where v_{F} is the velocity of electrons at the Fermi surface and l is the mean free path.

The present geometry is effectively one-dimensional. de Gennes therefore assumes that the kernel is a function of the arguments z, z' only. Equation (14.56) then simplifies to

$$2|\omega_n'|K_{\omega_n}(z, z') - D(z')\frac{\partial^2}{\partial z'^2}K_{\omega_n}(z, z') = 2\pi\Xi_0(z)\,\delta(z-z'). \tag{14.58}$$

Here D and Ξ_0 have the form of step functions; they are constant within each of the layers, but can change discontinuously at the surface of separation $z = 0$. Within a single layer, the solution is of the form

$$K_{\omega_n}(z, z') = C_1\exp\left(\frac{z-z'}{\xi_{\omega_n}}\right) + C_2\exp\left(\frac{z'-z}{\xi_{\omega_n}}\right), \tag{14.59}$$

where

$$\xi^2_{\omega_n} = \frac{D}{2|\omega_n'|}. \tag{14.60}$$

The coherence length ξ_0 is the maximum ξ_{ω_n}, i.e.

$$\xi_0 = \xi_{\bar{\omega}} = (\hbar D/2\pi \mathrm{k}T)^{\frac{1}{2}}. \tag{14.61}$$

To solve (14.58), two boundary conditions are needed. One condition follows on integrating (14.54) with respect to z' from $-\infty$ to ∞, and making use of the sum rule (de Gennes 1964 p. 229):

$$\int K_{\omega_n}(\mathbf{r}, \mathbf{r}')\,\mathrm{d}^3r' = \frac{\pi}{|\omega_n'|}\Xi_0(\mathbf{r}). \tag{14.62}$$

The boundary condition thus found is

$$\left\{D(z)\frac{\mathrm{d}}{\mathrm{d}z'}K_{\omega_n}(z, z')\right\}_{z'=+0} = \left\{D(z)\frac{\mathrm{d}}{\mathrm{d}z'}K_{\omega_n}(z, z')\right\}_{z'=-0}. \tag{14.63}$$

Exercise. Derive the boundary condition (14.63).

de Gennes derives the second boundary condition

$$\{K_{\omega_n}(z, z')/\Xi_0(z')\}_{z'=+0} = \{K_{\omega_n}(z, z')/\Xi_0(z')\}_{z'=-0} \quad (14.64)$$

by a symmetry argument. The boundary conditions obeyed by the gap function are found from (14.54):

$$\left.\begin{aligned} \Delta/\Xi_0 V \equiv \Delta/\mathscr{F} \quad & \text{continuous,} \\ \frac{D}{V}\left(\frac{\mathrm{d}\Delta}{\mathrm{d}z}\right) \quad & \text{continuous.} \end{aligned}\right\} \quad (14.65)$$

In the present geometry the thicknesses d_+, d_- of the layers $z > 0$, $z < 0$ are small compared with the coherence lengths. $K_{\omega_n}(z, z')$ is therefore constant except for discontinuous jumps when one of the arguments z or z' changes sign. From the pair of boundary conditions (14.63) and (14.64), it follows that

$$\frac{K_{\omega_n}(+,+)}{\Xi_0(+)} = \frac{K_{\omega_n}(+,-)}{\Xi_0(-)}; \quad \frac{K_{\omega_n}(+,-)}{\Xi_0(+)} = \frac{K_{\omega_n}(-,-)}{\Xi_0(-)}. \quad (14.66)$$

From the sum rule (14.62), there follows the further pair of relations:

$$\left.\begin{aligned} d_+K_{\omega_n}(+,+) + d_-K_{\omega_n}(+,-) &= \Xi_0(+)\frac{\pi}{|\omega_n'|}, \\ d_+K_{\omega_n}(-,+) + d_-K_{\omega_n}(-,-) &= \Xi_0(-)\frac{\pi}{|\omega_n'|}. \end{aligned}\right\} \quad (14.67)$$

Hence

$$\frac{K_{\omega_n}(+,-)}{\Xi_0(+)\Xi_0(-)} = \frac{K_{\omega_n}(+,+)}{\Xi_0(+)^2} = \frac{K_{\omega_n}(-,-)}{\Xi_0(-)^2}$$

$$= \frac{\pi}{|\omega_n'|\{\Xi_0(+)d_+ + \Xi_0(-)d_-\}}. \quad (14.68)$$

With the form (14.68) for the kernel, the integral equation (14.54) for the gap function reduces to the pair of algebraic equations:

$$\left.\begin{aligned}\Delta(+) &= \\ &\sum_{\omega_n'} V(+)\,\mathrm{k}T\frac{\pi}{\hbar|\omega_n'|}\frac{\Xi_0(+)^2 d_+\Delta(+)+\Xi_0(+)\Xi_0(-)d_-\Delta(-)}{\Xi_0(+)d_+ + \Xi_0(-)d_-},\\ \Delta(-) &= \\ &\sum_{\omega_n'} V(-)\,\mathrm{k}T\frac{\pi}{\hbar|\omega_n'|}\frac{\Xi_0(+)\Xi_0(-)d_+\Delta(+)+\Xi_0(-)^2 d_-\Delta(-)}{\Xi_0(+)d_+ + \Xi_0(-)d_-}.\end{aligned}\right\} \tag{14.69}$$

de Gennes assumes the frequency cut-off value $\bar{\omega}$ to be the same in the two layers, and makes the replacement (14.53). The condition for (14.69) to have a non-trivial solution is

$$\mathscr{F}_{\mathrm{eff}} \equiv \frac{V(+)\Xi_0(+)^2 d_+ + V(-)\Xi_0(-)^2 d_-}{\Xi_0(+)d_+ + \Xi_0(-)d_-} > 0. \tag{14.70}$$

The entire system has a *single* transition temperature, governed by the effective Fröhlich constant (14.70). This effective constant is a weighted mean of the Fröhlich constants† of the two layers. Note that if $\mathscr{F}(+)$ and $\mathscr{F}(-)$ are both positive (i.e. both layers have attractive electron–electron interactions) then (14.70) is always satisfied. The two-layer system will have a superconducting transition temperature somewhere between those of the two materials. On the other hand, if $\mathscr{F}(+)$ is positive and $\mathscr{F}(-)$ is negative, a superconducting transition temperature exists *only if* $d_- < d_+\mathscr{F}(+)\Xi_0(+)/\mathscr{F}(-)\Xi_0(-)$. There is some experimental evidence‡ that in lead–silver, tin–silver, lead–copper and tin–copper systems, the superconducting transistion temperature tends to zero very rapidly as the silver or copper layer approaches a critical thickness ~ 300Å. The experimental evidence is still not completely conclusive,

† Including the Coulomb part, cf. (11.32).
‡ Smith *et al.* (1961), Simmons and Douglass (1962), P. and R. Hilsch (1964).

since the surface oxide barrier plays an important role. However, the theoretical predictions of Silvert and Cooper (1966) agree well with the empirical curve of P. and R. Hilsch (1964). It is reasonable to conclude that $\mathscr{F}$ is negative for silver and copper.

Thus the study of the superconducting transition in two-layer systems gives a method, at least in principle, of measuring $\mathscr{F}$ for a metal which has not been observed to become superconducting. It can therefore be used to discriminate between true non-superconductors and metals whose transition temperatures are too low for present-day techniques.

Another interesting geometry for the proximity effect is the s-n-s sandwich. de Gennes shows that in this system a supercurrent can travel through the normal layer. The theory is quite similar to that of the Josephson current (§ 8.2), but it is found that the maximum supercurrent is proportional to $(T_c-T)^2$. This is in contrast to the (T_c-T) factor in the Josephson current (equation (8.12)).

CHAPTER 15

CRITERIA FOR SUPERCONDUCTIVITY

IT should be possible to predict from first principles whether or not a given material will be a superconductor. Indeed it should even be possible to calculate the transition temperature. In fact, however, the estimates of the strength of the electron–phonon interaction and of the density of states are not very accurate. Consequently the criteria for superconductivity remain rather ill-defined. It is not yet possible to give an unequivocal answer to the question: do other mechanisms for superconductivity, apart from the Fröhlich interaction, exist?

15.1. Theoretical criteria

Since the work of Fröhlich (1950), it has been recognized that the most important factor in determining whether a given material will be a superconductor is the value of $\mathscr{F}$. Fröhlich estimated the strength of the electron–phonon interaction from the measured resistivity of materials at room temperature. For several reasons, however, this does not give a reliable estimate. Umklapp processes play an important part in the scattering of electrons by phonons, and the structure of the density-of-states function $\Xi_0(\varepsilon)$ is complicated. Considering the difficulties, Fröhlich's criterion was fairly successful; the superconductors in his list tend to have high values of $\mathscr{F}$, and the non-superconductors low $\mathscr{F}$. However, the classification is not sharp.

In the BCS theory, the criterion is slightly different. The decisive factor is the sign of

$$\tilde{\mathscr{F}} \equiv \mathscr{F} - \mathscr{F}_{\text{Coul}}, \tag{15.1}$$

(p. 191). However, $\tilde{\mathscr{F}}$ is even more difficult to estimate theoretically for a real material than $\mathscr{F}$. To find $\tilde{\mathscr{F}}$ it is necessary to calculate the mean Coulomb matrix element $\mathscr{F}_{\text{Coul}}/\Xi_0$ *in the actual crystal lattice*. If the effects of the lattice are neglected, then the mean Coulomb interaction can be found by the method of Bardeen and Pines (1955). But, under these assumptions, $\mathscr{F}_{\text{Coul}}$ is nearly constant, and the criterion reduces to that of Fröhlich.

If $\mathscr{F}$ were accurately known, it should be possible to do more than predict whether a material will be a superconductor. From (12.47), the transition temperature should also be known. However, there is very little correlation between the observed transition temperature and Fröhlich's $\mathscr{F}$. This is hardly surprising, since the function $\exp(-1/\mathscr{F})$ depends very sensitively on its argument $\mathscr{F}$; a very small error in $\mathscr{F}$ will produce a large error in T_c. In practice, one has to interpret (12.47) phenomenologically. From the transition temperature, one can estimate $\mathscr{F}$, and use this value to correlate the transition temperature with other properties (e.g. the zero-temperature energy gap $2\Delta(0)$).

The derivation of (15.1) depends on the assumption that the Coulomb matrix element $\mathscr{F}_{\text{Coul}}/\Xi_0$ is similar in its general behaviour to the phonon-mediated matrix element $\mathscr{F}/\Xi_0$. In particular the BCS theory assumes that both interactions are constant, but are cut off at an energy $\eta_c = \hbar\bar{\omega}$ from the Fermi surface. Bogoliubov *et al.* (1959 Chapter 5) have questioned the validity of cutting off the Coulomb interaction at the same energy $\hbar\bar{\omega}$ as the phonon interaction .

In their calculation, Bogoliubov *et al.* assume that

$$\left.\begin{aligned}(V_{\text{phonon}})_{\mathbf{k},\mathbf{k}'} &= \text{const.}, > 0; \; |\eta(\mathbf{k})| < \hbar\bar{\omega},\; |\eta(\mathbf{k}')| < \hbar\bar{\omega}, \\ &= 0 \qquad\qquad \text{otherwise,}\end{aligned}\right\} \tag{15.2}$$

and that

$$\left.\begin{aligned}(V_{\text{Coul}})_{\mathbf{k},\mathbf{k}'} &= \text{const.}, < 0; \; |\eta(\mathbf{k})| < \eta_c,\; |\eta(\mathbf{k}')| < \eta_c, \\ &= 0 \qquad\qquad \text{otherwise.}\end{aligned}\right\} \tag{15.3}$$

Here η_c is assumed to be comparable with the Fermi energy μ. To simplify the calculation, they also take $\Xi_0(\eta)$ to be constant (even when $\eta \sim \mu$!). Under these assumptions the integral equation (12.24) has the solution for weak coupling:

$$\left.\begin{aligned} \Delta &= \Delta_1, \quad |\eta(\mathbf{k})| < \hbar\bar{\omega}, \\ &= \Delta_2, \quad \hbar\bar{\omega} < |\eta(\mathbf{k})| < \eta_c, \end{aligned}\right\} \tag{15.4}$$

where

$$\left.\begin{aligned} \Delta_1 &= (\mathscr{F} - \mathscr{F}_{\text{Coul}})\Delta_1 \ln(2\hbar\bar{\omega}/\Delta_1) - \mathscr{F}_{\text{Coul}}\Delta_2 \ln(\eta_c/\hbar\bar{\omega}), \\ \Delta_2 &= -\mathscr{F}_{\text{Coul}}\{\Delta_1 \ln(2\hbar\bar{\omega}/\Delta_1) - \Delta_2 \ln(\eta_c/\hbar\bar{\omega})\}. \end{aligned}\right\} \tag{15.5}$$

It is the energy gap $2\Delta_1$, for small $\eta(\mathbf{k})$, which is observed experimentally, and which should be related to the effective interaction by (11.27). Hence the effective interaction constant is

$$\bar{\bar{\mathscr{F}}} = \mathscr{F} - \bar{\mathscr{F}}_{\text{Coul}} = \mathscr{F} - \frac{\mathscr{F}_{\text{Coul}}}{1 + \mathscr{F}_{\text{Coul}} \ln(\eta_c/\hbar\bar{\omega})}. \tag{15.6}$$

Exercise. Assuming the interaction to be of the form of (15.2) and (15.3), show that the Bogoliubov integral equation (12.24) has the solution (15.4).

This result is surprising. The Coulomb interaction is rendered *less* effective by increasing its range in momentum space to include states far from the Fermi surface. The effective Coulomb interaction is reduced by the appearance of a negative amplitude for Cooper pairs in states with

$$\hbar\bar{\omega} < |\eta(\mathbf{k})| < \eta_c. \tag{15.7}$$

The scattering between Bloch states *within* the region (15.7) makes a positive contribution to the energy. But this positive part is outweighed by a negative contribution from the interaction between states in this region (15.7) and states in the BCS region $|\eta(\mathbf{k})| < \hbar\bar{\omega}$.

Morel and Anderson (1962) have criticized the use of an *instantaneous* electron–electron interaction in both the BCS and Bogoliubov theories. They point out that the Coulomb interaction propagates at the velocity of light, and is indeed instantaneous for all practical purposes. But, in contrast, the

Fröhlich interaction propagates at only the velocity of sound. This is slower than the Fermi velocity by about a factor 10^3. It is therefore important that the interaction potential (10.55) be replaced by a retarded potential. Morel and Anderson show how Gor'kov's gap equation (14.38) must be generalized to include retardation.

In the absence of Coulomb interaction one of the results of the Morel–Anderson analysis is that the Fröhlich parameter $\mathscr{F}$ is renormalized. $\mathscr{F}$ is no longer strictly proportional to the density of states at the Fermi surface. The expression found for $\mathscr{F}$ is

$$\mathscr{F} = \tfrac{1}{2}\left\{\frac{\kappa^2}{\kappa^2+\frac{3}{5}k_0^2}\right\}^2, \tag{15.8}$$

where κ is the Thomas–Fermi inverse screening length (10.18)† and k_0 is the Debye cut-off wave number. Note that, *however large the density of states*, $\mathscr{F}$ cannot exceed the value $\frac{1}{2}$.

When the screened Coulomb term (11.30) is included, Morel and Anderson find for the energy gap:

$$\Delta = 2\hbar\bar{\omega}\exp\left\{-\mathscr{F}+\frac{\mathscr{F}_{\text{Coul}}}{1+\mathscr{F}_{\text{Coul}}\ln(\mu/\hbar\bar{\omega})}\right\}. \tag{15.9}$$

This is identical with the expression of Bogoliubov *et al.*, with the Coulomb cut off η_c chosen to be the Fermi energy μ. It should be noted that the phonon frequency $\bar{\omega}$ appears *inside* the expression for the effective coupling constant. One of the consequences is that the form of the isotope effect is no longer given by (10.52) or (10.53). If the isotopic-mass dependence is written

$$T_c \propto M^{-a}, \tag{15.10}$$

then the value of a can be calculated from (15.9).

† Since the Coulomb interaction has been neglected, it is puzzling that the screening length appears in (15.8). In fact it appears as a result of substituting the expression of Bohm and Staver (1951) for the velocity of sound (see also Staver 1952, Bardeen and Pines 1955, Morel 1959). The Bohm–Staver velocity of sound is calculated from a plasma model, of charged electrons and heavy ions, interacting *only* through a Coulomb interaction. The screening length thus appears naturally. 'Neglecting Coulomb interactions' in deriving (15.8) really means neglecting only the short-range part of the Coulomb interaction.

Table 15.1 shows the calculated values of several of the above parameters, for a number of elements. Where possible the experimental values are given for comparison.

The agreement with experiment is not very good. In particular, the isotope-effect exponents are systematically too low. Moreover the effective Coulomb matrix elements are practically the same for all materials;

$$\bar{\mathscr{F}}_{\text{Coul}} = \mathscr{F}_{\text{Coul}}/\{1+\mathscr{F}_{\text{Coul}}\ln(\mu/\hbar\bar{\omega})\} \simeq 0{\cdot}1. \qquad (15.11)$$

As a result, the Morel–Anderson theory predicts that $\tilde{\bar{\mathscr{F}}}$ should be positive for all metals (including the alkali and noble metals). This result is in conflict with the evidence from the proximity effect (§ 14.5) that in copper and silver $\tilde{\mathscr{F}}$ is probably negative.

15.1.1 Superconducting semiconductors

It was remarked by Cohen (1964) that the BCS criterion (15.1) can be fulfilled in a semiconductor. Although the density of states Ξ_0 in a semiconductor is very much lower than in a metal, the electron–phonon interaction can be strong. The situation is particularly favourable in many-valley semiconductors. In these the density of states is comparatively large and the most important phonon modes are those associated with intervalley transitions. These phonons are of short wavelength, and their interaction with electrons is almost unscreened.

The theoretical predictions have been confirmed, with the discovery of a superconducting transition in self-doped germanium telluride (Hein *et al.* 1964), in reduced strontium titanate (Schooley *et al.* 1964) and in mixed barium–strontium titanates (Frederikse *et al.* 1966). In all cases the transition temperature was observed to vary with carrier concentration in the way predicted by Cohen. The materials are all extreme Type II, with $\kappa \sim 10^2$. Cupric sulphide, observed to become a superconductor by Meissner (1929), is presumably also in this category.

Element	Debye temp.[a] Θ(°K)	Transition temp. T_c(°K)	$\mathscr{F}$ (Morel–Anderson)[a]	$\mathscr{F}_{Coul}$ (Morel–Anderson)[a]	$\tilde{\mathscr{F}}$ (Morel–Anderson)[a]	$\tilde{\mathscr{F}}$ (experimental)	a (Morel–Anderson)	a (Garland)[b]	a (experimental)[b]
Na	160	<0·09[c]	0·25	0·12	0·13	<0·13			
K	100	<0·08[c]	0·25	0·12	0·13	<0·14			
Cu	343	<0·05[c]	0·20	0·10	0·10	<0·11			
Au	164	<0·006[c]	0·18	0·10	0·08	<0·09			
Mg	342	<0·05[c]	0·32	0·12	0·20	<0·11			
Ca	220	<0·35[c]	0·27	0·11	0·16	<0·15			
Zn	235	0·9[a]	0·25	0·09	0·16	0·18[a]	0·35[a]	0·40	0·45±·05
Cd	164	0·56[a]	0·23	0·09	0·14	0·175[a]	0·34[a]	0·37	0·50±·10
Hg	70	4·16[a]	0·37	0·10	0·27	0·35[a]	0·46[a]	0·47	0·50±·03
Al	375	1·2[a]	0·33	0·10	0·23	0·17[a]	0·34[a]	0·35	
In	109	3·4[a]	0·34	0·10	0·24	0·29[a]	0·44[a]		
Tl	100	2·4[a]	0·32	0·09	0·23	0·27[a]	0·43[a]	0·45	0·50±·10
Sn	195	3·75[a]	0·34	0·10	0·24	0·25[a]	0·42[a]	0·44	0·47±·02
Pb	96	7·22[a]	0·40	0·10	0·30	0·39[a]	0·47[a]	0·47	0·48±·01
Ti	430	0·4[a]	0·41	0·11	0·30	0·14[a]	0·25[a]	0·15	
Zr	265	0·55[a]	0·37	0·11	0·26	0·16[a]	0·30[a]	0·35	
V	338	4·9[a]	0·47	0·12	0·35	0·24[a]	0·41[a]	0·15	
Nb	320	8·8[a]	0·47	0·12	0·35	0·32[a]	0·45[a]		
Ta	230	4·4[a]	0·45	0·11	0·34	0·25[a]	0·42[a]	0·35	
Mo	360	0·92[c]	0·38	0·10	0·28	0·17[a]	0·3[b]	0·35	0·37±·07
U	200	1·1[a]	0·47	0·12	0·35	0·19[a]	0·36[a]		
Ru		0·47[c]					0·25[b]	0·00	0·00±·05
Os		0·71[c]					0·25[b]	0·10	0·10±·10

[a] From the table of Morel and Anderson (1962).
[b] From the table of Garland (1963).
[c] From the table of Roberts (1964).

15.2. Matthias's empirical rules

As a result of a careful study of the superconducting transition temperatures of a great number of metals and alloys, Matthias (1957) proposed a number of empirical rules governing the superconducting transition temperature.

(a) The transition temperature of an element depends in a regular fashion on its position in the periodic table. For the transition metals and the metals with complete *d*-shells, the functional dependence of T_c on the valency is quite different (Fig. 15.1). However, it is rather similar for successive periods

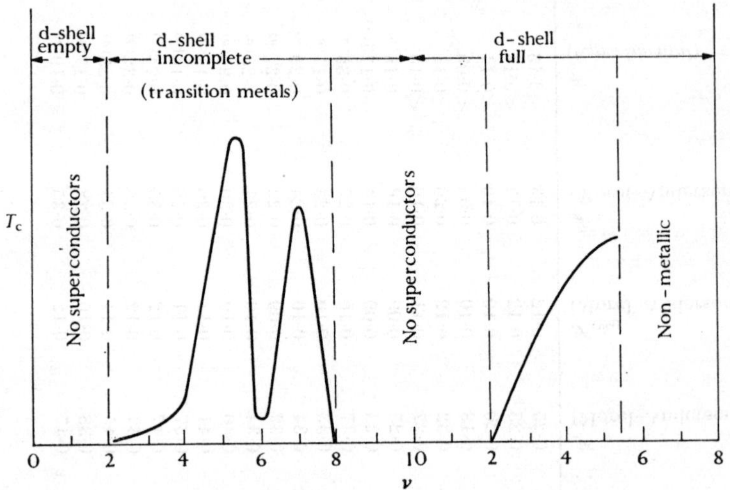

FIG. 15.1. Qualitative variation of the transition temperature T_c with the mean valency ν (Matthias 1957).

within either the transition or non-transition family. By the 'valency', Matthias means the *maximum* valency that the element in question can exhibit. In other words, the valency is the number of electrons outside a closed shell.

(b) For alloys of one transition metal with another, and for alloys of one non-transition metal with another, it is possible to define a mean valency ν (weighted according to the atomic fraction of the constituent metals). *The transition temperatures of alloys are a smooth function of their valency.* Pure metals appear in this picture merely as particular points on the smooth curve. However, the concept of a mean valency does not prove useful in describing alloys in which one of the components is a transition element and another is a non-transition element.

(c) For elements and alloys with closed d-shells, T_c rises from zero (or a very small value) at $\nu = 2$ to a maximum at about $\nu = 5$. It is not possible to extrapolate to $\nu > 5$, since elements in this series with $\nu \geqslant 5$ are non-metals, or at best semimetals. However, it is generally believed that the high-pressure metallic phases of these materials will be superconductors. A superconducting modification of bismuth is known to exist.

(d) In the transition series, T_c has maxima at $\nu = 5$ and $\nu = 7$, with a very sharp minimum at $\nu = 6$ (see Fig. 15.1). Matthias originally proposed the existence of a smaller maximum at $\nu = 3$, on account of the superconductivity of lanthanum. However (Kondo 1963, Hamilton and Jensen 1963), there are reasons for regarding lanthanum as a special case; none of the other lanthanide elements has been observed to become superconducting. For the most part intermetallic compounds behave like transition-metal alloys, showing the characteristic maxima at $\nu = 5$ and $\nu = 7$.

Considerable irregularities occur when one of the alloying materials is magnetic (in particular iron or manganese). In a few cases the mean valency concept works well enough. But more often the addition of only a small amount of iron has a quite disproportionate influence on the transition temperature. For example, the addition of a little iron to a molybdenum–rhenium alloy depresses T_c very drastically; iron in tungsten alloys behaves similarly. Indeed, Matthias suggests that superconductivity has not been observed in pure tungsten because available tungsten specimens did not have a sufficiently low concentration of magnetic impurities.

The effect of adding iron to niobium ($\nu = 5$) should be contrasted with the effect of iron on molybdenum or tungsten. In niobium systems, iron has an anomalously *small* effect on the transition temperature. The transition temperature of pure niobium is 9°K. If niobium is alloyed with ruthenium or osmium, the transition temperature is depressed to below 4°K at valency $\nu = 5{\cdot}3$. But a niobium-iron alloy with $\nu = 5{\cdot}3$ has its transition temperature at 6·8°K (Fig. 15.2).

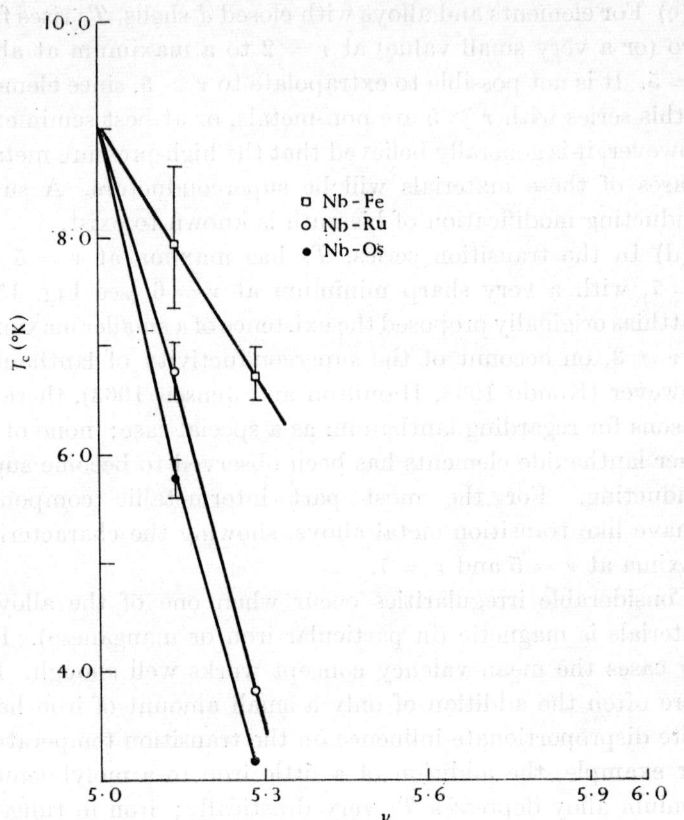

FIG. 15.2. Transition temperatures of dilute solutions of Fe, Ru, and Os in Nb (Geballe 1964).

These experiments have stimulated a great deal of further work. It has been widely conjectured that the anomalous behaviour of iron alloys is evidence for the existence of a non-phonon mechanism for superconductivity. Where iron acts to suppress superconductivity, it seems that it does so as a consequence of the existence of localized magnetic moments (Clogston *et al.* 1962*b*, Anderson 1961). These localized states presumably scatter electrons strongly, and may tend to break Cooper pairs. But the opposite type of anomaly is more difficult to interpret.

15.3. Other mechanisms for superconductivity

Historically, the BCS theory was based on the Bloch–Fröhlich electron–phonon interaction. The observation of the isotope effect in many superconductors provided strong evidence in support of this view of the nature of superconductivity. But the structure of the BCS theory is such that the details of the interaction mechanism are largely irrelevant to the problem. *Any* attractive interaction between Fermions will lead to Cooper pair formation. When the interactions are not of uniform sign, the problem is more complicated; many of the difficulties discussed in § 15.1 stem from the non-uniform sign of the electron–electron interaction in real metals. The present consensus of opinion is that a BCS condensation will occur in a system for which there are *some regions in* **k**-*space where the interactions are attractive.*

Kohn and Luttinger (1965) have predicted that even 'jellium' (cf. p. 168) will be a superconductor at low enough temperatures. This prediction follows from the detailed theory of the Coulomb screening. The modern theory of a Coulomb gas gives for the wave-vector-dependent dielectric constant (e.g. Pines 1961 p. 64)

$$\boldsymbol{\epsilon}(k) = 1+\frac{\kappa^2}{k^2}\left\{\frac{1}{2}+\frac{k_{\mathrm{F}}}{2k}\left(1-\frac{k^2}{4k_{\mathrm{F}}^2}\right)\ln\left|\frac{k-2k_{\mathrm{F}}}{k+2k_{\mathrm{F}}}\right|\right\}, \qquad (15.12)$$

instead of the Thomas–Fermi value $1+\kappa^2/k^2$ (cf. § 10.2). The expression (15.12) is not of uniform sign, and Kohn and Luttinger show the possibility of Cooper-pair formation in the regions of **k**-space for which the dielectric constant is negative.

'Superconducting' wave functions for the nuclei of heavy atoms have been extensively discussed (e.g. Bohr *et al.* 1958,

Baranger 1963). The Yukawa force between nucleons has a pairing part, and the methods of the BCS theory are applicable. Changes in the formalism are, of course, needed to take account of the relatively small number of particles comprising even the largest atomic nuclei; plane-wave states are obviously inappropriate! The pairing theory accounts in a very satisfactory way for the fact that the excitation spectrum shows an energy gap when the numbers of protons and neutrons are both even.

Another system for which a BCS condensation has been predicted is liquid helium-3 (Brueckner *et al.* 1960, Emery and Sessler 1960). For this system the internal structure of helium atoms can be disregarded, and the atoms treated as interacting Fermions. The interactions are, however, too strong for the BCS approach to be quantitatively reliable. Instead, one must start from the phenomenological Fermi liquid model (Landau 1956), and estimate the scattering lengths for scattering of one 'dressed' atom by another, in various angular-momentum eigenstates. It is found that the interaction is repulsive for s-scattering, and slightly attractive for p-scattering. For d-scattering it is quite strongly attractive. Thus the BCS condensation should occur into states in which the Cooper-pair wave function has d-symmetry. Since the atoms of helium are neutral, the condensed phase will be a superfluid, but not a superconductor. Emery and Sessler (1960) predicted a transition temperature of 0·08°K, but experiments down to below 0·005°K (Abel *et al.* 1965, Peshkov 1964) have failed to establish the existence of a condensed phase. The fact that the predicted transition temperature is too high by at least a factor 20, and is uncertain by probably a factor 10^2, is another illustration of the inadequacy of the present criteria for a BCS condensation.

There has been considerable discussion recently whether non-phonon mechanisms play a role in the superconductivity of any ordinary materials. The measured isotope-effect exponent a is significantly different from $\frac{1}{2}$ in several materials. In particular it is nearly zero in rhenium and osmium (Table 15.1). This fact was at first interpreted to mean that the phonon interaction was not responsible for the superconductivity of these elements. However the theory of Morel and Anderson (1962) predicts that a will differ from $\frac{1}{2}$ as a consequence of the time lag in phonon interactions. There is no quantitative agreement with experiment, but the evidence is insufficient to conclude that a new mechanism is operative.

Magnetic interactions have been widely discussed as a possible mechanism for the superconductivity of the transition metals. In these substances the Fermi surface is at an energy where two bands overlap. In one of these bands, the Bloch one-electron wave functions are built up mainly of superpositions of atomic s- and p-wave functions. But in the much narrower second band, the wave functions have predominantly d-wave nature. The density of states Ξ_0 varies in a rather complicated way with the number of electrons (Morin and Maita 1963). Figure 15.3 illustrates the variation of Ξ_0 with ν; note the peaks at $\nu = 5$ and $\nu = 7$. Pines (1958) has attempted to account for Matthias's rules on the basis of the variation of Ξ_0 with ν, but this programme has been only partly successful.

Two-band models with magnetic interactions have been discussed by, *inter alia*, Suhl *et al.* (1959), Kondo (1963), and Kuper *et al.* (1964).† Garland (1963) has studied the problem of superconductivity in two-band metals very carefully. He shows that the high effective mass of electrons in the d-band modifies the Coulomb screening drastically. After extensive numerical computations, Garland concludes that the phonon and screened-Coulomb interactions are sufficient to account for the transition temperatures and isotope-effect coefficients in nearly all cases. However, the question remains controversial.

† The last-mentioned paper is concerned with the effect of the very narrow f-band on superconductivity in lanthanum and uranium. When the f-band is partially full, the magnetic interaction in it is inimical to superconductivity. However, in the cases of lanthanum and uranium the f-band is empty but very near the Fermi surface. Under the magnetic interactions, Cooper pairs can be formed in the f-band. The model sheds some light on possible reasons for the departure of these two elements from Matthias's rules.

Finally, Little's (1964) proposal should be mentioned, that certain organic polymers should be superconducting at room temperature. These are materials with a long spine of alternate double and single bonds, and highly polarizable side chains

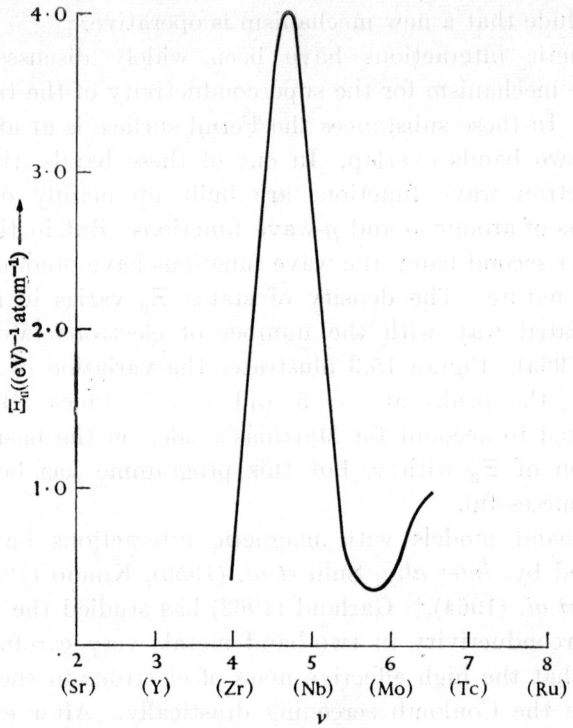

FIG. 15.3. Variation of the density of states Ξ_0 with mean valency ν (Morin and Maita 1963).

at regular intervals along the spine. One electron per spine atom is in a π-state, and exhibits quasimetallic properties in one dimension. Little proposes that the interaction with the side chains can lead to very strong formation of Cooper pairs. The proposal has been severely criticized by Ferrell (1964, 1966), Paulus (1966), and Kuper (1966)†.

† *Added in proof:* Bychkov *et al.* (1965, 1966) have investigated the possibility of a four-particle condensation. Kuper's objection may not be applicable to this model, but Ferrell's criticism would still appear to apply.

APPENDIX A

Free energy of a magnetic medium

FOR a small change $\delta\mathbf{B}$ in the magnetic induction, the change in the magnetic energy of a system is given by

$$\delta U_{\mathrm{m}} = \frac{1}{4\pi}\int \mathrm{d}\tau\, \mathbf{H}.\delta\mathbf{B}, \tag{A.1}$$ [9]

where the integral is to be taken over all space. Equation (A.1) may be rewritten[78]

$$\begin{aligned} 4\pi\delta U_{\mathrm{m}} = \int \mathrm{d}\tau\, \{&(\mathbf{H}-\mathbf{H}_{\mathrm{e}}).\delta\mathbf{B} &\text{(A.2a)}\\ &+\mathbf{H}_{\mathrm{e}}.\delta(\mathbf{B}-\mathbf{H}) &\text{(A.2b)}\\ &+\mathbf{H}_{\mathrm{e}}.\delta(\mathbf{H}-\mathbf{H}_{\mathrm{e}}) &\text{(A.2c)}\\ &+\mathbf{H}_{\mathrm{e}}.\delta\mathbf{H}_{\mathrm{e}}\}. &\text{(A.2d)} \end{aligned}$$

Here $\mathbf{H}_{\mathrm{e}}$ is the applied field, i.e. the value which the field would take in the absence of the magnetic medium but in the presence of the currents which generate the field. Introduce vector potentials $\mathbf{A}_{\mathrm{e}}$, $\mathbf{A}$, satisfying respectively

$$\mathbf{H}_{\mathrm{e}} = \operatorname{curl}\mathbf{A}_{\mathrm{e}}, \tag{A.3a}$$ [79]

$$\mathbf{B} = \operatorname{curl}\mathbf{A}. \tag{A.3b}$$

Then (A.2a) may be written

$$\int \mathrm{d}\tau\, (\mathbf{H}-\mathbf{H}_{\mathrm{e}}).\operatorname{curl}\delta\mathbf{A}$$
$$= \int \mathrm{d}\tau\, [\operatorname{div}\{(\mathbf{H}-\mathbf{H}_{\mathrm{e}})\wedge\mathbf{A}\} + \delta\mathbf{A}.\operatorname{curl}(\mathbf{H}-\mathbf{H}_{\mathrm{e}})]. \tag{A.4}$$

The divergence may be expressed as a surface integral over a surface at infinity, and vanishes. In the other term of (A.4) we note that

$$\operatorname{curl}\mathbf{H} = \operatorname{curl}\mathbf{H}_{\mathrm{e}} = 4\pi\mathbf{J}/c, \tag{A.5}$$ [26][30]

since, by definition, both **H** and $\mathbf{H}_e$ are produced by the same distribution of currents **J**. Hence (A.2a) is zero. If we write (A.2c) as

$$\int d\tau\,(\text{curl}\,\mathbf{A}_e)\,.\,\delta(\mathbf{H}-\mathbf{H}_e), \tag{A.6}$$

it may similarly be shown to vanish. Hence only (A.2b) and (A.2d) survive:[80]

$$\delta U_m = \frac{1}{4\pi}\int d\tau\,\{\mathbf{H}_e\,.\,\delta(\mathbf{B}-\mathbf{H})+\mathbf{H}_e\,.\,\delta\mathbf{H}_e\}, \tag{A.7}$$

$$= \mathbf{H}_e\,.\,\delta\mathbf{M}+\frac{1}{4\pi}\int d\tau\,\mathbf{H}_e\,.\,\delta\mathbf{H}_e \tag{A.8}$$

when $\mathbf{H}_e$ is uniform in the region occupied by the medium. For a fuller account, see Landau and Lifshitz (1960 §§ 30, 31).

Equation (A.8) is equivalent to (2.4), apart from the term[(10),(11)]

$$\frac{1}{4\pi}\int d\tau\,\mathbf{H}_e\,.\,\delta\mathbf{H}_e = \frac{1}{8\pi}\int d\tau\,\delta(H_e^2),$$

which represents the field energy stored in the magnetic field $\mathbf{H}_e$ (and which is independent of the magnetic properties of the system).

From (2.6) and (2.7), the magnetic Gibbs function satisfies

$$\delta G = -\mathbf{M}\,.\,\delta\mathbf{H}_e - p\delta\Omega - S\delta T. \tag{A.9}$$

The mixed second derivative $(\partial^2 G/\partial T\partial\mathbf{H}_e)_\Omega$ may be written in the two equivalent forms

$$\left[\frac{\partial}{\partial\mathbf{H}_e}\left(\frac{\partial G}{\partial T}\right)\right]_\Omega = -\left(\frac{\partial S}{\partial\mathbf{H}_e}\right)_{T,\Omega}, \tag{A.10}$$

$$\left[\frac{\partial}{\partial T}\left(\frac{\partial G}{\partial\mathbf{H}_e}\right)\right]_\Omega = -\left(\frac{\partial\mathbf{M}}{\partial T}\right)_{\Omega,\mathbf{H}_e}. \tag{A.11}$$

Hence the Maxwell relation used in (6.5) follows:

$$\left(\frac{\partial S}{\partial \mathbf{H}_{\mathrm{e}}}\right)_{T,\Omega} = \left(\frac{\partial \mathbf{M}}{\partial T}\right)_{\Omega,\mathbf{H}_{\mathrm{e}}}. \tag{A.12}$$

In the above expressions, the vector operator $\partial/\partial\mathbf{H}_{\mathrm{e}}$ is to be interpreted as the vector whose components are the differential operators $\partial/\partial(H_{\mathrm{e}})_x$, $\partial/\partial(H_{\mathrm{e}})_y$, $\partial/\partial(H_{\mathrm{e}})_z$.

APPENDIX B

Alternative derivation of the upper critical field

THE following alternative derivation of equation (6.56) (Josephson 1965*b*) is illuminating. For small Ψ, the non-linear term in (6.24) may be neglected, to give[26]

$$\frac{1}{2m_s}\left(-i\hbar\nabla-\frac{\varepsilon}{c}\mathbf{A}\right)^2\Psi=-\alpha\Psi. \tag{B.1}$$

This equation is formally identical with the Schrödinger equation for a particle of charge ε and mass m_s in a magnetic field; $-\alpha$ plays the role of the energy eigenvalue. It follows immediately that H_{c2} is the largest magnetic field compatible with the energy $-\alpha$.

However, for a given magnetic field $\mathbf{H}$, the lowest eigenstate has energy

$$\tfrac{1}{2}\hbar\omega_c\equiv\tfrac{1}{2}\hbar\frac{\varepsilon H}{m_s c}. \qquad \text{(B.2)}^{(81)}$$

Here ω_c is the cyclotron frequency; see e.g. Mott and Jones (1936 p. 202). Hence

$$-\alpha=\tfrac{1}{2}\frac{\hbar\varepsilon H_{c2}}{m_s c}; \qquad \text{(B.3)}^{(81)}$$

or finally, using (6.26), (6.27), (6.37), and (6.38),

$$H_{c2}=\kappa\sqrt{2}\,H_c. \tag{B.4}$$

APPENDIX C

The Schrödinger and Heisenberg pictures

In the usual Schrödinger picture of quantum mechanics, physically observable quantities are represented by time-independent operators. Thus, for example, the momentum operator for a single particle is given by

$$\hat{\mathbf{p}} = -\mathrm{i}\hbar\nabla. \tag{C.1}$$

Stationary states of the system are eigenstates of the energy operator

$$\hat{E} = \mathrm{i}\hbar\partial/\partial t. \tag{C.2}$$

From the time-dependent Schrödinger† equation

$$\hat{H}\psi^{\mathrm{S}}(\mathbf{r}, t) = \hat{E}\psi^{\mathrm{S}}(\mathbf{r}, t), \tag{C.3}$$

they are also eigenstates of the Hamiltonian operator $\hat{H}$. The eigenvalue condition

$$\hat{E}\psi^{\mathrm{S}} = E\psi^{\mathrm{S}} \tag{C.4}$$

gives

$$\psi^{\mathrm{S}}(\mathbf{r}, t) = \psi^{\mathrm{S}}(\mathbf{r}, 0)\mathrm{e}^{-\mathrm{i}Et/\hbar}. \tag{C.5}$$

Thus the stationary states have constant modulus, but their phase is time dependent.

To pass to the Heisenberg picture, we perform the unitary transformation

$$\hat{O}^{\mathrm{H}} \equiv \mathrm{e}^{\mathrm{i}\hat{H}t/\hbar}\,\hat{O}\,\mathrm{e}^{-\mathrm{i}\hat{H}t/\hbar}, \tag{C.6}$$

$$\phi^{\mathrm{H}} \equiv \mathrm{e}^{\mathrm{i}\hat{H}t/\hbar}\phi^{\mathrm{S}}, \tag{C.7}$$

† Superscripts S and H will be used to refer to states in the Schrödinger and Heisenberg pictures respectively. Schrödinger operators are designated $\hat{O}$, and Heisenberg operators $\hat{O}^{\mathrm{H}}$. $\hat{H}$ is the Hamiltonian in the Schrödinger picture. However, we are usually interested in systems for which the Hamiltonian is independent of the time. It will then be independent of time even in the Heisenberg picture, so that $\hat{H}^{\mathrm{H}} = \hat{H}$.

where $\hat{O}$ is any operator and ϕ^{s} any state in the Schrödinger picture. The new picture is designed to bring stationary states to rest. To see this, let ϕ^{s} in (C.7) have the form (C.5). Since, by (C.3), (C.5) is an eigenstate of $\hat{H}$ with eigenvalue E, it follows that

$$\psi^{\mathrm{H}}(\mathbf{r}, t) = \psi^{\mathrm{S}}(\mathbf{r}, 0). \tag{C.8}$$

The generalization to systems with many particles and with spins is trivial.

The transformation has removed the time dependence from the stationary states of the system, at the expense of making the operators time-dependent. The equation of motion of a Heisenberg operator is found by differentiating (C.6):

$$\frac{\mathrm{d}}{\mathrm{d}t}(\hat{O}^{\mathrm{H}}) = \frac{1}{\mathrm{i}\hbar}[\hat{O}^{\mathrm{H}}, \hat{H}]. \tag{C.9}$$

The (m, n)th matrix element of $\hat{O}^{\mathrm{H}}$ is

$$(\phi_m^{\mathrm{H}}|\hat{O}^{\mathrm{H}}|\phi_n^{\mathrm{H}}) = (\phi_m^{\mathrm{H}}|\hat{O}|\phi_n^{\mathrm{H}})\mathrm{e}^{\mathrm{i}(E_m - E_n)t/\hbar}, \tag{C.10}$$

i.e. the matrix elements of observables oscillate at the Bohr frequencies $(E_m - E_n)/\hbar$. For a fuller account of the relation between the Schrödinger and Heisenberg pictures, see Dirac (1958 pp. 108–18).

APPENDIX D

Compensation of dangerous diagrams

Bogoliubov (1958) has given a criterion for the choice of the parameters $u_{\mathbf{k}}$, $v_{\mathbf{k}}$ which may be applied directly to the Bloch–Fröhlich interaction Hamiltonian.

Applying the transformation (12.16) to the grand Hamiltonian (cf. (10.32), (12.1))

$$\hat{\mathscr{H}} = \sum_{k} (\varepsilon(\mathbf{k})-\mu)\hat{a}_{k}^{+}\hat{a}_{k} + \sum_{\mathbf{q}} \hbar\omega_{\mathbf{q}}\hat{b}_{\mathbf{q}}^{+}\hat{b}_{\mathbf{q}} + \mathrm{i}\sum_{k,\mathbf{q}} D_{\mathbf{q}}(\hat{a}_{k-\mathbf{q}}^{+}\hat{a}_{k}\hat{b}_{\mathbf{q}}^{+} - \hat{a}_{k}^{+}\hat{a}_{k-\mathbf{q}}\hat{b}_{\mathbf{q}}) \tag{D.1}$$

gives

$$\hat{\mathscr{H}}^{\mathrm{T}} = \hat{\mathscr{H}}_{0} + \hat{\mathscr{H}}_{1} + \hat{\mathscr{H}}_{2} + \hat{\mathscr{H}}_{3}, \tag{D.2}$$

where

$$\hat{\mathscr{H}}_{0} = 2\sum_{k} (\varepsilon(\mathbf{k}) - \mu)v_{\mathbf{k}}^{2} + \sum_{\mathbf{q}} \hbar\omega_{\mathbf{q}}\hat{b}_{\mathbf{q}}^{+}\hat{b}_{\mathbf{q}} + \sum_{k} (\varepsilon(\mathbf{k})-\mu)(u_{\mathbf{k}}^{2}-v_{\mathbf{k}}^{2})\hat{\alpha}_{k}^{+}\hat{\alpha}_{k}, \tag{D.3}$$

$$\hat{\mathscr{H}}_{1} = \sum_{k} (\varepsilon(\mathbf{k})-\mu)u_{\mathbf{k}}v_{\mathbf{k}}(\hat{\alpha}_{k}^{+}\hat{\alpha}_{-k}^{+} + \hat{\alpha}_{-k}\hat{\alpha}_{k}), \tag{D.4}$$

$$\hat{\mathscr{H}}_{2} = \mathrm{i}\sum_{\mathbf{q},k} \{D_{\mathbf{q}}\sigma_{\mathbf{k}}(u_{\mathbf{k}}v_{\mathbf{k}+\mathbf{q}}\hat{\alpha}_{k}^{+}\hat{\alpha}_{-k-\mathbf{q}}^{+} + u_{\mathbf{k}+\mathbf{q}}v_{\mathbf{k}}\hat{\alpha}_{k+\mathbf{q}}\hat{\alpha}_{-k})\hat{b}_{\mathbf{q}}^{+} - \text{h.c.}\}, \tag{D.5}$$

and

$$\hat{\mathscr{H}}_{3} = \mathrm{i}\sum_{\mathbf{q},k} \{D_{\mathbf{q}}(u_{\mathbf{k}}u_{\mathbf{k}+\mathbf{q}} - v_{\mathbf{k}}v_{\mathbf{k}+\mathbf{q}})\hat{\alpha}_{k}^{+}\hat{\alpha}_{k+\mathbf{q}}^{+}\hat{b}_{\mathbf{q}}^{+} - \text{h.c.}\}. \tag{D.6}$$

Here h.c. represents the Hermitian conjugate of the preceding expression.

The programme is to regard $\hat{\mathscr{H}}_{0}$ as an unperturbed Hamiltonian, and to treat $\hat{\mathscr{H}}_{1} + \hat{\mathscr{H}}_{2} + \hat{\mathscr{H}}_{3}$ by perturbation theory. However, there are divergence difficulties associated with the vanishing of energy denominators. Bogoliubov classifies as 'dangerous' those processes which, in a given order of perturbation theory, can lead to a vanishing denominator. He shows that a process in which the number of phonons changes is never dangerous. Indeed, before transformation the Hamiltonian has no dangerous processes below fourth order.

After the transformation, dangerous processes occur even in first order, since $\hat{\mathscr{H}}_1$ creates two Bogoliubons without any change in the number of phonons. On the other hand $\hat{\mathscr{H}}_3$ creates or absorbs a phonon, conserving the number of Bogoliubons, and $\hat{\mathscr{H}}_2$ creates or absorbs two Bogoliubons, with the emission or absorption of a phonon. Thus neither $\hat{\mathscr{H}}_2$ nor $\hat{\mathscr{H}}_3$ is dangerous in first order. The structure of these operators is illustrated in Fig. D.1. In Fig. D.2, a dangerous second-order process is illustrated. In it, $\hat{\mathscr{H}}_2$ creates two Bogoliubons and a phonon. Then $\hat{\mathscr{H}}_3$ absorbs the phonon and scatters one of the Bogoliubons. The result of these two processes is equivalent to the single process $\hat{\mathscr{H}}_1$ of Fig. D.1(a).

Bogoliubov's method consists in choosing the coefficients $u_{\mathbf{k}}$, $v_{\mathbf{k}}$, in such a way that these two dangerous processes cancel each other. Since the vacuum state of $\hat{\mathscr{H}}_0$ is the BCS ground state $|\Phi_0\rangle$, the compensation condition is

$$\{\hat{\mathscr{H}}_1-\hat{\mathscr{H}}_3(\hat{\mathscr{H}}_0-E_0)^{-1}\hat{\mathscr{H}}_2\}|\Phi_0\rangle = 0, \tag{D.7}$$

i.e.

$$2(\varepsilon(\mathbf{k})-\mu)u_{\mathbf{k}}v_{\mathbf{k}} = \sum_{\mathbf{q}} \frac{D_{\mathbf{q}}^2\{u_{\mathbf{k}+\mathbf{q}}v_{\mathbf{k}+\mathbf{q}}(u_{\mathbf{k}}^2-v_{\mathbf{k}}^2) - u_{\mathbf{k}}v_{\mathbf{k}}(u_{\mathbf{k}+\mathbf{q}}^2-v_{\mathbf{k}-\mathbf{q}}^2)\}}{E(\mathbf{k})+E(\mathbf{k}+\mathbf{q})+\hbar\omega_{\mathbf{q}}}, \tag{D.8}$$

where

$$E(\mathbf{k}) = (\varepsilon(\mathbf{k})-\mu)(u_{\mathbf{k}}^2-v_{\mathbf{k}}^2) = \eta(\mathbf{k})\,(u_{\mathbf{k}}^2-v_{\mathbf{k}}^2). \tag{D.9}$$

With the additional definitions

$$\left.\begin{aligned} \tilde{E}(\mathbf{k}) &= \tilde{\eta}(\mathbf{k})\,(u_{\mathbf{k}}^2-v_{\mathbf{k}}^2), \\ \tilde{\eta}(\mathbf{k}) &= \eta(\mathbf{k})-\sum_{\mathbf{q}} D_{\mathbf{q}}^2 \frac{u_{\mathbf{k}+\mathbf{q}}^2-v_{\mathbf{k}+\mathbf{q}}^2}{E(\mathbf{k})+E(\mathbf{k}+\mathbf{q})+\hbar\omega_{\mathbf{q}}}, \end{aligned}\right\} \tag{D.10}$$

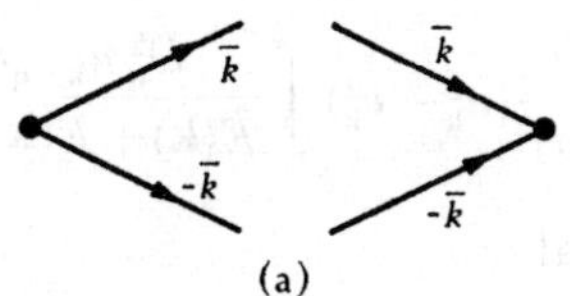

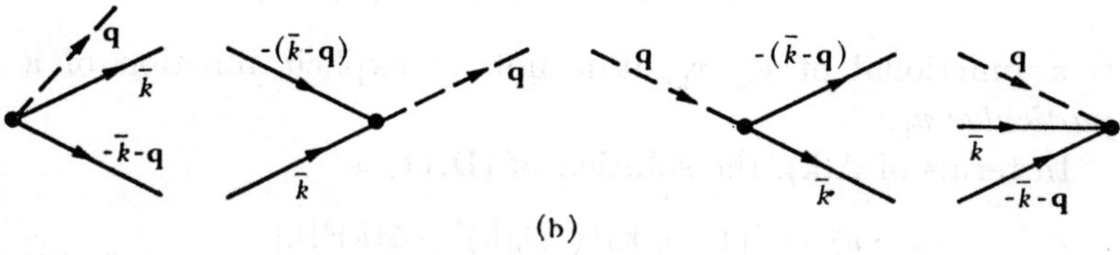

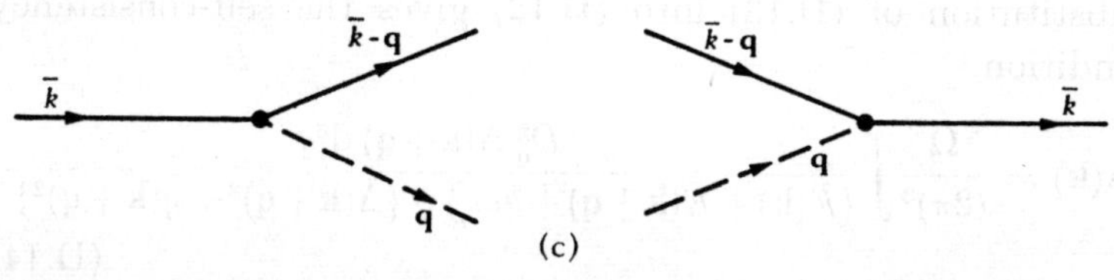

FIG. D.1. Feynman vertices described by the Bogoliubov Hamiltonian. (a) Diagrams belonging to $\hat{\mathscr{H}}_1$. (b) Diagrams belonging to $\hat{\mathscr{H}}_2$. (c) Diagrams belonging to $\hat{\mathscr{H}}_3$.

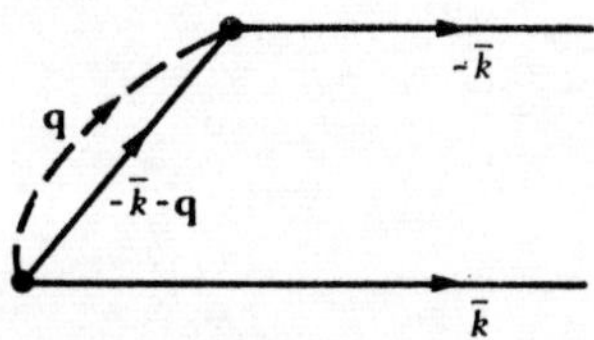

FIG. D.2. A dangerous second-order process, produced by the successive operation of $\hat{\mathscr{H}}_2$ and $\hat{\mathscr{H}}_3$ (Figs. D.1(b) and D.1(c)).

equation (D.8) may be rewritten

$$\tilde{\eta}\, u_{\mathbf{k}} v_{\mathbf{k}} = \frac{\Omega}{(2\pi)^3}(u_{\mathbf{k}}^2 - v_{\mathbf{k}}^2) \int \frac{D_{\mathbf{q}}^2\, u_{\mathbf{k}+\mathbf{q}} v_{\mathbf{k}+\mathbf{q}}\, \mathrm{d}^3 q}{\tilde{E}(\mathbf{k}) + \tilde{E}(\mathbf{k}+\mathbf{q}) + \hbar\omega_{\mathbf{q}}}. \tag{D.11}$$

Although the integral

$$\Delta(\mathbf{k}) = \frac{\Omega}{(2\pi)^3} \int \frac{D_{\mathbf{q}}^2\, u_{\mathbf{k}+\mathbf{q}} v_{\mathbf{k}+\mathbf{q}}\, \mathrm{d}^3 q}{\tilde{E}(\mathbf{k}) + \tilde{E}(\mathbf{k}+\mathbf{q}) + \hbar\omega_{\mathbf{q}}} \tag{D.12}$$

is a functional of $u_{\mathbf{k}}$, $v_{\mathbf{k}}$, it is not an explicit function of a *particular* $u_{\mathbf{k}}$.

In terms of $\Delta(\mathbf{k})$, the solution of (D.11) is

$$\left.\begin{aligned} u_{\mathbf{k}}^2 &= \tfrac{1}{2}[1 \pm \tilde{\eta}(\mathbf{k})/\surd\{\tilde{\eta}(\mathbf{k})^2 + \Delta(\mathbf{k})^2\}], \\ v_{\mathbf{k}}^2 &= \tfrac{1}{2}[1 \mp \tilde{\eta}(\mathbf{k})/\surd\{\tilde{\eta}(\mathbf{k})^2 + \Delta(\mathbf{k})^2\}]. \end{aligned}\right\} \tag{D.13}$$

Substitution of (D.13) into (D 12) gives the self-consistency condition

$$\Delta(\mathbf{k}) = \frac{\Omega}{(2\pi)^3} \int \frac{D_{\mathbf{q}}^2\, \Delta(\mathbf{k}+\mathbf{q})\, \mathrm{d}^3 q}{\{\tilde{E}(\mathbf{k}) + \tilde{E}(\mathbf{k}+\mathbf{q}) + \hbar\omega_{\mathbf{q}}\}\surd\{\Delta(\mathbf{k}+\mathbf{q})^2 + \eta(\mathbf{k}+\mathbf{q})^2\}}. \tag{D.14}$$

Although (D.14) is more complicated in form than the BCS condition (11.25), Bogoliubov has shown that it has a similar solution to (11.25) in the limit of weak coupling:

$$\Delta = \hbar\bar{\omega}\exp(-1/\mathscr{F}). \tag{D.15}$$

APPENDIX E

The function $\lim_{\theta\to+0} 1/(x+\mathrm{i}\theta)$.

LET $f(x)$ be an arbitrary function of x, continuous at $x = 0$. Then

$$\int_{-\infty}^{\infty} f(x)x^{-1}\,\mathrm{d}x \tag{E.1}$$

is not uniquely defined, on account of the pole at $x = 0$. The introduction of an infinitesimal positive imaginary part $\mathrm{i}\theta$ of x makes (E.1) unique; it is equivalent to integrating along the contour $ABCDE$ (Fig. E.1). The integral

$$\int_{ABCDE} f(x)x^{-1}\,\mathrm{d}x \tag{E.2}$$

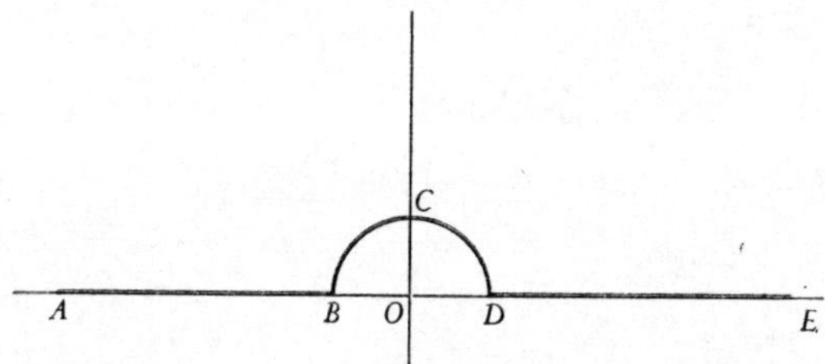

FIG. E.1. Integration contour for equation (E.2).

may be written as a sum of two parts. One is the principal value

$$\mathscr{P}\int_{-\infty}^{\infty} f(x)x^{-1}\,\mathrm{d}x, \tag{E.3}$$

arising from the open contour AB, DE. The second part is the integral along the semicircle BCD; in the limit $\theta \to 0$, this is clearly

$$-f(0)\int_0^{\pi} \mathrm{i}\,\mathrm{d}\phi = -\mathrm{i}\pi f(0). \tag{E.4}$$

Hence

$$\lim_{\theta\to+0}\int_{-\infty}^{\infty} f(x)(x+\mathrm{i}\theta)^{-1}\,\mathrm{d}x = \mathscr{P}\int_{-\infty}^{\infty} f(x)x^{-1}\,\mathrm{d}x - \mathrm{i}\pi f(0)$$

$$= \int_{-\infty}^{\infty} f(x)\{\mathscr{P}(x^{-1}) - \mathrm{i}\pi\delta(x)\}\,\mathrm{d}x. \tag{E.5}$$

Since (E.5) is valid for *any* function $f(x)$ (with only the restriction that $f(x)$ must be analytic in the neighbourhood of $x = 0$), we may write symbolically

$$\lim_{\theta\to+0}(x+\mathrm{i}\theta)^{-1} = \mathscr{P}(x^{-1}) - \mathrm{i}\pi\delta(x), \tag{E.6}$$

which is equation (14.19).

SUGGESTED FURTHER READING

Theoretical

BLATT, J. M., 1964. *Theory of Superconductivity*, Academic Press, New York.
The main emphasis is on the 'Bose-condensation' aspects of the microscopic theory, treated particularly from the point of view of Schafroth, Butler, and Blatt (1957).

DE GENNES, P. G., 1966. *Superconductivity of Metals and Alloys*, Benjamin, New York.
Especially useful for its account of the Ginzburg–Landau equations and their application to Type II materials and tunnelling phenomena.

RICKAYZEN, G., 1965. *Theory of Superconductivity*, Interscience, New York.
A comprehensive account of the theory. Strongly recommended.

SCHRIEFFER, J. R., 1964. *Theory of Superconductivity*, Benjamin, New York.
A course of introductory lecture notes on the microscopic theory. It provides a very readable introduction to the use of Green-function techniques in superconductivity.

Experimental

LIVINGSTON, J. D. and SCHADLER, H. W., 1964. The effect of metallurgical variables on superconducting properties, *Progress in Materials Science* (Ed. B. Chalmers), Vol. 12, p. 183, Pergamon Press, Oxford.
An extensive review article, summarizing much experimental work. It also contains a lengthy bibliography.

LYNTON, E. A., 1962. *Superconductivity*, Methuen, London.
An introductory and largely descriptive account.

ROBERTS, B. W., 1964. Superconducting materials and some of their properties, *Progress in Cryogenics* (Ed. K. Mendelssohn), Vol. 4, p. 161, Heywood, London.
An extensive compilation of data, otherwise available only highly dispersed in the literature. Much of the information is tabulated.

BIBLIOGRAPHY

ABEL, W. R., ANDERSON, A. C., BLACK, W. C., and WHEATLEY, J. C., 1965. *Phys. Rev. Lett.* **14**, 129.

ABRIKOSOV, A. A., 1957. *Zh. éksp. teor. Fiz.* **32**, 1442 (Translation: *Soviet Phys. JETP* **5**, 1174).

ADKINS, C. J., 1964. *Rev. mod. Phys.* **36**, 211.

ALLEN, W. D., DAWTON, R. H., BÄR, M., MENDELSSOHN, K., and OLSEN, J. L., 1950*a*. *Nature, Lond.* **166**, 1071.

—— —— LOCK, J. M., PIPPARD, A. B., and SHOENBERG, D., 1950*b*. *Nature, Lond.* **166**, 1071.

ANDERSON, P. W., 1958. *Phys. Rev.* **112**, 1900.

—— 1959. *Phys. Rev. Lett.* **3**, 325.

—— 1961. *Phys. Rev.* **124**, 41.

—— 1962. *Phys. Rev. Lett.* **9**, 309.

—— and DAYEM, A., 1964. *Phys. Rev. Lett.* **13**, 195.

—— and KIM, Y. B., 1964. *Rev. mod. Phys.* **36**, 39.

—— and ROWELL, J. M., 1963. *Phys. Rev. Lett.* **10**, 230.

—— WERTHAMER, N. R. and LUTTINGER, J. M., 1965. *Phys. Rev.* **138**, A 1157.

ANDREW, E. R., 1948. *Proc. R. Soc.* A **194**, 98.

ANDROES, G. M. and KNIGHT, W. D., 1961. *Phys. Rev.* **121**, 779.

ARKADIEV, V., 1947. *Nature, Lond.* **160**, 330.

ATKINS, K. R., 1951. *Liquid Helium*, p. 33, Cambridge University Press.

BAIRD, D. C., and MUKHERJEE, B., 1967. *Phys. Lett.* In press.

BARANGER, M., 1963. 1962 *Cargèse Lectures in Theoretical Physics* (Ed. M. Lévy), Vol. 1, Benjamin, New York.

BARDEEN, J., 1937. *Phys. Rev.* **52**, 688.

—— 1950. *Phys. Rev.* **80**, 567.

—— 1951. *Rev. mod. Phys.* **23**, 261.

—— 1954. *Phys. Rev.* **94**, 554.

—— 1956. *Handbuch der Physik/Encyclopedia of Physics* (Ed. S. Flügge), Vol. 15, p. 274, Springer, Berlin.

—— 1961. *Phys. Rev. Lett.* **7**, 162.

—— COOPER, L. N., and SCHRIEFFER, J. R., 1957. *Phys. Rev.* **106**, 162; *Phys. Rev.* **108**, 1175.

—— and PINES, D., 1955. *Phys. Rev.* **99**, 1140.

—— RICKAYZEN, G., and TEWORDT, L., 1959. *Phys. Rev.* **113**, 982.

—— and SCHRIEFFER, J. R., 1961. *Progress in Low Temperature Physics*, (Ed. C. J. Gorter), Vol. 3, p. 212, North Holland, Amsterdam.

—— and STEPHEN, M. J., 1965. *Phys. Rev.* **140**, A 1197.

BEAN, C. P., 1962. *Phys. Rev. Lett.* **8**, 250.

BECKER, R., HELLER, G., and SAUTER, F. 1933. *Z. Phys.* **85**, 772.

BERNARDES, N., 1957. *Phys. Rev.* **107**, 354.

BLATT, J. M., 1960. *Prog. theor. Phys., Osaka* **23**, 447.

—— 1964. *Theory of Superconductivity*, Academic Press, New York.

BLOCH, F., 1928. *Z. Phys.* **52**, 555.

BOGOLIUBOV, N. N., 1947. *J. Phys., Moscow.* **11**, 23.

—— 1958. *Nuovo Cim.* (Ser. X). **7**, 794; *Zh. éksp. teor. Fiz.* **34**, 58, 73 (Translation: *Soviet Phys. JETP* **7**, 41, 51).

—— TOLMACHEV, V. V. and SHIRKOV, D. V., 1959. *A New Method in the Theory of Superconductivity*, Akad. Nauk. SSSR., Moscow (1958). (Translation: Consultants Bureau, Inc., New York.)

Bohm, D., 1949. *Phys. Rev.* **75**, 502.
—— and Pines, D., 1953. *Phys. Rev.* **92**, 609.
—— and Staver, T., 1951. *Phys. Rev.* **84**, 826. See also Staver.
Bohr, A., Mottelson, B. R., and Pines, D., 1958. *Phys. Rev.* **110**, 936.
Borcherds, P. H., Gough, C. E., Vinen, W. F., and Warren, A. C., 1964. *Phil. Mag.* (Ser. viii) **10**, 349.
Born, M. and Cheng, K. C., 1948. *Nature, Lond.* **161**, 968, 1017; *J. Phys. Radium, Paris.* **9**, 249.
Brenig, W., 1961. *Phys. Rev. Lett.* **7**, 337.
Brout, R. H, 1965. *Phase Transitions*, Benjamin, New York.
Brueckner, K. A., Soda, T., Anderson, P. W., and Morel, P., 1960. *Phys. Rev.* **118**, 1442.
Buck, D. A., 1956. *Proc. Inst. Radio Engrs* **44**, 482.
Buckingham, M. J., 1957. *Nuovo Cim.* (Ser. X) **5**, 1763.
Bychkov, Yu. A., Gor'kov, L. P., and Dzyaloshinskiĭ, I. E., 1965. *Zh. éksp. teor. Fiz. Pis'ma v Redaktsiyu* **2**, 146 (Translation: *JETP Lett.* **2**, 92.
—— —— —— 1966. *Zh. éksp. teor. Fiz.* **50**, 738 (Translation: *Soviet Phys. JETP* **23**, 489.)
Byers, N. and Yang, C. N., 1961. *Phys. Rev. Lett.* **7**, 46.
Campbell, A. M., Evetts, J. E., and Dew-Hughes, D., 1964. *Phil. Mag.* (Ser. viii) **10**, 333.
Casimir, H. B. G., 1953. *Physica* **19**, 764.
Charles, R. J. and Harrison, W. A., 1963. *Phys. Rev. Lett.* **11**, 75.
Clogston, A. M., Gossard, A. C., Jaccarino, V., and Yafet, Y., 1962*a*. *Phys. Rev. Lett.* **9**, 262.
—— Matthias, B. T., Peter, M., Williams, H. J., Corenzwit, E., and Sherwood, R. C., 1962*b*. *Phys. Rev.* **125**, 541.
Cohen, M. L., 1964. *Phys. Rev.* **134**, A 511; *Rev. mod. Phys.* **36**, 240.
Cooper, L. N., 1956. *Phys. Rev.* **104**, 1189.
—— 1961. *Phys. Rev. Lett.* **6**, 689.
—— 1962. *Phys. Rev. Lett.* **8**, 367.
Corak, W. S., Goodman, B. B., Satterthwaite, C. B., and Wexler, A., 1956. *Phys. Rev.* **102**, 656.
—— and Sattherthwaite, C. B., 1956. *Phys. Rev.* **102**, 662.
Courant, R. and Hilbert, D., 1953. *Methods of Mathematical Physics*, Vol. I, pp. 351–388, Interscience, New York.
Cribier, D., Jacrot, B., Rao, L. M., and Farnoux, B., 1964. *Phys. Lett.* **9**, 106.
Crowe, J. W., 1957. *IBM Journal of Research*, p. 294.
Deaver, B. S. Jr. and Fairbank, W. M., 1961. *Phys. Rev. Lett.* **7**, 43.
Désirant, M. and Shoenberg, D., 1948. *Proc. R. Soc.* A **194**, 63.
Dheer, P. N., 1961. *Proc. R. Soc.* A **260**, 333.
Dirac, P. A. M., 1958. *The Principles of Quantum Mechanics*, 4th edn., Clarendon Press, Oxford.
Doll, R. and Näbauer, M., 1961. *Phys. Rev. Lett.* **7**, 51.
Ehrenfest, P., 1933. *Proc. K. ned. Acad. Wet.* Science section **36**, 153.
Einstein, A., 1917. *Phys. Z.* **18**, 121.
Emery, V. J. and Sessler, A. M., 1960. *Phys. Rev.* **119**, 43.
Essman, U. and Träuble, H., 1967. *Phys. Lett.* **24** A, 526.
Evans, W. A. B. and Rickayzen, G., 1964. *Proc. phys. Soc.* **83**, 311.
Evetts, J. E., Campbell, A. M., and Dew-Hughes, D., 1964. *Phil. Mag.* (Ser. viii) **10**, 339.

FABER, T. E., 1952. *Proc. R. Soc.* A **214**, 392.
—— 1954. *Proc. R. Soc.* A **223**, 174.
—— 1957. *Proc. R. Soc.* A **241**, 531.
—— 1958. *Proc. R. Soc.* A **248**, 460.
—— and PIPPARD, A. B., 1955. *Proc. R. Soc.* A **231**, 336.
FERRELL, R. A., 1959. *Phys. Rev. Lett.* **3**, 262.
—— 1964. *Phys. Rev. Lett.* **13**, 330.
—— 1966. *LT 10* conference contribution. In press.
—— and GLOVER, R. E. III, 1958. *Phys. Rev.* **109**, 1398.
—— and PRANGE, R. E., 1963. *Phys. Rev. Lett.* **10**, 479.
FEYNMAN, R. P., 1955. *Prog. in Low Temperature Physics* (Ed. C. J. Gorter), Vol. I, p. 17, North Holland, Amsterdam.
FREDERIKSE, H. P. R., SCHOOLEY, J. F., THURBER, W. R., PFEIFFER, E., and HOSLER, W. R., 1966. *Phys. Rev. Lett.* **16**, 579.
FRIEDEL, J., DE GENNES, P. G., and MATRICON, J., 1963. *Appl. Phys. Lett.* **2**, 119.
FRÖHLICH, H., 1950. *Phys. Rev.* **79**, 845.
—— 1952. *Proc. R. Soc.* A **215**, 291.
—— 1954. *Proc. R. Soc.* A **223**, 296.
—— 1966. *Proc. phys. Soc.* **87**, 330.
—— PELZER, H., and ZIENAU, S., 1950. *Phil. Mag.* (Ser. vii) **41**, 221.
GALKIN, A. A., KAN, YA. S., and LAZAREV, B. G., 1957. *Zh. éksp. teor. Fiz.* **32**, 1582 (Translation: *Soviet Phys. JETP* **5**, 1292).
GARLAND, J. W. Jr., 1963. *Phys. Rev. Lett.* **11**, 111.
GEBALLE, T. H., 1964. *Rev. mod. Phys.* **36**, 134.
DE GENNES, P. G., 1963. *Phys. Lett.* **5**, 22.
—— 1964. *Rev. mod. Phys.* **36**, 225.
—— 1966. *Superconductivity of Metals and Alloys*, pp. 240–4, Benjamin, New York.
—— and GUYON, E. 1963. *Phys. Lett.* **3**, 168.
—— and MATRICON, J., 1964. *Rev. mod. Phys.* **36**, 45
GIAEVER, I., 1960. *Phys. Rev. Lett.* **5**, 147, 464.
—— HART, H. R. Jr., and MEGERLE, K., 1962. *Phys. Rev.* **126**, 941.
—— and MEGERLE, K., 1961. *Phys. Rev.* **122**, 1101.
GINSBERG, D. M. and TINKHAM, M., 1960. *Phys. Rev.* **118**, 990.
GINZBURG, V. L., 1944. *J. Phys., Moscow* **8**, 148.
—— 1946. *Zh. éksp. teor. Fiz.* **14**, 134.
—— 1952. *Usp. fiz. Nauk.* **48**, 25 (German Translation: 1953, *Fortschr. Phys.* **1**, 101).
—— and LANDAU, L. D., 1950. *Zh. éksp. teor. Fiz.* **20**, 1064.
GLOVER, R. E., III, and TINKHAM, M., 1957. *Phys. Rev.* **108**, 243.
GOODMAN, B. B., 1953. *Proc. phys. Soc.* A **66**, 217.
—— 1961. *Phys. Rev. Lett.* **6**, 597.
—— 1966. *Rep. Prog. Phys.* **29** (2), 445.
GOR'KOV, L. P., 1958. *Zh. éksp. teor. Fiz.* **34**, 735 (Translation: *Soviet Phys. JETP* **7**, 505).
—— 1959. *Zh. éksp. teor. Fiz.* **36**, 1918 (Translation: *Soviet Phys. JETP* **9**, 1364.)
GORTER, C. J. and CASIMIR, H. B. G., 1934. *Phys. Z.* **35**, 963.
HAMILTON, D. C. and JENSEN, M. A., 1963. *Phys. Rev. Lett.* **11**, 205.
HAMMOND, R. H. and KELLY, G. M., 1964. *Rev. mod. Phys.* **36**, 185.
HEBEL, L. C. and SLICHTER, C. P., 1959. *Phys. Rev.* **113**, 1504.
HEIN, R. A., GIBSON, J. W., MAZELSKY, R., MILLER, R. C., and HULM, J. K., 1964. *Phys. Rev. Lett.* **12**, 320.

Heisenberg, W., 1947. *Z. Naturf.* **2a**, 185.
—— 1948. *Z. Naturf.* **3a**, 65; *Two Lectures*, Lecture 2, Cambridge University Press.
Hilsch, P. and Hilsch, R., 1964. *Z. Phys.* **180**, 10.
Huang, K., 1951. *Proc. phys. Soc.* A **64**, 867.
Jaklevic, R. C., Lambe, J. J., Silver, A. H., and Mercereau, J. E., 1964. *Phys. Rev. Lett.* **12**, 159, 274.
Jeans, Sir James, 1925. *The Mathematical Theory of Electricity and Magnetism*, 5th edn., pp. 152–3, Cambridge University Press.
Josephson, B. D., 1962. *Phys. Lett.* **1**, 251.
—— 1964. *Rev. mod. Phys.* **36**, 216.
—— 1965 (*a*). *Adv. Phys.* **14**, 419.
—— 1965 (*b*). Private communication.
Kadanoff, L. P. and Baym, G., 1962. *Quantum Statistical Mechanics*, Benjamin, New York.
Keesom, W. H., 1924. *Rapp. et Disc. 4e Congr. Phys. Solvay*, p. 288.
Kim, Y. B., Hempstead, C. F., and Strnad, A. R., 1962. *Phys. Rev. Lett.* **9**, 306.
—— —— —— 1963. *Phys. Rev.* **131**, 2486.
—— —— —— 1965. *Phys. Rev.* **139**, A 1163.
Kleiner, W. H., Roth, L. M. and Autler, S. H., 1964. *Phys. Rev.* **133** A, 1226.
Knight, W. D., 1949. *Phys. Rev.* **76**, 1259.
Kohn, W. and Luttinger, J. M., 1965. *Phys. Rev. Lett.* **15**, 524.
Kondo, J., 1963. *Prog. theor. Phys., Osaka* **29**, 1
Koppe, H., 1947. *Annln Phys.* **1**, 405.
—— 1950. *Ergebn. exakt. Naturw.* **23**, 283.
Kramers, H. A., 1927. *Atti Congresso dei Fisici, Como*, p. 545.
Kronig, R. de L., 1926. *J. opt. Soc. Am.* **12**, 547.
—— and Penney, W. G., 1931. *Proc. R. Soc.* A **130**, 499.
Kunzler, J. E., 1961. *Rev. mod. Phys.* **33**, 501.
—— Bueler, E., Hsu, F. S. L., Matthias, B. T., and Wahl, C., 1961. *J. appl. Phys.* **32**, 325.
—— —— —— and Wernick, J. H., 1961. *Phys. Rev. Lett.* **6**, 89.
Kuper, C. G., 1952. *Phil. Mag.* (Ser. vii) **43**, 1264.
—— 1954. Ph.D. thesis, Cambridge University.
—— 1966. *Phys. Rev.* **150**, 189.
—— Jensen, M. A., and Hamilton, D. C., 1964. *Phys. Rev.* **134**, A15.
Landau, L. D., 1937 *a*. *Zh. éksp. teor. Fiz.* **7**, 19.
—— 1937 *b*. *Phys. Z. SowjUn.* **11**, 129
—— 1941. *J. Phys., Moscow* **5**, 71.
—— 1943. *J. Phys., Moscow* **7**, 99.
—— 1956. *Zh. éksp. teor. Fiz.* **30**, 1058 (Translation: 1957 *Soviet Phys. JETP* **3**, 920).
—— and Lifshitz, E. M., 1959. *Statistical Physics*, Pergamon Press, London.
—— —— 1960. *Electrodynamics of Continuous Media*, Pergamon Press, London.
von Laue, M., 1949. *Theorie der Supraleitung*, 2nd edn, Springer Verlag, Berlin (Translation: *Theory of Superconductivity*, Academic Press, New York, 1952).
Leech, J. W., 1958. *Classical Mechanics*, p. 49. Methuen, London.
Lifshitz, E. M. and Sharvin, Yu. V., 1951. *Dokl. Akad. Nauk. SSSR* **79**, 783.

LITTLE, W. A., 1964. *Phys. Rev.* **134,** A 1416.
LIVINGSTON, J. D., 1963. *Phys. Rev.* **129,** 1943.
LONDON, F., 1936. *Physica* **3,** 450.
—— 1937. *Une Conception Nouvelle de la Supraconductibilite,* pp. 66–68 (Actualités Sci. et Ind. No. 458), Hermann et Cie, Paris.
—— 1950. *Superfluids,* Vol. I: *Macroscopic Theory of Superconductivity,* Wiley, New York.
—— and LONDON, H., 1935. *Physica* **2,** 341.
LONDON, H., 1935. *Proc. R. Soc.* A **152,** 650.
—— 1940. *Proc. R. Soc.* A **176,** 522.
LYNTON, E. A., 1962. *Superconductivity,* p. 3, Methuen, London.
MANDL, F., 1959. *Introduction to Quantum Field Theory,* Interscience, New York.
MATRICON, J., 1964. *Phys. Lett.* **9,** 289.
MATTHIAS, B. T., 1957. *Progress in Low Temperature Physics* (Ed. C. J. Gorter), Vol. 2, p. 138, North Holland, Amsterdam.
—— GEBALLE, T. H., and COMPTON, V. B., 1963. *Rev. mod. Phys.* **35,** 1.
—— —— LONGINOTTI, L. D., CORENZWIT, E., HULL, G. W., WILLENS, R. H., and MAITA, J. P., 1967. *Science* **156,** 645.
MEISSNER, W., 1929. *Z. Phys.* **58,** 570.
—— and OCHSENFELD, R., 1933. *Naturwissenschaften* **21,** 787.
MENDELSSOHN, K., 1935. *Proc. R. Soc.* A **152,** 34.
MESHKOVSKY, A. and SHALNIKOV, A., 1947. *J. Phys., Moscow* **11,** 1; *Zh. éksp. teor. Fiz.* **17,** 851.
MILNE-THOMSON, L. M., 1955. *Theoretical Hydrodynamics,* 3rd edn p. 67, Macmillan, London.
MOREL, P., 1959. *Physics Chem. Solids* **10,** 277.
—— and ANDERSON, P. W., 1962. *Phys. Rev.* **125,** 1263.
MORIN, F. J. and MAITA, J. P., 1963. *Phys. Rev.* **129,** 1115.
MORSE, R. W. and BOHM, H. V., 1957. *Phys. Rev.* **108,** 1094.
MOTT, N. F., 1936. *Proc. Camb. phil. Soc. math. phys. Sci.* **32,** 228.
—— and JONES, H., 1936. *The Theory of the Properties of Metals and Alloys,* Clarendon Press, Oxford.
MÜHLSCHLEGEL, B., 1959. *Z. Phys.* **155,** 313.
NAKAJIMA, S., 1953. *Busseiron kenk-yu* **65,** 116.
NAMBU, Y., 1960. *Phys. Rev.* **117,** 648.
NOER, R. J. and KNIGHT, W. D., 1964. *Rev. mod. Phys.* **36,** 177.
NORDHEIM, L., 1931. *Annln Phys.* **9,** 607.
—— 1934. *Actualités Scientifiques et Industrielles,* No. 131, Hermann et Cie, Paris.
NOZIÈRES, P. and PINES, D., 1958. *Nuovo Cim.* (Ser. X) **9,** 470.
ONNES, H. KAMERLINGH, 1908. *Proc. K. ned. Acad. Wet.* Science Section **11,** 168.
—— 1911. *Communs. phys. Lab. Univ. Leiden* No. 120b, 122b, 124c.
—— 1914. *Communs. phys. Lab. Univ. Leiden* No. 140b, 141b.
—— and TUYN, W., 1923. *Communs. phys. Lab. Univ. Leiden,* No. 160a, b, 167a.
ONSAGER, L., 1944. *Phys. Rev.* **65,** 117.
—— 1951. Private communication.
—— 1961. *Phys. Rev. Lett.* **7,** 50.
PARMENTER, R. H., 1963. *Phys. Rev.* **132,** 2490.
PAULUS, K. F. G., 1966. *Mol. Phys.* **10,** 381.
PEIERLS, R. E., 1930. *Annln Phys.* **4,** 121.
—— 1932. *Annln Phys.* **12,** 154.

PEIERLS 1936. *Proc. R. Soc.* A **155**, 613.
PESHKOV, V. P., 1964. *Zh. éksp. teor. Fiz.* **46**, 1510 (Translation: *Soviet Phys. JETP* **19**, 1023).
PINES, D., 1958. *Phys. Rev.* **109**, 280.
—— 1961. *The Many-Body Problem*, pp. 55–65, Benjamin, New York.
PIPPARD, A. B., 1947. *Proc. R. Soc.* A **191**, 370, 385, 399.
—— 1950 *a*. *Proc. R. Soc.* A **203**, 98.
—— 1950 *b*. *Proc. R. Soc.* A **203**, 210.
—— 1950 *c*. *Phil. Mag.* (Ser. vii) **41**, 243.
—— 1951. *Proc. Camb. phil. Soc. math. phys. Sci.* **47**, 617.
—— 1953. *Proc. R. Soc.* A **216**, 547.
REIF, F., 1957. *Phys. Rev.* **106**, 208.
—— and WOOLF, M. A., 1962. *Phys. Rev. Lett.* **9**, 315.
REUTER, G. E. H. and SONDHEIMER, E. H., 1948. *Proc. R. Soc.* A **195**, **336**.
REYNOLDS, C. A., SERIN, B., WRIGHT, W. H., and NESBITT, L. B., 1950. *Phys. Rev.* **78**, 487.
RICKAYZEN, G., 1958. *Phys. Rev.* **111**, 817.
—— 1959. *Phys. Rev.* **115**, 795.
—— 1965. *Theory of Superconductivity*, Interscience, New York.
ROBERTS, B. W., 1964. *Progress in Cryagenics* (Ed. K. Mendelssohn), Vol. 4, p. 161, Heywood, London.
SAINT–JAMES, D. and DE GENNES, P. G., 1963. *Phys. Lett.* **7**, 306.
SCHAFROTH, M. R., 1955. *Phys. Rev.* **100**, 463.
—— 1958. *Phys. Rev.* **111**, 72.
—— 1960. *Solid State Physics* (Ed. F. Seitz and D. Turnbull), Vol. 10, p. 293, Academic Press, New York.
—— and BLATT, J. M., 1956. *Nuovo Cim.* (Ser. X) **4**, 786.
—— BUTLER, S. T., and BLATT, J. M., 1957. *Helv. phys. Acta.* **30**, 93.
SCHIFF, L. I., 1955. *Quantum Mechanics*, 2nd edn, p. 199, McGraw-Hill, New York.
SCHOOLEY, J. F., HOSLER, W. R., and COHEN, M. L., 1964. *Phys. Rev. Lett.* **12**, **474**.
SCHRIEFFER, J. R., 1959. *Phys. Rev. Lett.* **3**, **323**.
—— 1964. *Theory of Superconductivity*, Benjamin, New York.
—— SCALAPINO, D. J., and WILKINS, J. W., 1963. *Phys. Rev. Lett.* **10**, **336**.
—— and WILKINS, J. W., 1963. *Phys. Rev. Lett.* **10**, 17.
SCOTT, R. B., 1948. *J. Res. natn. Bur. Stand.* **41**, 581.
SEWELL, G. L., 1959. *Proc. phys. Soc.* **74**, 340.
SHAPIRO, S., JANUS, A. R., and HOLLY, S., 1964. *Rev. mod. Phys.* **36**, 223.
SHOENBERG, D., 1940. *Proc. R. Soc.* A 157, 49.
—— 1952. *Superconductivity*, 2nd edn, pp. 20–22, Cambridge University Press.
SHUBNIKOV, L. W. and ALEKSEYEVSKY, N. E., 1936. *Nature, Lond.* **138**, 804.
SILSBEE, F. B., 1916. *J. Wash. Acad. Sci.* **6**, 597.
SILVERT, W. and COOPER, L. N., 1966. *Phys. Rev.* **141**, 336.
SIMMONS, W. A. and DOUGLASS, D. H. Jr., 1962. *Phys. Rev. Lett.* **9**, 153.
SLATER, J. C., 1929. *Phys. Rev.* **34**, 1293.
SMITH, P. H., SHAPIRO, S., MILES, J. L., and NICOL, J., 1961. *Phys. Rev. Lett.* **6**, 686.
SOMMERFELD, A., 1928. *Z. Phys.* **47**, 1.
—— and BETHE, H. A., 1933. *Handbuch der Physik*, 2e Aufl., Vol. 24, (2), p. 333. Springer, Berlin.
STAVER, T., 1952. Ph.D. thesis, Princeton University.

STEPHEN, M. J. and SUHL, H., 1964. *Phys. Rev. Lett.* **13**, 797.
SUHL, H., MATTHIAS, B. T., and WALKER, L. R., 1959. *Phys. Rev. Lett.* **3**, 552.
TAYLOR, A. W. B., 1961. *Proc. phys. Soc.* **78**, 1372.
THOULESS, D. J., 1960, *Ann. Phys.* **10**, 553.
TITCHMARSH, E. C., 1948. *Introduction to the Theory of Fourier Integrals*, 2nd edn, p. 128, Clarendon Press, Oxford.
TUYN, W. and ONNES, H. K., 1926. *Communs. phys. Lab. Univ. Leiden* 174a.
VALATIN, J. G., 1958. *Nuovo Cim.* (Ser. X) **7**, 843.
VOLGER, J., STAAS, F. A., and VAN VIJFEIJKEN, A. G., 1964. *Phys. Lett.* **9**, 303.
WALDRAM, J. R., 1964. *Adv. Phys.* **13**, 1.
WENTZEL, G., 1951. *Phys. Rev.* **83**, 168.
WHITTAKER, E. T., and WATSON, G. N., 1927. *A Course of Modern Analysis*, 4th edn, p. 464, Cambridge University Press.
WILLIAMS, D. L., 1962. *Proc. phys. Soc.* **79**, 594.
WILSON, A. H., 1953. *The Theory of Metals*, 2nd edn, Cambridge University Press.
YOSIDA, K., 1958. *Phys. Rev.* **110**, 769.

AUTHOR INDEX

SUBJECT INDEX